Handbook on Biological Warfare Preparedness

Handbook on Biological Warfare Preparedness

Edited by

S.J.S. Flora

Vidhu Pachauri

ACADEMIC PRESS
An imprint of Elsevier

Academic Press is an imprint of Elsevier
125 London Wall, London EC2Y 5AS, United Kingdom
525 B Street, Suite 1650, San Diego, CA 92101, United States
50 Hampshire Street, 5th Floor, Cambridge, MA 02139, United States
The Boulevard, Langford Lane, Kidlington, Oxford OX5 1GB, United Kingdom

Notices
Knowledge and best practice in this field are constantly changing. As new research and experience
broaden our understanding, changes in research methods, professional practices, or medical treatment
may become necessary.

Practitioners and researchers must always rely on their own experience and knowledge in evaluating
and using any information, methods, compounds, or experiments described herein. In using such
information or methods they should be mindful of their own safety and the safety of others, including
parties for whom they have a professional responsibility.

To the fullest extent of the law, neither the Publisher nor the authors, contributors, or editors, assume
any liability for any injury and/or damage to persons or property as a matter of products liability,
negligence or otherwise, or from any use or operation of any methods, products, instructions, or ideas
contained in the material herein.

Library of Congress Cataloging-in-Publication Data
A catalog record for this book is available from the Library of Congress

British Library Cataloguing-in-Publication Data
A catalogue record for this book is available from the British Library

ISBN: 978-0-12-812026-2

For information on all Academic Press publications
visit our website at https://www.elsevier.com/books-and-journals

Publisher: Andre Gerhard Wolff
Acquisition Editor: Kattie Washington
Editorial Project Manager: Ana Claudia A. Garcia
Production Project Manager: Punithavathy Govindaradjane
Cover Designer: Mark Rogers

Typeset by SPi Global, India

Dedicated to my mother Paramjeet

Contents

**CHAPTER 5   Advance detection technologies for select
biothreat agents** **83**
M.M. Parida, Paban Kumar Dash and Jyoti Shukla

**CHAPTER 6   Microfluidics application for detection of
biological warfare agents** **103**
Bhairab Mondal, N. Bhavanashri, S.P. Mounika,
Deepika Tuteja, Kunti Tandi and H. Soniya

Anju Tripathi, Kshirod Sathua, Vidhu Pachauri,
and S.J.S. Flora

Contributors

Tawseef Ahmad
Department of Biotechnology, Punjabi University, Patiala, India

A.S.B. Bhaskar
Division of Pharmacology and Toxicology, Defence Research and Development Establishment, Gwalior, India

N. Bhavanashri
Shankaranarayana Life Sciences, Bengaluru, India

Mannan Boopathi
Defence Research and Development Establishment, DRDO, Gwalior, India

Paban Kumar Dash
Division of Virology, Defence Research and Development Establishment (DRDE), Defence Research and Development Organization, Ministry of Defence, Gwalior, India

S.J.S. Flora
National Institute of Pharmaceutical Education and Research-Raebareli, Lucknow, India

Gaganjot Gupta
Department of Biotechnology, Punjabi University, Patiala, India

Baljinder Kaur
Department of Biotechnology, Punjabi University, Patiala, India

Kewal Krishan
Department of Defence and Strategic Studies, Punjabi University, Patiala, India

Anoop Kumar
National Institute of Pharmaceutical Education and Research-Raebareli, Lucknow, India

Chacha D. Mangu
National Institute for Medical Research, Mbeya Medical Research Center, Mbeya, Tanzania

Bhairab Mondal
Shankaranarayana Life Sciences, Bengaluru, India

S.P. Mounika
Shankaranarayana Life Sciences, Bengaluru, India

V. Nagaraajan
VN Neurocare Center, Madurai, India

Vidhu Pachauri
National Institute of Pharmaceutical Education and Research-Raebareli, Lucknow, India

Archna Panghal
National Institute of Pharmaceutical Education and Research-Raebareli,
Lucknow, India

M.M. Parida
Division of Virology, Defence Research and Development Establishment (DRDE),
Defence Research and Development Organization, Ministry of Defence, Gwalior, India

Jayant Patwa
National Institute of Pharmaceutical Education and Research-Raebareli,
Lucknow, India

Vipin K. Rastogi
U.S. Army Futures Command—Combat Capabilities Development Command,
Chemical Biological Center, Edgewood, MD, United States

Bhavana Sant
Division of Pharmacology and Toxicology, Defence Research and Development
Establishment, Gwalior, India

Kshirod Sathua
National Institute of Pharmaceutical Education and Research-Raebareli,
Lucknow, India

Anshula Sharma
Department of Biotechnology, Punjabi University, Patiala, India

Jyoti Shukla
Division of Virology, Defence Research and Development Establishment (DRDE),
Defence Research and Development Organization, Ministry of Defence, Gwalior, India

Virendra V. Singh
Defence Research and Development Establishment, DRDO, Gwalior, India

Beer Singh
Defence Research and Development Establishment, DRDO, Gwalior, India

H. Soniya
Shankaranarayana Life Sciences, Bengaluru, India

Kunti Tandi
Shankaranarayana Life Sciences, Bengaluru, India

Vikas B. Thakare
Defence Research and Development Establishment, DRDO, Gwalior, India

Duraipandian Thavaselvam
Defence Research and Development Establishment, DRDO, Gwalior, India

Anju Tripathi
National Institute of Pharmaceutical Education and Research-Raebareli,
Lucknow, India

Deepika Tuteja
Shankaranarayana Life Sciences, Bengaluru, India

Lalena Wallace
DTRA, CB Research Center of Excellence Division, APG, Edgewood, MD,
United States

Preface

Bioterrorism is the intentional and deliberate release of biological agents (bacteria, viruses, or toxins) to cause mass illness or death of people and animals. It is said that if the 20th century was the century of physics, the 21st century will be the century of biology. Bioterrorism is defined as a planned or destructive use of biological agents such as viruses, bacteria, fungi, or toxins produced from living organisms. The main aim of bioterrorism is to harm people, animals, and plants by causing death, so as to achieve political or social destruction. Due to rapid increase in the technical skills of terrorists and fast growing research in the field of molecular biology and biotechnology, the risk of bioterrorism is increasing day by day.

Anthrax was used during the First World War by Germany to infect the mules and horses of enemies. Cases of deliberate attacks using anthrax were reported in the United States in 2001. Letters laced with infectious anthrax were delivered to news media offices and the US Congress. Biological agents are relatively easy and inexpensive to obtain, can be easily disseminated, and can cause widespread fear and panic. Risk of massive destruction in the form of life is too high. Exposure to minute quantities of a biological agent may go unnoticed, yet ultimately be the cause of disease and death. They don't work immediately. The incubation period of a microbial agent can be days or weeks. Bio-weapons cannot be kept as a military asset, as there are important limitations, like they cannot differentiate between foes and friends. Thus such weapons may be used only to create mass panic. Three types of agents basically are used on the basis of ability and extent of damage that can be caused: (i)Category A: high-priority agents causing high mortality and mass impact, such as triggering panic in local populations; (ii) Category B: moderate-priority agents causing relatively less damage; and (iii) Category C: low-priority agents; these are emerging pathogens, which are readily available and can thus be easily mutated or engineered. Thus, making a strong public health system is prerequisite to effectively handle the threat. For this, the various components of the public health system, such as surveillance, assessment, medical management, information and education, etc. need to be made stronger. There is also a need to make the national stockpile of drugs readily available in case of an incidence. The most fundamental steps include the need for the spread of awareness, readiness with drugs and medicines, and readiness with decontamination procedures.

The primary objective of this book, "Handbook on Biological Warfare Preparedness," is to offer a comprehensive text/reference source on the subject for research veterinary toxicologists, students, teachers, clinicians, and environmentalists. The volume is organized with a total of 14 chapters, in order to cover as many topics as possible. Although the book is heavily focused on describing various bacterial and viral agents, their detection and medical management, it has many novel chapters on timely topics, such as new generation synthetic agents, emerging threat, advanced detection technologies for bio-threat agents, environmental sampling and decontamination of infrastructure contaminated with biological warfare agents, biological and toxin warfare convention: current status and future prospects, microfluidics application for the detection of biological warfare agents and toxins as biological warfare agents, etc. Several chapters provide the latest information on problems related to detection, prevention, protection, de-contamination, and medical management. A significant part of the book is devoted to providing details of bacterial and viral biological warfare agents, their categories, routes of infection, and detection techniques. Finally, the later part of the book emphasizes some new developments in bio-threat detection technologies, prevention, therapeutic measures, genome information of BW agents, and finally a brief chapter on planning for protection of civilians against bio-terrorism.

In the past few years, scientists from many parts of the world have realized the need for a simple handbook that can provide detailed coverage of biological warfare agents and our preparedness. This book can be used by students of graduate and postgraduate level, scientists, and may also be useful to the first responders who train first responders in military and para-military set up. This book mainly concentrates on an introduction to different bacterial and viral agents and other emerging threats like toxins. In addition, we have provided details about the detection, protection, and de-contamination of these agents and this information is what the first responders to any biological emergency look for in a handbook. As indicated above, we have a chapter on the environmental detection of these agents and the newer instrumental, molecular, and immunological test systems that are presently available for use. Our idea was to make this an easy handbook that will provide the latest information to scientists, researchers in the area, and to some extent, persons who train the first responders in disaster management as well as in military and para-military organizations.

We have identified experts to write each of these chapters with the latest and updated information that is required for scientists and graduate students. This book addresses global problems by offering useful information and practical solutions. A standalone chapter is provided on every topic, with major references for further reading. This book represents the collective wisdom of more than 20 authors, and offers a unique text/reference source for those involved in research in the area of biological warfare. I must confess that some of the experts in the area initially agreed to contribute chapters for this book but could not fulfill their commitments for reasons best known to them. We still, however, managed to get some highly qualified and well-experienced authors on the subject to write some of the chapters. I am also thankful to some young researchers at our institute who may not be experts in the field, but volunteered to write some of the chapters.

S.J.S. Flora

Acknowledgments

I could not begin this book without placing a word of acknowledgment and appreciation to all those who directly or indirectly contributed to it. I must thus express my gratitude to all my colleagues, mentors, and students, with whom I learned so much. Among these I am especially grateful to Dr D.T. Selvam, who first introduced me to the world of Biological Warfare many years ago when I was at the Defence Research and Development Establishment, Gwalior, India. We started this book together but unfortunately he could not continue with me when I took this new assignment at the National Institute of Pharmaceutical Education and Research, Raebareli, Lucknow, India. I also thank Dr Vidhu Pachauri, a former graduate student in my laboratory, for agreeing to assist me at the final stages of some editorial work. I would like to acknowledge the work of the contributors, many of whom were contacted at a very late stage with very little time to complete their task. Their efforts made compiling and editing this book fairly easy.

I owe a debt of gratitude as well to the editors at Elsevier: Ana Claudia Abad Garcia, Editorial Project Manager and Kattie Washington, Acquisitions Editor. Their vision and faith in the idea of a comprehensive Biological Warfare Handbook got this project off the ground and kept it running. This book would not have been physically possible without the tireless work and extraordinary skills of Ana Claudia and her team. I could not possibly acknowledge her enough. We had some very testing and frustrating times in last few years to keep our date with the deadline for the book. Thank you, Ana.

Kshirod, Jayant, Anoop, and Anju wrote material for this book. More importantly, however, they supported my efforts in this project wholeheartedly. They are the trusted colleagues in this institute.

Acknowledgments would hardly be complete without recognizing those who truly worked behind the scenes on this book. I mean of course the family who supported my time away from them as I worked. My wife, Gurpyari, frequently gave me advice, and showed me a great deal of patience. Our daughter, Preeti, and son, Ujjwal, often helped in proofreading my work and cheered me up when I was able to take breaks from the computer. Those will be fond memories.

S.J.S. Flora

Biological warfare agents: History and modern-day relevance

1

S.J.S. Flora

National Institute of Pharmaceutical Education and Research-Raebareli, Lucknow, India

History of biological warfare agents

Infectious diseases and war are always interlinked. Even without a precise understanding of how diseases were spread, it was understood early that dead animals or humans could cause disease. There are some accounts of biological warfare in the form of poison on arrows, and polluting wells and other source of water by opposing armies. However, apart from some rare well-documented events (Eitzen and Takafuji, 1997) it is often very difficult for historians and microbiologists to understand natural epidemics from alleged biological attacks, because little information is available for times before the advent of modern microbiology and the passage of time may also have distorted the reality of the past. Evidence from Persian, Greek, and Roman literature suggests the use of animal cadavers to contaminate wells and other sources of water around 300 BC, and in 400 BC, Scythian archers infecting their arrows by dipping them in decomposing bodies or in blood mixed with manure. In 190 BC during the Battle of Eurymed, Hannibal won a naval victory over King Eumenes II of Pergamon by firing earthen vessels full of venomous snakes onto the enemy ships. In the 12th century AD, during the battle of Tortona, Barbarossa used the bodies of dead soldiers to poison wells. During the siege of Caffa, a well-fortified Genoese-controlled seaport (now Feodosia, Ukraine), in 1346, the attacking Tartar force experienced an epidemic of plague (Wheelis, 2002). The Tartars threw the cadavers of their deceased into the city, and it is believed that this has initiated a plague epidemic in the city. The outbreak of plague forced the retreat of the Genoese forces. The plague pandemic known as the Black Death spread throughout Europe, the Near East, and North Africa in the 14th century and was probably the most devastating public health disaster in recorded history. The ultimate origin of the plague remains uncertain: several countries in the Far East, China, Mongolia, India, and Central Asia have reported these instances in the past (Rauw, 2012). The biological warfare attack in Caffa is one of the terrible consequences in history when diseases were used as weapons.

During the battle between Russian and Swedish forces at Reval in Estonia in 1710, catapulted plague cadavers were used. In the 18th century, during the French and Indian War, British forces in North America gave blankets from smallpox patients to the Native Americans to transmit the disease to the immunologically naïve tribes. In 1863, a Confederate surgeon was arrested and charged with attempting to import yellow fever-infected clothes into northern parts of the United States during the Civil War (Hunsicker, 2006). Biological warfare became more sophisticated against both animals and humans during the 1900s. The conception of Koch's postulates and the development of modern microbiology during the 19th century made possible the isolation and production of stocks of specific pathogens (Robertson and Robertson, 1995). During World War I (WWI), some reports have suggested that the Germans developed horses and cattle inoculated with disease-producing bacteria, such as *Bacillus anthracis* (anthrax) and *Pseudomonas pseudomallei* (glanders), and shipped them to the United States, Russia, and other countries (Hugh-Jones); although Germany denied these reports. In 1924, a subcommittee of the Temporary Mixed Commission of the League of Nations found no hard evidence that bacteriological weapons had been employed in war. On June 17, 1925, the "Protocol for the Prohibition of the Use in War of Asphyxiating, Poisonous or Other Gases and of Bacteriological Methods of Warfare," commonly called the Geneva Protocol of 1925, was signed. Because viruses were not differentiated from bacteria at that time, they were not specifically mentioned in the protocol. A total of 108 nations, including the five permanent members of the United Nations (UN) Security Council, eventually signed the agreement. However, the Geneva Protocol did not address verification or compliance, making it a "toothless" and less meaningful document. Several countries that were parties to the Geneva Protocol of 1925 began to develop biological weapons soon after its ratification. These countries included Belgium, Canada, France, Great Britain, Italy, the Netherlands, Poland, Japan, and the Soviet Union. The United States did not ratify the Geneva Protocol until 1975 (Riedel, 2004).

Biological agents to cause destruction have been used since 6th century BC, wherein it was the poisoning of wells and water supplies that was most common. The very first deliberate act of spreading disease occurred in 1346 at the Siege of Caffa (Feodosia, Ukraine), when a Tartar army disposed its plague-ridden dead over the walls of the besieged city (Wheelis, 2002). Later, in 1763, during the French and Indian war, an English general intentionally distributed blankets contaminated with small pox scabs to the Native Americans loyal to the French, which caused a huge epidemic to kill the tribal people. The Germans also has their own bioweapon program for WWI, in which they purposely infected horses and other transport animals with disease causing microbes of anthrax and glanders (Frischknecht, 2003). After WWI, the Geneva Protocol was signed banning biological weapons, which all countries in attendance signed except Japan.

Before the start of WWII, the Japanese produced bioweapons agents to cause anthrax, plaque, cholera, and shigellosis, for which field experiments were done on Chinese prisoners of war and civilians, which led to several thousand deaths (Barras and Greub, 2014). One such experiment included the use of ceramic bomblets containing plague-infected fleas and grain on Chinese cities including Nanking, which attracted rats bitten by fleas contaminated with plague causing microorganisms.

When reports were spread about Japan's bioweapons program, a research program was launched by President Roosevelt on biological agents in 1941. George W. Merck of Merck Pharmaceuticals was named head of the Army's Chemical Warfare Service, and Camp Detrick, Frederick, MD, was developed into a site for biological weapons research and development.

In 1946, the US officially announced its participation in research related to bioweapons. In 1969, the World Health Organization issued a report that described the volatility of biological weapons, on the basis of which, later that year, President Nixon shut down the US offensive biological warfare program and limited biological weapons research to defensive purposes only (Frischknecht, 2003). In 1972, the Biological Weapons Convention Treaty, which called for all countries to destroy their stocks of bioweapons, was signed by 103 nations including the former Soviet Union (Riedel, 2004). Laboratory acquired small pox infections made the WHO recommend merging all the variola virus stocks, which had to be maintained only by United States and Russia (McFadden, 2010). In 1979, *Bacillus anthracis* spores were accidentally released at a bioweapons research facility in Sverdlovsk, Union of Soviet Socialist Republics (USSR), which resulted in 68 deaths due to inhalational of anthrax. In the fall of 2001, the US experienced bioterrorism when the members of the Rajneesh cult in Oregon experimented with various bioweapon prior to their 1984 act of bioterrorism. Seeking to influence the outcome of upcoming municipal elections, the cult deliberately contaminated salad bars of local restaurants with *Salmonella typhimurium*, which caused illness in over 700 people.

In 1992 in Minnesota, the Members of Minnesota Patriots Council, a militia group, planned to kill local authorities with ricin, a potent toxin obtained from castor beans. Another bioterrorism act occurred in the fall of 2001, when *Bacillus anthracis* spores were placed in at least seven envelopes passing through US mail facilities in Florida, Washington DC, New York, and New Jersey, leading to 22 confirmed and suspected cases of anthrax, of which five were fatal. Table 1 summarizes the historical incidences of biological warfare.

Biological warfare agents

Biological warfare agents (BWAs) may be defined as a class of living biological agents used with an intention of creating a state of war by causing disease to humans, plants, and animals. It is used to kill or harm other life forms. Biological agents include infectious agents such as bacteria, viruses, fungi, and insects. In the main, chemical and microorganism toxins are used as warfare agents. These represent two of the three forms of weapons mass destruction: chemical, biological, and nuclear. A chemical weapon uses chemicals to kill living organisms; biological weapons make use of living organisms to kill; whereas nuclear weapons are atomic and hydrogen bombs with explosive yields. Biological weapons are one of the most researched and used ways of causing havoc. They can be delivered in various ways, such as through the air by aerosol sprays, which disperse in the

Table 1 Examples of biological warfare agents used during the past millennium (Barras and Greub, 2014).

Year	Event
14th century BC	The Hittites send rams infected with tularemia to their enemies
4th century BC	According to Herodotus, Scythian archers infect their arrows by dipping them into decomposing cadavers
600 BC	Solon uses the purgative herb hellebore during the siege of Krissa
1155	Emperor Barbarossa poisons water wells with human bodies, Tortona, Italy
1346	Mongols catapult bodies of plague victims over the city walls of Caffa, Crimean Peninsula (now Feodosia, Ukraine)
1422	Lithuanian army hurls manure made of infected victims into the town of Carolstein (Bohemia)
1495	Spanish mix wine with blood of leprosy patients to sell to their French foes, Naples (Italy)
1650	Polish fire saliva from rabid dogs towards their enemies
1675	First deal between German and French forces not to use "poison bullets"
1710	Russian army catapult plague cadavers over the Swedish troops in Reval (Estonia)
1763	British officers distribute blankets from smallpox hospital to Native Americans
1797	The Napoleonic armies flood the plains around Mantua (Italy), to enhance the spread of malaria among the enemy
1863	Confederates sell clothing from yellow fever and smallpox patients to Union troops, United States
WWI	German and French agents use glanders and anthrax
WWII	Japan uses plague, anthrax, and other diseases; several other countries experiment with and develop biological weapons programs
1980–88	Iraq uses mustard gas, sarin, and tabun against Iran and ethnic groups inside Iraq during the Persian Gulf War
1995	Aum Shinrikyo uses sarin gas in the Tokyo subway system

air to cause illness in the respiratory tract. They can also be delivered in the form of explosives, and through food and water by contaminating bodies of water with large amounts of disease causing organisms. A biological warfare attack can be defined at a situation when a large number of people are detected with uncommon symptoms, dead animals with uneven medical findings. Protective measures should be used against bioweapons such as the high-efficiency particulate air (HEPA) filter masks used for tuberculosis exposure, which filter out most biological warfare particles delivered through the air, and providing antibiotic treatment for immunization of officials.

Characteristics of biological weapons

Major classes of living organisms that infect living hosts may depend upon the interactions between the host and the biological agent. These interactions may depend upon the individual, e.g., immune response, and nutritional and health status, and the environmental conditions to which the host is exposed, such as sanitation, water quality, etc. (Rauw, 2012). Some of the important characteristic features by which they can be classified are as follows (Table 2):

1. **Virulence:** The relative disease causing ability of a microorganism, which may differ from species to species.
2. **Infectivity:** The ability of the agent to enter, survive, and multiply inside the host, and what rate of infection it causes.
3. **Incubation period:** The time span between first exposure of infective agent and the first appearance of symptoms in the host body.
4. **Lethality:** The death causing ability of an organism. It varies from organism to organism and the virulence factor present in them.
5. **Mode of transmission:** How an organism can be transferred in the environment: by vector or without vector.

Table 2 Various type of warfare agents and their definition.

Characteristics	Chemical warfare agents	Biological warfare agents	Nuclear warfare agents
Definition	Use of chemicals made by humans for destruction of life	Use of living biological agents to kill living organisms	Use of nuclear weapons for destruction of life
History	Not used in ancient time period	Has been used for centuries	First used near the end of World War II
Target	Only those exposed to them	Can also affect nontargeted individuals	Can also affect nontargeted individuals
Cost	More expensive	Less expensive	More expensive
Rate of action	Almost immediately	Takes time	Slow to fast
Victims	Fewer individuals are affected with the same dose of biological weapon	More individuals are affected with the same dose of biological weapon	Large number of victims
Reason for death	Poisoning	Disease	Genetic defects and mutations
Production and storage area	Much bigger space	Less space	Very large space
Uses	Used for targeted death	Used for mass extinction	Used for mass extinction and causing defects in future generations

Advantages and disadvantages of biological agents

Use of biological warfare agents has both positive and negative aspects, as stated below in Table 3. As an advantage, the agents should be easier and simple to produce, so that minimum time is required for production. Cost effectiveness is another important aspect that should be taken into account when developing a warfare agent. The agent should be a potent killer, able to cause mass destruction with very high transmission rate, and have no available treatment (Thavaselvam and Vijayaraghavan, 2010). The disadvantages of these agents include the difficulty in delivery and problems in protecting the workers from the agents.

Characteristics of an ideal biological warfare agent

In order to consider a biological warfare agent as an ideal one, it should possess the following important features (Fig. 1). It should be cost effective, should harm the

Table 3 Advantages and disadvantages of biological warfare agents.

	Advantages of biological warfare agents	Disadvantages of biological warfare agents
1.	May be easier and faster to produce	Difficulty of protecting workers at every stage until final delivery
2.	Cost effective	Difficulty in delivery
3.	High motility and morbidity	Difficulty in maintaining quality control
4.	High person to person transmission	Difficult to control once released
5.	Difficult to treat	Difficult to store

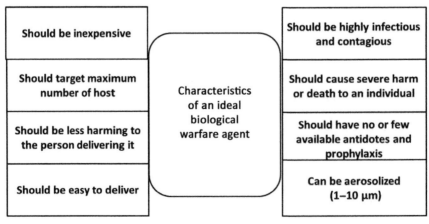

FIG. 1

Characteristics of an ideal biological warfare agent.

maximum individuals, and should target the maximum number of hosts. Its delivery should be easy and should be aerosolized.

Classification of biological warfare agents according to the Center for Disease Control

According to the Centers for Disease Control and Prevention, Atlanta, United States (CDC) the Biological agents can be classified into A, B, and C, on the basis of priority from highest to lowest based on their virulence rate and rate of transmission, as shown in Fig. 2 (Sofaer et al., 1999; Perry, 2006).

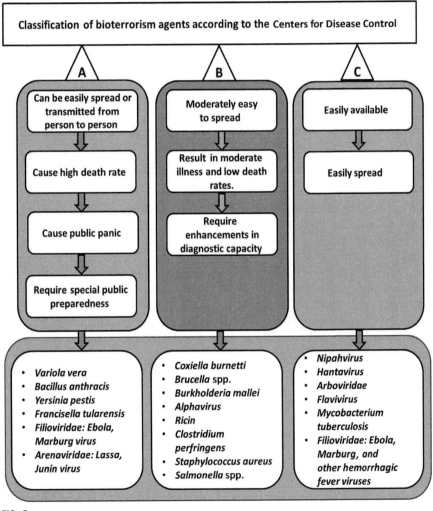

FIG. 2

Classification of biological agents according to CDC.

Present-day relevance of the agents

Since the 1980s, bioterrorism has been a very common problem. Several countries have been offensive in using these biological weapons. In 1985, Iraq used anthrax, botulinum toxin, and aflatoxin in the Persian Gulf War (Zilinskas, 1997). Another incidence of biowar seen in the recent past was the incident in 1984 wherein followers of the Bhagwan Shree Rajneesh contaminated restaurant salad bars in Oregon with *Salmonella*, which infected around 750 people. Another incidence was in 1993, where a Japanese sect of the Aum Shinrikyo cult attempted an aerosolized release of anthrax from building tops in Tokyo (Takahashi et al., 2004). In 1986, Tamil guerrillas operating in Sri Lanka poisoned tea with potassium cyanide in an effort to cripple the Sri Lankan tea export industry. In 1993, Iran allegedly deliberately contaminated the water supplies of the forces from United States and Europe to contaminate the water of Galilee with an unspecified biological agent (Sofaer et al., 1999). Two members of the Minnesota Militia group, in 1995, were found to possess ricin for revenge against local government officials (Johnson, 2012). In 1996, an Ohio man attempted to obtain bubonic plague cultures through the mail (Mishra and Trikamji, 2014). Later, in 2001, anthrax was delivered by mail to US media and government offices resulting in five deaths. In 2002, terrorists were arrested in Manchester, England, for the production of ricin with the plan to attack the Russian embassy with the toxin (Mishra and Trikamji, 2014).

Detection

Biological agents that can be used during war and for terrorism should be detected accurately and reliably so that damage and further spread of these agents can be controlled. The challenge in the development of detection systems is to make instruments or methods sensitive enough to detect the concentration at which microorganism can cause disease in humans, and also confirm the presence of these agents in varied matrices. Other than this, detection system should be portable, user friendly, and able to be utilized for the detection of multiple agents. The various types of sample, such as human clinical samples like blood, sputum, urine, stool, cerebrospinal fluid, powdery material, food, air, and water samples need to be examined. There are various detection protocols, for instance, biochemical test based assays, antibody based assays, and nucleic acid based assay, with limited range of scale that can be successfully use at the time of an emergency situation (Thavaselvam and Vijayaraghavan, 2010). BWAs are difficult to detect in real time because they can produce disease at very low concentrations and can be embedded in the naturally occurring environment. These agents can be detected by extensive analysis and when they cause disease in a host. The conventional diagnostic test (double layer agar and polymerase chain reaction) and analytical laboratory methods, such as high performance liquid chromatography or enzyme linked immunosorbent assay, have some drawbacks. Therefore, the development of methods and systems to detect BWAs in real time is urgently required.

Currently, research has been prompted to develop miniaturized systems that incorporate multiple functions into a single chip, known as "lab on chip," which represents the alternative method used for detection of BWAs in real time. Various sensor based systems are used for detection of BWAs. These are acoustic wave sensor, immunosensors, and biosensors with microfluidics (Weingart et al., 2012; Dudak and Boyaci, 2014; Matatagui et al., 2013). A Love-wave immunosensor combined with a micro fluidic chip can detect BWA samples in a dynamic mode and has characteristics like fast response time and competence to differentiate different BWAs (Matatagui et al., 2014). Surface plasmon resonance technique has been explored for the on-site detection of biological warfare. The technique is highly sensitive and capable of real time detection (Trzaskowski and Ciach, 2017). The multianalyte biosensor, Bead ARray Counter, uses magnetic microbeads, DNA hybridization, and giant magneto resistive sensors to detect and identify biological warfare agents (Edelstein et al., 2000). An electrochemical detection platform for parallel and sensitive on-site identification of multiple biothreat agents has been developed (Pöhlmann et al., 2017).

Nowadays, nanomaterials are employed in almost every research field to obtain better results or to overcome the disadvantages associated with conventional techniques/methods. Nanomaterials are used in biological recognition mechanisms and detection methods like colorimetric, photoluminescence, electrochemical, and plasmon resonance (Syed, 2014; Liu et al., 2012; Saha et al., 2012). Nanomaterial bioconjugates are used as probes for detecting bioagents; quantum dots-antibody conjugates have been developed for detecting protein toxin and viruses, for example, ricin, shiga toxin, etc. (Boeneman Gemmill et al., 2013; Goldman et al., 2004); magnetic nanoparticles have been tested for detection of bioagents that use surface plasmon resonance; and nuclear magnetic resonance (Kamikawa et al., 2012; Koh et al., 2008) for signal transduction, carbon nanotubes, silica nanoparticle, gold nanomaterials, nanowires, and nanobarcodes are different forms of nanomaterials that have been studied for the detection of bioagents.

Conclusion

BWAs are microbes that are too small to be visible to the human eye. This property of BWAs facilitates their use in war, and by terrorists to cause sickness, fear, and death in a large number of victims. The development of BWAs as weapons should be prevented and if such a situation occurs, detection and medical facilities should be applied quickly. Traditional techniques possess excellent sensitivity and specificity for the detection of BWAs, but have slow response times, require sample preparation and laboratory condition for implementation, and are expensive. Laboratory conditions and instrument access may not be feasible in affected areas. Nanomaterials have been used in the development of nanosensors to reduce the assay time of immunogenic assays and can increase the detection limit of the BWAs. Therefore, portable, highly sensitive, specific, and reliable methods need to be developed for the detection of BWAs at the point of release.

References

Barras, V., Greub, G., 2014. History of biological warfare and bioterrorism. Clin. Microbiol. Infect. 20 (6), 497–502.

Boeneman Gemmill, K., Deschamps, J.R., Delehanty, J.B., Susumu, K., Stewart, M.H., Glaven, R.H., et al., 2013. Optimizing protein coordination to quantum dots with designer peptidyl linkers. Bioconjug. Chem. 24 (2), 269–281.

Dudak, F.C., Boyaci, İ.H., 2014. Peptide-based surface plasmon resonance biosensor for detection of staphylococcal enterotoxin b. Food Anal. Methods 7 (2), 506–511.

Edelstein, R., Tamanaha, C., Sheehan, P., Miller, M., Baselt, D., Whitman, L., et al., 2000. The BARC biosensor applied to the detection of biological warfare agents. Biosens. Bioelectron. 14 (10), 805–813.

Eitzen, E.M., Takafuji, E.T., 1997. Historical overview of biological warfare. In: Sidell, F.R., Takafuji, E.T., Franz, D.R. (Eds.), Medical Aspects of Chemical and Biological Warfare. Office of the Surgeon General, Department of the Army, Walter Reed Army Medical Center, 415–423.

Frischknecht, F., 2003. The history of biological warfare. EMBO Rep. 4 (6S), S47–S52.

Goldman, E.R., Clapp, A.R., Anderson, G.P., Uyeda, H.T., Mauro, J.M., Medintz, I.L., et al., 2004. Multiplexed toxin analysis using four colors of quantum dot fluororeagents. Anal. Chem. 76 (3), 684–688.

Hunsicker, A., 2006. Understanding International Counter Terrorism: A Professional's Guide to the Operational Art. Universal-Publishers.

Johnson, D., 2012. Right Wing Resurgence: How a Domestic Terrorist Threat Is Being Ignored. Rowman & Littlefield.

Kamikawa, T.L., Mikolajczyk, M.G., Kennedy, M., Zhong, L., Zhang, P., Setterington, E.B., et al., 2012. Pandemic influenza detection by electrically active magnetic nanoparticles and surface Plasmon resonance. IEEE Trans. Nanotechnol. 11 (1), 88–96.

Koh, I., Hong, R., Weissleder, R., Josephson, L., 2008. Sensitive NMR sensors detect antibodies to influenza. Angew. Chem. Int. Ed. 47 (22), 4119–4121.

Liu, Y., Dong, X., Chen, P., 2012. Biological and chemical sensors based on graphene materials. Chem. Soc. Rev. 41 (6), 2283–2307.

Matatagui, D., Moynet, D., Fernández, M., Fontecha, J., Esquivel, J., Gràcia, I., et al., 2013. Detection of bacteriophages in dynamic mode using a Love-wave immunosensor with microfluidics technology. Sensors Actuators B Chem. 185, 218–224.

Matatagui, D., Fontecha, J.L., Fernández, M.J., Gràcia, I., Cané, C., Santos, J.P., et al., 2014. Love-wave sensors combined with microfluidics for fast detection of biological warfare agents. Sensors 14 (7), 12658–12669.

McFadden, G., 2010. Killing a killer: what next for smallpox? PLoS Pathog. 6 (1), e1000727.

Mishra, S., Trikamji, B., 2014. Historical and preventive aspect of biological warfare. Int. J. Health Syst. Disaster Manage. 2 (4), 204.

Perry, W.B., 2006. Biological weapons: an introduction for surgeons. Surg. Clin. North Am. 86 (3), 649–663.

Pöhlmann, C., Bellanger, L., Drevinek, M., Elßner, T., 2017. Multiplex detection of biothreat agents using an automated electrochemical ELISA platform. Procedia Technol. 27, 104–105.

Rauw, W.M., 2012. Immune response from a resource allocation perspective. Front. Genet. 3, 1–14.

Riedel, S., 2004. Biological warfare and bioterrorism: a historical review. Proc. (Baylor Univ. Med. Cent.) 17 (4), 400.

Robertson, A.G., Robertson, L.J., 1995. From asps to allegations: biological warfare in history. Mil. Med. 160 (8), 369–373.

Saha, K., Agasti, S.S., Kim, C., Li, X., Rotello, V.M., 2012. Gold nanoparticles in chemical and biological sensing. Chem. Rev. 112 (5), 2739–2779.

Sofaer, A.D., Wilson, G.D., Dell, S.D., 1999. The New Terror: Facing the Threat of Biological and Chemical Weapons. Hoover Institution, Stanford, CA.

Syed, M.A., 2014. Advances in nanodiagnostic techniques for microbial agents. Biosens. Bioelectron. 51, 391–400.

Takahashi, H., Keim, P., Kaufmann, A.F., Keys, C., Smith, K.L., Taniguchi, K., et al., 2004. Bacillus anthracis bioterrorism incident, Kameido, Tokyo, 1993. Emerg. Infect. Dis. 10 (1), 117.

Thavaselvam, D., Vijayaraghavan, R., 2010. Biological warfare agents. J. Pharm. Bioallied Sci. 2 (3), 179.

Trzaskowski, M., Ciach, T., 2017. SPR system for on-site detection of biological warfare. Curr. Anal. Chem. 13 (2), 144–149.

Weingart, O.G., Gao, H., Crevoisier, F., Heitger, F., Avondet, M.-A., Sigrist, H., 2012. A bioanalytical platform for simultaneous detection and quantification of biological toxins. Sensors 12 (2), 2324–2339.

Wheelis, M., 2002. Biological warfare at the 1346 siege of Caffa. Emerg. Infect. Dis. 8 (9), 971.

Zilinskas, R.A., 1997. Iraq's biological weapons: the past as future? JAMA 278 (5), 418–424.

Bacterial biological warfare agents

2

Kshirod Sathua, S.J.S. Flora

National Institute of Pharmaceutical Education and Research-Raebareli, Lucknow, India

Introduction

There has been an intimate connection between mankind and microbial diversity since the primitive era. This association in nature has been more symbiotic than pathogenic for humans. However, it is also true that with changing times, the pathogenicity of the microbial population is considered to have become more virulent and resistant, especially in view of interfering with nature and the catastrophic activity of humans. Hence, in recent times, the relationships between mankind and microbial diversity have become more pathogenic, especially with the emergence of new pathogens as causative agents. At the same time, the most terrible version of this relationship is exploiting microbial populations by humans for self-destruction, also termed "bioterrorism." Thus, unfortunately, advancement in the field of bacteriology from the 19th century onwards has given both constructive and devastating directions to the microbiological era (Zilinskas, 2017). Since that time, many scientists from different nations and military research organizations have employed various bacteria as bioweapons. The situation became more concerning and devastating when different militant organizations adapted some infectious bacteria as bioweapons for terrorizing the human population. In this way, bacterial biological warfare agents came into force and have become a great threat for mankind and civilizations.

On the other hand, throughout history, human beings have had an intimate relationship with animals in various aspects like food, clothing, etc., and that is the prime reason for transmission of zoonotic diseases from livestock to humans. Bioterrorists have taken advantage of this relationship and have targeted various devastating zoonotic diseases like anthrax and plague (Arun Kumar et al., 2011).

The practice of using bacteria as warfare agents is not new; it started before the advent of the microbial era. The first ever evidence of a biological attack is believed to date back to the 14th century BC, when the Hittites sent tularemia-infected rams to their enemies in order to weaken them. However, the "Black Death" (bubonic plague) in the mid-14th century is considered one of most dangerous health disasters of all time. Rat fleas living on the wild black rat are thought to have been the prime carrier

Handbook on Biological Warfare Preparedness. https://doi.org/10.1016/B978-0-12-812026-2.00002-5

Table 1 Bacterial biological warfare before the microbiology era.

Year	Incident	Outcome
14th century BC	Hittites were sending *Francisella tularensis* infected rams	Causes tularemia via infected rams
4th century BC	Infected arrows used by Scythians	Death of many people infected with *Clostridium perfringens*, *Clostridium tetani*, and snake venom
1343	Black Death in Eurasia	75–200 million people died

of the plague, responsible for transmission of *Yersinia pestis*. It is estimated that one-quarter to one-third of the European population died during this time. Although there is little documentary evidence of plague's later use in bacterial warfare, victims have testified to its use by Japan as a weapon in World War II (Barras and Greub, 2014). A brief list of bacterial biological warfare before the microbiology era is given in Table 1.

The use of infectious bacteria as predominant warfare agents started in the 19th century when advancements in the field of bacteriology were made. It became a serious concern during World Wars I and II, when many developed nations like Germany, Japan, France, and the United States vigorously started using *Bacillus anthracis*, *Vibrio cholerae*, *Y. pestis*, *Shigella* species, *Burkholderia mallei*, etc., tularemia, anthrax, brucellosis, and botulism toxins in weaponized form during that period. The "Anthrax letter" was another horrible example of bacteriological attack, in September 2001 (Zacchia and Schmitt, 2018). The lists of bacterial biological warfare in modern microbiology era are briefly given in Table 2.

Since antiquity, Middle Ages to modern and contemporary periods show the main indicator of bacterial biological threats for socioeconomic concern and also show the needfulness of clinical microbiologists for detection and identification as well as solution tools for bacterial biowarfare agents.

The ubiquitous existence of bacteria in nature makes them unique and a prime target for bioterrorists as their bioweapon. The unique way of their spreadable nature like aerosolized droplets gives best opportunity to some infectious bacteria to enter in unintended population. Cost-effectiveness, easy application, difficulty in detection, and lack of concern and advancement in the field of bacteriology make bacteria unique among biological warfare agents (D'Arcangelis, 2016).

The various bacterial warfare agents include *Bacillus anthracis*, *Y. pestis*, *Brucella suis*, *Coxiella burnetii*, *Francisella tularensis*, *Burkholderia pseudomallei*, *Burkholderia mallei*, *Salmonella typhimurium*, *Chlamydophila psittaci*, *Rickettsiaceae*, *Shigella* spp., *Vibrio cholera*, etc. Based upon disseminated ability, mortality rate, actions required for public health preparedness, and capability of causing public panic, the US Centers for Diseases Control and Prevention (CDC) and National Institute of Allergy and Infectious Diseases (NIAID) classify biological warfare agents into three categories (Cenciarelli et al., 2014):

Table 2 Bacterial biological warfare in the modern microbiology era.

Year	Incident	Outcome
1914–18	*Bacillus anthracis* and *Burkholderia mallei* infected animal feed	Infecting enemies with anthrax and glanders
1932	Japanese dropped plague-infected fleas, and infected food and clothing by aircraft into areas of nonoccupied China	Many people died, including several Japanese soldiers
1940	Japanese Air Force applied ceramic bombs in Ningbo, which contained fleas carrying bubonic plague	Roughly 400,000 people may have died
1942	US War Research Service engaged with testing of anthrax spores, brucellosis, and botulism toxins	Lack of plentiful evidence available of biological warfare attack
1947–91	Cold war era biological agents	
1979	Anthrax outbreak in Sverdlovsk town	
1984	Contaminating food by use of *Salmonella typhimurium*	751 Cases been detected and 45 hospitalized
1990–91	Iraq started research on *Bacillus anthracis* under leadership of Saddam Hussein	
2001	Anthrax letter (letters containing *Bacillus anthracis*) sent to government officials in New York	22 People infected and five people died of anthrax

Category A: Agents that are easily transmitted among the population, result in high mortality, and are considered a major public health concern, for which special action is required.

Category B: Agents that are moderately easily transmitted, result in moderate morbidity and low mortality rates, and require specific disease surveillance.

Category C: Agents that are easy to produce and transmitted because of their availability. They are also linked to major health impacts.

A list of various bacterial biological warfare agents, their category, routes of infection, and detection techniques are given in Table 3. Most of them are transmitted from animals to humans and they are responsible for the occurrence of various devastating zoonotic diseases like anthrax, plague, brucellosis, Q fever, tularaemia, melioidosis, glanders, etc. (Pal et al., 2017).

Black Death victims and anthrax outbreaks taught us that there is a need for immediate detection tools for identification of bacterial biothreat agents. Ideally, biological detecting systems should be able to detect very low concentrations of microbes/toxins, and be capable of detecting in various matrices and multiple threat agents simultaneously. Some advanced techniques for detection of bacterial biological include: (1) biochemical assay; (2) bioluminescence-based detection; (3) antigen- and antibody-based detection techniques; (4) nucleic acid-based identification;

Table 3 Lists of bacterial biological warfare agents, categories, routes of infection, and detection techniques.

Disease	Causative agent	Categories	Target route of infection as warfare agents	Incubation period (days)	Diagnosis techniques
Anthrax	Bacillus anthracis	A	Aerosol	1–5	Culture, serology, ELISA, PCR
Plague	Yersinia pestis	A	Aerosol	2–3	Culture, serology, ELISA, PCR
Tularemia	Francisella tularensis	A	Skin when handling infected animal tissue	3–5	Culture, serology, ELISA, PCR
Brucellosis	Brucella suis, Brucella melitensis, Brucella abortus	B	Aerosol	5–60	Culture, serology, ELISA, PCR
Glanders	Burkholderia mallei	B	Aerosol	3–7	Culture, serology, ELISA, PCR
Melioidosis	Burkholderia pseudomallei	B	Aerosol	3–7	Culture, serology, ELISA, PCR
Q fever	Coxiella burnetii	B	Dust or aerosols	2–42	Serology, ELISA
Botulinum toxin	Clostridium botulinum	A	Contaminated food	1–5	ELISA, mouse inoculation for toxin detection

and (5) sensor-based detection technique (Thavaselvam and Vijayaraghavan, 2010). The most recent reports suggest use of antibodies like monoclonal antibodies. Recombinant antibodies used with a biosensor for detection of biological warfare agents are under development (Mobed et al., 2019).

To overcome the problems of bacterial weaponization and to know the pathogenic nature of a bacterial biological agent, many biological weapon programs have been established over time. Some nations like the Union of Soviet Socialist Republic (USSR), Japan (Unit 731), the United States, and Iraq actively participated in the establishment of various biological weapon programs (Zilinskas, 2017). In April 2013, 170 countries signed a treaty—the Biological Weapons Convention (BWC)—for prevention of biological attacks (Haider, 2018). Many countries have started research into protection against biological warfare agents. According to customary international humanitarian law and various international treaties, use of biological weapons is prohibited and is considered a war crime (Henckaerts and Doswald-Beck, 2005). Although most of the bioweapon programs and treaties were made for self-benefits yet, it helped minimizing illegal and unethical practice of bacteria being used as warfare agents.

Using bacterial agents as bioweapons is considered a serious concern for society globally, having a negative impact on human lives as well as socioeconomic development. Humanity should be the biggest concern for social well-being, irrespective of anything. Indeed one must not forget the history of largest biological weapons accidents, like anthrax outbreaks and plague victims, which have taught us how bacteria played a significant role in destruction of a civilization (Dols, 2019; Jernigan et al., 2002). Although many steps have been taken to limit the use of bacterial biological agents, such protocols are still in the initial stages in terms of strengthening their implementation. It must always be remembered that biological weapons are considered not only as a warfare agent, but also as a threat for human beings' survival in the near future (Tournier et al., 2019). Hence, despite the research and production, political involvement, and self-benefiting treaties, there must be complete elimination of the practice of bacterial warfare.

Warfare agents

The term "warfare agents" denotes exposure to chemical, nuclear, and biological agents during military service, which potentially cause harm to a large number of humans or human-made structures like buildings, natural structures like mountains, or the biosphere. The main purpose behind the use of warfare agents is to terrorize enemies. Warfare agents are broadly categorized in four types: chemical, biological, radiological, and nuclear, abbreviated as CBRN (Fountain, 2018). The threat of use of these agents by militants, either individually or in combination formats, such as biological and chemical warfare, may make it a difficult challenge to counter.

Biological warfare agents as a preferred weapon of destruction

Biological warfare agents can be easy to use, cost-effective, and, most importantly, difficult to detect. The cost of biological warfare agents is much lower, even less than 0.05% than that of conventional warfare agents (Beeching et al., 2002; Danzig and Berkowsky, 1997). Moreover, very simple technology is needed for its production. In recent times, biological warfare agents have been expected to become the first choice for bioterrorists, as they can easily escape before detection. Bioterrorists are therefore now focusing on biological warfare.

Biological warfare agents

The use of infectious biological threat agents or toxins from a biological source of origin as an act of war, aiming to debilitate living beings, leads to severe disruption to economic and societal infrastructure. The type of biological warfare is generally defined from its source, such as bacterial, viral, fungal, or even insect. Nowadays, biological warfare agents are considered as one of the most emerging and terrible weapons among NBC (nuclear warfare, biological warfare, and chemical warfare) as they are easy to use, cost-effective, and difficult to detect, but can target an entire population (Thavaselvam and Vijayaraghavan, 2010).

Bacterial biological warfare agents are unique

Bacteria are ubiquitous in the environment and are thus a unique and prime source for biological warfare agents. Possible transmission from aerosolized droplets, as in the case of *Y. pestis*, makes bacterial agents spreadable even in unintended populations. Despite the fact that modern microbiologists have been focused regarding advancement in the field of bacteriology since the 19th century, at the same time people started using various infectious bacteria to terrorize civilization in various ways, and in this way, bacterial biological agents came into force as predominant warfare agents. Hence cost-effectiveness, easy use, difficulty to detect, and lack of concern and advancement in the field of bacteriology have made bacteria unique among biological warfare agents (Arun Kumar et al., 2011; Hassani et al., 2004).

During the 19th century, when modern microbiology saw great advancements, ironically it was the same time when the use of infectious bacteria to terrorize populations was first observed and predominantly established as a warfare agent. Hence, the lack of concern at that time, along with simultaneous advancements in the field of bacteriology, made bacteria a preferred and unique choice among biological warfare agents.

Historical aspects

Although use of bacterial biological warfare agents has been believed to have started in the 14th century BC, its practice began much earlier. Hittites were sending tularemia-infected rams to their enemies to weaken them in the 14th century BC, which was assumed to be the first evidence of bacterial bioweaponization (Hirschmann, 2018; Trevisanato, 2007). Later on, in the 4th century BC, Scythian archers were reported to have used infected arrows, which were thought to be composed of dangerous bacterial disease causing agents like *Clostridium perfringens* and *Clostridium tetani*, and snake venom.

In the mid-14th century, the Black Death (bubonic plague) was considered to be the most devastating pandemic health disaster in human history (Dols, 2019). The Black Death was assumed to originate in 1343 in Central Asia during traveling to reach Crimea. Wild black rat and oriental rat fleas were thought to be the prime carrier of plague by transmission of *Y. pestis*. Eurasia was affected by this plague, and it has been estimated that 75–200 million people died—roughly one-quarter to one-third of the European population at that time (Barras and Greub, 2014; Wheelis, 2002). Although there is little evidence to go on, eastern countries like China and India are also believed to have been severely affected by plague during that period. According to the view of Gabriele de' Mussi, the Black Death was assumed to reach from Crimea to Europe (Wheelis, 2002). Although there was a controversy in the statement of Mussi, it was believed to be the most successful and terrible biological attack ever (Wheelis, 1999). This victims provided a lead that bacterial warfare agents for terrorizing people. Later Japan used plague as a weapon in World War II (Harris, 2002), and the Soviet stockpiles of *Y. pestis* prepared for use in an all-out war further proving that plague remains a genuine problem for modern arms control even six and a half centuries later (Inglesby et al., 2000).

Due to advancements in the field of bacteriology in the 19th century, people started using various bacteria to terrorize civilization in various ways, and bacterial biological agents thus came into force as predominant warfare agents. During World War I, nations like Germany and France were using *Bacillus anthracis* and *Burkholderia mallei* in infected animal feed to infect their enemies (Geissler and van Courtland Moon, 1999; Robertson and Robertson, 1995). Later the use of biological weapons was strictly prohibited by the Geneva Protocol of 1925, yet anthrax and glanders remained a threat among population (Baxter and Buergenthal, 1970). Countries like France, the United Kingdom, Italy, Canada, Belgium, Poland, and the Soviet Union adopted the 1925 treaties and instead of misusing these agents, they started research and production on biological weapons; however, nations like the United States did not follow the Geneva Protocol until 1975 (Barras and Greub, 2014).

For advancement of bioweapon research, with the interest of the Japanese government, the Army Epidemic Prevention Research Laboratory (Unit 731) was set up in 1932. Japanese scientists infected prisoners with organisms like *Bacillus anthracis*, *V. cholerae*, *Y. pestis*, *Shigella* species, and *Francisella tularensis*, and without medical treatment, studied various effects of these diseases (Barenblatt, 2004).

The research of Unit 731 led the Japanese army to conduct large-scale trials of biological weapons, such as the development of bioweapon bombs for spreading pathogens, the infection of reservoirs and wells with deadly pathogens (composed of *Bacillus anthracis*, *V. cholerae*, *Y. pestis*, *Shigella* species, and *Salmonella* species), and the dropping of plague-infected fleas, and infected food and clothing, by aircraft into areas of China that were not occupied by Japanese soldiers. It is believed that many people died as a result of these attacks, including several Japanese soldiers (Harris, 2002).

The practice of using bacterial warfare agents was more generalized during the onset of World War II. Tularemia, anthrax, brucellosis, and botulism toxins are some examples of terrible bacteria-borne diseases that were weaponized during that period. Bacterial warfare became even more deadly when, in 1940, the Japanese Air Force applied ceramic bombs containing fleas carrying bubonic plague in Ningbo (Barenblatt, 2004). Many of these operations failed due to inefficient delivery systems (Williams and Wallace, 1989); however, roughly 400,000 people may have died. During the same time, the United States also feared possible biological warfare agent development by the Germans, and hence pushed its allies to perform bioweapon research (Barras and Greub, 2014; Carus, 2001). In 1942, the US government set up the US War Research Service, which has been engaged with the development of facilities for scientific research against mass production and testing of many dangerous bacteria-borne disease like anthrax spores, brucellosis, botulism toxins, etc. (Covert, 2000; Guillemin, 2006). Although there is little evidence available of biological warfare attacks, it is believed that many nations used biological agents during the Cold War era in the years 1947–91. After the ineffectiveness of the 1925 Geneva agreements and under the intervention of the WHO, in 1972, more than 100 nations, including the United States, United Kingdom, and Soviet governments, signed another treaty—the Convention on the Prohibition of the Development, Production, and Stockpiling of Bacteriological and Toxin Weapons, and on their Destruction—known as BWC (Wheelis and Rózsa, 2009).

In April 1979, there was an accidental outbreak of anthrax in the Soviet town of Sverdlovsk. According to the view of Western specialists, some potentially deadly spores were released into the environment during the transferring of anthrax bacteria into containers by Soviet workers. Although the incident was announced as unintentional by Soviet officials, later on, in 1992, the matter become more controversial when President Boris Yelstin admitted officially that "our military developments were the cause" (Meselson et al., 1994; Noah et al., 2002).

In 1984, in Dalles, Oregon, *Salmonella typhimurium* was used to contaminate food, as a result of which 751 cases were detected, of which 45 were hospitalized. After World War II, these have been considered as instances of bioterrorism (Carus, 2001).

Undoubtedly the existence of BWC has been influential in the minimization of development and production of bacteriological and toxin weapons; however, during the Gulf War, some nations, like Iraq under the leadership of Saddam Hussein, started research on the deadly bacteria *Bacillus anthracis*.

The "anthrax letter" was another horrible example of bacteriological attack, in September 2001. Several letters containing *Bacillus anthracis* were sent to government officials or journalists in New York, resulting in 22 people becoming infected and five people dying of anthrax.

The above historical evidence of BW during antiquity, Middle Ages, modern and contemporary periods shows the main indicator of bacterial biological threats. It also showed the necessity for clinical microbiologists to detect and prevent bacterial biowarfare agents. So far, 182 countries have ratified the treaty made by the Soviet Union's major biological warfare agency, "Biopreparat" (Handelman and Alibek, 2014). Apart from 9 nations that are still suspected, the remaining nations are not believed to be involved in offensive warfare programs.

Bacterial biological warfare agents
Lists of bacterial biological warfare agents

Various lethal bacteria such as *Bacillus anthracis*, *Y. pestis*, *Brucella suis*, *Coxiella burnetii*, *Francisella tularensis*, *Burkholderia pseudomallei*, *Burkholderia mallei*, *Salmonella typhimurium*, *Chlamydophila psittaci*, *Rickettsiaceae*, *Shigella* spp., and *Vibrio cholera* are commonly used biological warfare agents. A brief description of some of the important bacterial biological warfare agents is given as follows.

Bacillus anthracis

In the fashion world there has been an increasing demand for animal products like wool, leathers, hides, hair products, etc. This is one of the prime targets of bioterrorists for transmission of deadly transmissible disease like anthrax. The endospore-forming, nonmotile, anaerobic gram-positive bacteria, *Bacillus anthracis*, is the main causative factor for this zoonotic disease. It is transmitted through inhalation or consumption of contaminated animal products, such as cutaneous anthrax through handling of contaminated wool, or using infected animal products such as leather, goat hair products, etc. Contaminated wool or leather is the reason for cutaneous anthrax. The bacterial spore enters through abraded skin during handling of the abovementioned contaminated products and causes infection by cutaneous anthrax (Sridhar and Chandrashekhar, 1991). Consumption of preinfected contaminated meat is the reason for gastrointestinal anthrax (Singh et al., 2011). Initially there is acute inflammation in the intestinal tract characterized by nausea, vomiting, loss of appetite, and fever, which may be followed by severe diarrhea and vomiting blood.

Case study 1: The US Postal Service sent a letter containing *Bacillus anthracis* spores in 2001, leading to 12 people becoming affected by cutaneous anthrax and 11 people being affected by inhalation anthrax (Calfee et al., 2011).

Yersinia pestis

One of the common hazardous bacteria affecting human beings, *Y. pestis*, is transmitted via flea bites and causes plague—it even sometimes causes fatal bubonic plague

(Achtman et al., 1999; Pechous et al., 2016). The clinical symptoms of plague include fever, chills, weakness, headache, tender lymph nodes, etc. Other types of plague include pneumonic plague, infected via respiratory droplets from infected individuals or animals. Pneumonic plague is the most likely target for bioterrorists in the form of aerosolized *Y. pestis*, which may cause sudden severe pneumonia and sepsis (Pechous et al., 2016).

Case study 1: In the 14th century, one-third of the population of Western Europe died of plague. At that time it was named the "Black Death." Plague has killed and terrorized many people across many countries over the course of history.

Brucella suis

One of the neglected zoonotic bacterial diseases is caused by *Brucella suis*. It is generally found in livestock like cattle, buffalo, goats, horses, and pigs, particularly in areas where multiple livestock farming systems are present (Smith and Bhimji, 2019; Smits and Kadri, 2005). It is a facultative intracellular, gram-negative pathogen, transmitted to the human population via dairy products. General symptoms of brucellosis include fever, body pain, especially muscle and joint pain, splenomegaly, etc. Being a neglected disease nowadays, it is an emerging target for the bioterrorist (Franc et al., 2018). Even an aerosol form of brucellosis is presently considered to be a dangerous biothreat as it causes respiratory brucellosis (Araj, 2010).

Coxiella burnetii

Coxiella burnetii is a gram-negative, aerobic pathogen, which causes Q fever. Common modes of transmission to human beings are via inhalation or tick bite, and major symptoms include fever, pneumonia, hepatitis, etc. The spores of *Coxiella burnetii* are resistant to adverse environmental stimuli such as heat, pressure, and even certain antiseptics. Due to the resistant nature in response to adverse environmental stimuli, and spreading nature in aerosolized form, it is considered a most desirable biowarfare agent for bioterrorists. Both United States and Soviet biological arsenals developed *Coxiella burnetii* as biological warfare agents (Gronvall, 2017; Rotz et al., 2002).

Francisella tularensis

Francisella tularensis is a gram-negative coccobacillus that causes tularemia. It is also called rabbit fever, as it acts as a primary pathogen for rabbits and squirrels. Humans become infected when bitten by an infected tick, by handling infected rabbits, or by eating undercooked meat. Common symptoms of tularemia include high fever, septicemia, toxemia, and some intestinal infection even including typhoid. It was considered among the top six agents having great adverse public health impacts (Rega et al., 2017).

Case studies: In the past, *Francisella tularensis* was an important subject of military research in the Soviet Union, Japan, and the United States, and was considered among the top six bioterrorism agents at that time (Johansson et al., 2004).

Burkholderia pseudomallei

Burkholderia pseudomallei is a saprophytic rod-shaped bacteria that causes emerging endemic diseases like melioidosis. This disease is commonly seen in Southeast Asia and Northern Australia (Cheng and Currie, 2005). It has been considered as a biological warfare agent in the past. The US Centers for Disease Control and Prevention classified it as a category B biothreat agent (Williamson et al., 2018).

Burkholderia mallei

Burkholderia mallei is an obligate pathogen that generally infects animals like horses, mules, and donkeys. It causes disease like glanders (Kettle and Wernery, 2016). Although humans are rarely affected by *Burkholderia mallei* via handling of infected animals, it is still prominently seen in areas like Northern Africa, Asia, and the Mediterranean (Neubauer et al., 2005). The US Centers for Disease Control and Prevention also classified it as a category B biothreat agent.

Case study 1: *Burkholderia mallei* was used to infect Russian horses and mules on the Eastern Front during World Wars I and II.

Case study 2: The Japanese used *Burkholderia mallei* to infect animals and humans at Pinfang Institute in China during World War I (Franz et al., 1997a; Rath, 2002).

Reason behind rising bacterial bioterrorism

Although bacterial bioterrorism has been used since antiquity and the Middle Ages, bioterrorism activities increased from the 19th century onwards. Beside cost-effectiveness and easy applicability, an important and unique feature for the rise of bacterial bioterrorism is that bacteria-borne diseases like anthrax can be controlled and created easily. On the other hand, airborne viral infectious diseases like smallpox have no such confined boundary of spreadability, and may affect a user's home country. Further, bacterial bioterrorism can be done for as little as $2500 using readily available laboratory equipment. It can also be suitably modified so that it will be effective only in a narrow environmental range; hence, only the target may be affected adversely (D'Arcangelis, 2016).

Besides the infectivity and virulence abilities of weaponized bacteria, another key factor is its sustainability. For example, agents like *Bacillus anthracis* initially form hard spores that are well dispersed in aerosol form, hence primary infection is common; however, secondary infection is rare, due to the lower chance of transmission from one person to another. Infection of pulmonary anthrax, on the other hand, primarily starts with influenza-like symptoms, progressing to dangerous hemorrhagic mediastinitis, and finally becomes fatal within 3–7 days.

Furthermore, due to available effective and efficient delivery systems and sustained high infectivity and virulence capacity even after a prolonged period of storage, biological warfare agents of bacterial origin are the most popular choice and the prime reason why there has been an exponential rise in bacterial bioterrorism (Arun Kumar et al., 2011; Hassani et al., 2004).

Identifying the signs of biological attack

Although it is difficult to identify the signs of biological attack, common epidemiological evidence or clues may be given, based upon the following signs (Dorner et al., 2016; Song et al., 2005; Treadwell et al., 2003):

1. Sudden onset of disease due to uncommon agents whose epidemiological explanations are lacking.
2. Onset of rare, genetically engineered strain.
3. High morbidity and mortality rates due to similar types of symptoms.
4. Sudden and unusual presentation of a disease (e.g., anthrax and plague).
5. Unusual geographic or seasonal distribution (e.g., tularemia due to unusual geographic distribution and influenza in summer season).
6. Sudden, unexpectedly increase in incidents (e.g., tularemia and plague).
7. Transmission of disease through aerosols, food, and water sources.
8. No illness presented in people who were/are not exposed to "common ventilation systems (have separate closed ventilation systems) when illness is seen in persons in close proximity who have a common ventilation system."
9. Without any other cause, different diseases coexisting in the same patient.
10. Respiratory disease due to inhalation of pathogen may unusually affect the large groups within a population.
11. Atypical illness that may unusually affect particular age groups (e.g., measles outbreak in adults).
12. Unusual pattern of death or illness.
13. Similar genetic pattern of infected agents that are identified from different sources.
14. Similar types of illness noticed simultaneously.
15. Unexpected and unexplainable disease or death in large numbers.

Detection techniques for bacterial biological warfare agents

The terrible suffering for victims of the Black Death and anthrax outbreaks taught us that there is a need for immediate detection tools for identification of bacterial biothreat agents (Lim et al., 2005). Ideally, biological detecting systems should be: (1) able to detect very low concentrations of microbes and/or toxins; (2) capable of detection in various matrices; (3) portable; (4) user friendly, and (5) capable of detecting multiple threat agents simultaneously.

Although many advanced detection techniques are available, so far none of them satisfies all these criteria (Lim et al., 2005; Walt and Franz, 2000). Some advanced techniques for detection of bacterial biological agents include: (1) biochemical test-based assays; (2) bioluminescence-based detection; (3) antigen- and antibody-based detection systems; (4) nucleic acid-based detection; (5) and sensor-based detection systems.

In biochemical test-based assays, identification of various bacteria is done after culturing them on a routine basis in a microbiological laboratory. Identification of various lethal bacterial warfare agents, like *Bacillus anthracis*, *Y. pestis*, *Burkholderia* sp., and *Brucella* sp., can be achieved using this technique. Although this technique is cheap and highly reliable, the major disadvantage is that it is time consuming and needs a pure culture (Thavaselvam and Vijayaraghavan, 2010).

The enzyme-based detection of bacterial biological agents is done by using cost-effective bioluminescence-based detection techniques (Lee and Deininger, 2004). In antigen- and antibody-based detection systems, antigen- and antibody-based immunoassays are used for detection of bacterial warfare agents. It is a highly sensitive technique, used for the detection of anthrax, plague, botulism, brucellosis, glanders, and melioidosis (Andreotti et al., 2003; Coudron et al., 2019). Currently immunochromatographic (ICT) line assays, in the form of single use and disposable tests, have been used for rapid detection of bacterial biological agents (King et al., 2003).

The nucleic acid-based detection system is one of the most advanced detection systems for detection of bacterial warfare agents. Amplification of specific regions of the genome by using PCR and real-time or Q PCR-based assays is considered to be a rapid and sensitive detection system, making it a preferred method of detection (Matero, 2017; Pal et al., 2018). In recent times, several Q-PCR kits have been developed for the detection of bacterial biothreat agents like anthrax.

Sensor-based detection systems are the best detection systems of all as they involve integration of biochemical, immunologic, and nucleic acid-based techniques. This technique is unique in that it detects specific toxins produced during the growth of bacteria, product formed due to an enzymatic activity linked with microbial metabolism, and multiple analytes. Recently, antibodies like monoclonal antibodies and recombinant antibodies used with biosensors for detection of biological warfare agents have been under development (Aberl and Kosslinger, 1998; Mobed et al., 2019).

Establishment of bacterial biological weapon programs
Biological warfare program of the USSR

The biological warfare program of the USSR is considered to be the oldest and longest biological warfare program in history (Leitenberg et al., 2012). The Soviet Military Chemical Agency set up a small biological warfare laboratory in Moscow in 1925, which eventually became the Scientific Research Institute of Health. Dr. Yakov Fishman, Director of the Soviet Military Chemical Agency, was the pathfinder of this biological warfare program at that time. In 1928, a laboratory progress report in four parts was submitted by Fishman, of which the second part contains "an assessment of the potential uses of bacteria for the purposes of warfare and sabotage" (Bojtzov and Geissler, 1999). It has been clearly described by Ginsburg with the help of dangerous anthrax causing pathogen *Bacillus anthracis*, as its spores are virulent and sustain adverse environmental pressure. During that time, they also investigated the warfare potential of other bacterial agents such as *V. cholerae* and *Y. pestis* (Handelman and Alibek, 2014).

In 1945, during World War II, the USSR obtained some important information regarding the virulence of *Y. pestis* strains when they had destroyed Japanese BW facilities; this valuable information helped them to potentiate their weaponization efforts (Bojtzov and Geissler, 1999; Ouagrham-Gormley et al., 2006). In this way, the Soviets assessed the pathogenic potentials of some bacteria, including *Bacillus anthracis*, *Y. pestis*, *Burkholderia mallei*, *Coxiella burnetii*, and *Francisella tularensis*, of which most cause zoonotic disease. Biopreparat is considered to be the second generation of BW organization, created by the USSR at that time and code named "Ferment." Although later on Ferment started investigation of viral warfare agents, initially it was focused on some bacterial biological agents like *Bacillus anthracis*, *Burkholderia mallei*, *Francisella tularensis*, and *Y. pestis*.

This provides conclusive evidence that, before 1992, the USSR directly or indirectly engaged with weaponized bacterial agents like *Bacillus anthracis*, *Brucella* species, *Burkholderia mallei*, *Burkholderia pseudomallei*, *Coxiella burnetii*, *Francisella tularensis*, *Legionella pneumophila*, and *Y. pestis* by developing several biological programs (Domaradskij and Orent, 2006).

Japanese biological warfare program

Unit 731 was one of the premier biological warfare organizations of Japan, established in 1930. Unit 731 was started as a warfare program, with a special focus on plague. Eventually it was also engaged with the weapons utility of some bacterial pathogens that cause "plague, cholera, gas gangrene, anthrax, typhoid, and paratyphoid." However, the highest priority was to weaponize *Bacillus anthracis* and *Y. pestis*. Two groups of victims in particular were affected by the inhuman activities of Unit 731; one group were affected due to being subjected to infected pathogens using various routes (Tsuchiya, 2011), and the other group was affected by exposure to *Y. pestis* via fleas (Zilinskas, 2017).

United States biological warfare program

To defend against the German BW program, in 1942, the United States started its own BW program with the United Kingdom and Canada. For researching possible BW agents, the US BW program screened 39 microbials and toxins from which *Bacillus anthracis* and *Y. pestis* were selected as high-priority lethal weaponized agents, based on their virulent potency (King and Guillemin, 2019). However, *Bacillus anthracis*, being the most lethal zoonotic pathogen, was selected as the first preferable weaponized agent. Before 1969, apart from *Bacillus anthracis*, other bacterial biological agents like *Brucella* species, *Coxiella burnetii*, and *Francisella tularensis* were also considered to be preferable biological weaponized agents by the United States (Franz et al., 1997b; Hay, 1999).

Iraq's biological warfare program

Iraq's biological warfare program started during Operation Desert Storm, in 1991. During that time, Iraqi scientists found five bacterial strains with BW potency. Of these, two nonpathogenic bacterial strains, *Bacillus subtilis* and *B. thuringiensis*, were used for testing purposes and two pathogenic bacteria, *Bacillus anthracis* and *Clostridium perfringens*, were also investigated as research targets. In addition to these, Iraq successfully started to prepare bombs such as the 250-lb (LD-250) and 400-lb (R-400), which had a capacity of 60–85 L of biological solution. Of the LD-250 and R-400, the R-400 was considered to be the most suitable for delivery of a biological weapon. Even in 1991, many R-400 bombs were filled with dangerous bacterial agents like *Bacillus anthracis* and bacteria-based toxins like botulinum toxin. Although the Iraqis never used their biological weapons, they always used zoonotic bacterial pathogens like *Bacillus anthracis* for achievement of their research targets (Duelfer, 2016; Trevan, 2016).

Impact of bacterial bioweaponization for society

Practicing weaponization of bacterial biological agents is considered a serious concern all over the world. It has a negative impact on society. Advancements in the field of bacteriology and biotechnology have undoubtedly produced many benefits for society; however, use of lethal bacteria as a weaponizing agent has adverse impacts for a nation in terms of its social and economic development. Although many steps have been taken to limit the use of bacterial biological warfare for the well-being of society by establishing many bioweapons conventions, these protocols are still in the initial stages of implementation (Harigel, 2001).

Despite a number of treaties having been signed by the end of World War II, among them most treaties are bilateral or multilateral, but universal treaties are very rare. These treaties, when evaluated, have several limitations; in addition, many have been made for self-benefits only. The prime goal behind making treaties should be complete elimination of bacterial warfare practice.

One should not forget the history of largest biological weapons accident like "anthrax outbreak" and "plague victims," which has been taught us how bacteria played a terrible role in the destruction of a civilization both socially as well as economically (Morea et al., 2018).

It must be remembered that biological weapons are considered not only as warfare agents but also as a threat to the survival of human beings in the near future. Hence, humanity should be the biggest concern for social well-being, irrespective of anything.

Conclusion

Human beings are quite advanced in the field of bacteriological research, but using bacteria as warfare agents is the most terrible and dangerous sign in terms of socioeconomic degeneration. The nature of any bioterrorist event involving bacterial biological warfare agents is highly unpredictable and difficult to prevent. Hence, it is a serious concern and rapid and accurate detection systems must be developed for survival against them. Currently we are living in a modern era and as per the biological warfare point of view, we are more concerned than our predecessors. Hence, humanity is the biggest concern for social well-being, irrespective of anything. Lastly, for complete eradication of this inhuman weapon across the globe, political community, defense sectors, and international security agencies should be working closely together to make ethical decisions to safeguard the very existence of the human race.

References

Aberl, F., Kosslinger, C., 1998. Biosensor-based methods in clinical diagnosis. Methods Mol. Med. 13, 503–517.

Achtman, M., Zurth, K., Morelli, G., Torrea, G., Guiyoule, A., Carniel, E., 1999. Yersinia pestis, the cause of plague, is a recently emerged clone of Yersinia pseudotuberculosis. [Research Support, Non-U S Gov't] Proc. Natl. Acad. Sci. U. S. A. 96 (24), 14043–14048.

Andreotti, P.E., Ludwig, G.V., Peruski, A.H., Tuite, J.J., Morse, S.S., Peruski Jr., L.F., 2003. Immunoassay of infectious agents. BioTechniques 35 (4), 850–859.

Araj, G.F., 2010. Update on laboratory diagnosis of human brucellosis. [Review] Int. J. Antimicrob. Agents 36 (1), 9.

Arun Kumar, R., Nishanth, T., Ravi Teja, Y., Sathish Kumar, D., 2011. Biothreats—bacterial warfare agents. J. Bioterr. Biodef. 2 (112), 2.

Barenblatt, D., 2004. A Plague Upon Humanity: The Secret Genocide of Axis Japan's Germ Warfare Operation. Edn. HarperCollins, New York.

Barras, V., Greub, G., 2014. History of biological warfare and bioterrorism. Clin. Microbiol. Infect. 20 (6), 497–502.

Baxter, R.R., Buergenthal, T., 1970. Legal aspects of the Geneva Protocol of 1925. Am. J. Int. Law 64 (5), 853–879.

Beeching, N.J., Dance, D.A., Miller, A.R., Spencer, R.C., 2002. Biological warfare and bioterrorism. BMJ 324 (7333), 336–339.

Bojtzov, V., Geissler, E., 1999. Military biology in the USSR. 1920–1945.

Calfee, M., Choi, Y., Rogers, J., Kelly, T., Willenberg, Z., Riggs, K., 2011. Lab-scale assessment to support remediation of outdoor surfaces contaminated with Bacillus anthracis spores. J. Bioterr. Biodef. 2 (3), 25–26.

Carus, W.S., 2001. Bioterrorism and Biocrimes: The Illicit Use of Biological Agents Since 1900. National Defense University, Washington, DC.

Cenciarelli, O., Pietropaoli, S., Gabbarini, V., Carestia, M., D'Amico, F., Malizia, A., et al., 2014. Use of non-pathogenic biological agents as biological warfare simulants for the development of a stand-off detection system. J. Microb. Biochem. Technol. 6, 375–380.

Cheng, A.C., Currie, B.J., 2005. Melioidosis: epidemiology, pathophysiology, and management. Clin. Microbiol. Rev. 18 (2), 383–416.

Coudron, L., McDonnell, M.B., Munro, I., McCluskey, D.K., Johnston, I.D., Tan, C.K., et al., 2019. Fully integrated digital microfluidics platform for automated immunoassay; a versatile tool for rapid, specific detection of a wide range of pathogens. Biosens. Bioelectron. 128, 52–60.

Covert, N.M., 2000. Cutting Edge: A History of Fort Detrick, Maryland. Public Affairs Office, Headquarters US Army Garrison.

Danzig, R., Berkowsky, P.B., 1997. Why should we be concerned about biological warfare? JAMA 278 (5), 431–432.

D'Arcangelis, G., 2016. Defending White Scientific Masculinity: the FBI, the media and profiling tactics during the post-9/11 anthrax investigation. Int. Fem. J. Polit. 18 (1), 119–138.

Dols, M.W., 2019. The Black Death in the Middle East. vol. 5354. Princeton University Press.

Domaradskij, I.V., Orent, W., 2006. Achievements of the Soviet biological weapons programme and implications for the future. Rev. Sci. Tech. 25 (1), 153.

Dorner, B.G., Zeleny, R., Harju, K., Hennekinne, J.-A., Vanninen, P., Schimmel, H., et al., 2016. Biological toxins of potential bioterrorism risk: current status of detection and identification technology. TrAC Trends Anal. Chem. 85, 89–102.

Duelfer, C., 2016. WMD elimination in Iraq, 2003. Nonprolif. Rev. 23 (1–2), 163–184.

Fountain, A.W., 2018. Chemical, biological, radiological, nuclear, and explosive threats: an introduction. In: Handbook of Security Science. Springer, 1–6.

Franc, K., Krecek, R., Häsler, B., Arenas-Gamboa, A., 2018. Brucellosis remains a neglected disease in the developing world: a call for interdisciplinary action. BMC Public Health 18 (1), 125.

Franz, D.R., Jahrling, P.B., Friedlander, A.M., McClain, D.J., Hoover, D.L., Bryne, W.R., et al., 1997a. Clinical recognition and management of patients exposed to biological warfare agents. JAMA 278 (5), 399–411.

Franz, D.R., Parrott, C.D., Takafuji, E.T., 1997b. The US biological warfare and biological defense programs, first ed In: Medical Aspects of Chemical and Biological Warfare. 425–436. United States Government Printing.

Geissler, E., van Courtland Moon, J.E., 1999. Biological and Toxin Weapons: Research, Development and Use From the Middle Ages to 1945. Oxford University Press.

Gronvall, G.K., 2017. Biodefense in the 21st Century. American Association for the Advancement of Science.

Guillemin, J., 2006. Scientists and the history of biological weapons: a brief historical overview of the development of biological weapons in the twentieth century. EMBO Rep. 7 (1S), S45–S49.

Haider, N., 2018. Chemical and biological weapons conventions: orienting to emerging challenges through a cooperative approach. In: Enhancing CBRNE Safety & Security: Proceedings of the SICC 2017 Conference. Springer, pp. 253–260.

Handelman, S., Alibek, K., 2014. Biohazard: The Chilling True Story of the Largest Covert Biological Weapons Program in the World—Told From the Inside by the Man Who Ran It. Delta.

Harigel, G.G., 2001. Chemical and Biological Weapons: Use in Warfare, Impact on Society and Environment. Carnegie Endowment for International Peace.

Harris, S.H., 2002. Factories of Death: Japanese Biological Warfare, 1932–1945, and the American Cover-Up. Psychology Press.

Hassani, M., Patel, M.C., Pirofski, L.-A., 2004. Vaccines for the prevention of diseases caused by potential bioweapons. Clin. Immunol. 111 (1), 1–15.

Hay, A., 1999. A magic sword or a big itch: an historical look at the United States biological weapons programme. Med. Confl. Surviv. 15 (3), 215–234.

Henckaerts, J.-M., Doswald-Beck, L., 2005. Customary International Humanitarian Law. vol. 1. Cambridge University Press.

Hirschmann, J., 2018. From squirrels to biological weapons: the early history of tularemia. Am. J. Med. Sci. 356 (4), 319–328.

Inglesby, T.V., Dennis, D.T., Henderson, D.A., Bartlett, J.G., Ascher, M.S., Eitzen, E., et al., 2000. Plague as a biological weapon: medical and public health management. JAMA 283 (17), 2281–2290.

Jernigan, D.B., Raghunathan, P.L., Bell, B.P., Brechner, R., Bresnitz, E.A., Butler, J.C., et al., 2002. Investigation of bioterrorism-related anthrax, United States, 2001: epidemiologic findings. Emerg. Infect. Dis. 8 (10), 1019.

Johansson, A., Farlow, J., Larsson, P., Dukerich, M., Chambers, E., Byström, M., et al., 2004. Worldwide genetic relationships among Francisella tularensis isolates determined by multiple-locus variable-number tandem repeat analysis. J. Bacteriol. 186 (17), 5808–5818.

Kettle, A., Wernery, U., 2016. Glanders and the risk for its introduction through the international movement of horses. Equine Vet. J. 48 (5), 654–658.

King, W., Guillemin, J., 2019. The price of alliance: Anglo-American intelligence cooperation and Imperial Japan's criminal biological warfare programme, 1944–1947. Intell. Natl Secur. 34 (2), 263–277.

King, D., Luna, V., Cannons, A., Cattani, J., Amuso, P., 2003. Performance assessment of three commercial assays for direct detection of Bacillus anthracis spores. J. Clin. Microbiol. 41 (7), 3454–3455.

Lee, J., Deininger, R.A., 2004. A rapid screening method for the detection of viable spores in powder using bioluminescence. Lumin. J. Biol. Chem. Lumin. 19 (4), 209–211.

Leitenberg, M., Zilinskas, R.A., Kuhn, J.H., 2012. The Soviet Biological Weapons Program: A History. Harvard University Press.

Lim, D.V., Simpson, J.M., Kearns, E.A., Kramer, M.F., 2005. Current and developing technologies for monitoring agents of bioterrorism and biowarfare. Clin. Microbiol. Rev. 18 (4), 583–607.

Matero, P.H., 2017. Identification of bacterial biothreat agents and pathogens by rapid molecular amplification methods. Doctoral Dissertation, University of Helsinki, Faculty of Medicine, Finland.

Meselson, M., Guillemin, J., Hugh-Jones, M., Langmuir, A., Popova, I., Shelokov, A., et al., 1994. The Sverdlovsk anthrax outbreak of 1979. Science 266 (5188), 1202–1208.

Mobed, A., Baradaran, B., de la Guardia, M., Agazadeh, M., Hasanzadeh, M., Rezaee, M.A., et al., 2019. Advances in detection of fastidious bacteria: from microscopic observation to molecular biosensors. TrAC Trends Anal. Chem. 113, 157–171.

Morea, D., Poggi, L.A., Tranquilli, V., 2018. Economic impact of biological incidents: a literature review. In: Enhancing CBRNE Safety & Security: Proceedings of the SICC 2017 Conference. Springer, pp. 291–297.

Neubauer, H., Sprague, L., Zacharia, R., Tomaso, H., Al Dahouk, S., Wernery, R., et al., 2005. Serodiagnosis of Burkholderia mallei infections in horses: state-of-the-art and perspectives. J. Vet. Med. B 52 (5), 201–205.

Noah, D.L., Huebner, K.D., Darling, R.G., Waeckerle, J.F., 2002. The history and threat of biological warfare and terrorism. Emerg. Med. Clin. North Am. 20 (2), 255–271.

Ouagrham-Gormley, S.B., Melikishvili, A., Zilinskas, R.A., 2006. The Soviet anti-plague system: an introduction. Crit. Rev. Microbiol. 32 (1), 15–17.

Pal, M., Tsegaye, M., Girzaw, F., Bedada, H., Godishala, V., Kandi, V., 2017. An overview on biological weapons and bioterrorism. Am. J. Biomed. Res. 5, 24–34.

Pal, V., Saxena, A., Singh, S., Goel, A., Kumar, J., Parida, M., et al., 2018. Development of a real-time loop-mediated isothermal amplification assay for detection of Burkholderia mallei. Transbound. Emerg. Dis. 65 (1), e32–e39.

Pechous, R.D., Sivaraman, V., Stasulli, N.M., Goldman, W.E., 2016. Pneumonic plague: the darker side of Yersinia pestis. Trends Microbiol. 24 (3), 190–197.

Rath, J., 2002. Biological weapons, war crimes, and WWI. Science 296 (5571), 1235–1237.

Rega, P., Guinness, M., McMahon, C., 2017. Tularemia—a review with concern for bioterrorism. Med. Res. Arch. 5 (8), 1–15.

Robertson, A.G., Robertson, L.J., 1995. From asps to allegations: biological warfare in history. Mil. Med. 160 (8), 369–373.

Rotz, L.D., Khan, A.S., Lillibridge, S.R., Ostroff, S.M., Hughes, J.M., 2002. Public health assessment of potential biological terrorism agents. Emerg. Infect. Dis. 8 (2), 225.

Singh, R.K., Sudhakar, A., Lokeshwar, B., 2011. From normal cells to malignancy: distinct role of pro-inflammatory factors and cellular redox mechanism in human malignancy. J. Cancer Sci. Ther. 3 (4), 70–75.

Smith, M.E., Bhimji, S.S., 2019. Brucellosis. In: StatPearls [Internet]. StatPearls Publishing, Treasure Island, FL.

Smits, H.L., Kadri, S.M., 2005. Brucellosis in India: a deceptive infectious disease. Indian J. Med. Res. 122 (5), 375.

Song, L., Ahn, S., Walt, D.R., 2005. Detecting biological warfare agents. Emerg. Infect. Dis. 11 (10), 1629.

Sridhar, S., Chandrashekhar, P., 1991. Cutaneous anthrax with secondary infection. Indian J. Dermatol. Venereol. Leprol. 57 (1), 38.

Thavaselvam, D., Vijayaraghavan, R., 2010. Biological warfare agents. J. Pharm. Bioallied Sci. 2 (3), 179.

Tournier, J.-N., Peyrefitte, C.N., Biot, F., Merens, A., Simon, F., 2019. The threat of bioterrorism. Lancet Infect. Dis. 19 (1), 18–19.

Treadwell, T.A., Koo, D., Kuker, K., Khan, A.S., 2003. Epidemiologic clues to bioterrorism. Public Health Rep. 118 (2), 92.

Trevan, T., 2016. The Iraqi biological warfare program. In: Biological Threats in the 21st Century: The Politics, People, Science and Historical Roots. World Scientific, pp. 113–129.

Trevisanato, S.I., 2007. The 'Hittite plague', an epidemic of tularemia and the first record of biological warfare. Med. Hypotheses 69 (6), 1371–1374.

Tsuchiya, T., 2011. The imperial Japanese experiments in China. In: The Oxford Textbook of Clinical Research Ethics. Oxford University Press, UK, 31–45.

Walt, D.R., Franz, D.R., 2000. Peer Reviewed: Biological Warfare Detection. ACS Publications.

Wheelis, M., 1999. Biological warfare before 1914. In: Geissler, E., Moon, J.E.V.C. (Eds.), Biological and Toxin Weapons: Research, Development and Use from the Middle Ages to 1945. 2003 ed. SIPRI Chemical & Biological Warfare Studies, vol. 18. Oxford University Press, Oxford.

Wheelis, M., 2002. Biological warfare at the 1346 siege of Caffa. Emerg. Infect. Dis. 8 (9), 971.

Wheelis, M., Rózsa, L., 2009. Deadly Cultures: Biological Weapons Since 1945. Harvard University Press.

Williams, P., Wallace, D., 1989. Unit 731: Japan's Secret Biological Warfare in World War II. Free Press New York.

Williamson, C.H., Wagner, D.M., Keim, P., Sahl, J.W., 2018. Developing inclusivity and exclusivity panels for testing diagnostic and detection tools targeting Burkholderia pseudomallei, the causative agent of melioidosis. J. AOAC Int. 101 (6), 1920–1926.

Zacchia, N.A., Schmitt, K., 2018. Medical spending for the 2001 anthrax letter attacks. Disaster Med. Public Health Prep. 12, 1–8.

Zilinskas, R., 2017. A brief history of biological weapons programmes and the use of animal pathogens as biological warfare agents. Rev. Sci. Tech. 36 (2), 415–422.

Toxins as biological warfare agents

A.S.B. Bhaskar, Bhavana Sant

Division of Pharmacology and Toxicology, Defence Research and Development Establishment,
Gwalior, India

Introduction

Natural toxins or biotoxins are substances produced by one organism that elicit their toxic action on another organism. Toxins are extremely poisonous products of the metabolism of living organisms like bacteria, plants, animals, and fungi. Biotoxins are individual chemical compounds of natural origin. These are biologically active chemical compounds or compounds produced by a specific chemical mechanism in a living organism. Chemically, they are a wide variety of complex structures like proteins, cyclic peptides, alkaloids, etc. After knowing the structure, toxins can be prepared by chemical synthesis in desired quantities. Some toxins can also be prepared by biotechnological procedures like cloning and expression. Toxins, from a pharmacological and toxicological perspective, can be considered chemical weapons (Anderson, 2012a). Comparison of physicochemical and functional characteristics place toxins between chemical warfare agents and biological warfare agents. There are several differences between toxins and traditional chemical warfare agents. When compared, toxins have a higher molecular weight, most of them are odorless and not dermally active, and most of them produce immune responses in the host. Toxins are easy to use via inhalation route in the form of aerosols. Their toxicity potential is much higher than highly toxic chemical agents like sarin. As a comparison, the lethal inhalation concentration (LCt50) of a botulinum toxin aerosol, on average, is 1000 times more toxic than sarin vapor. Toxic substances based weapons have a long history of several thousand years. It is associated with traditional methods of hunting, including the use of poisoned arrows, water, or the fumigation of animals with toxic combustion products. All of these hunting forms involving toxic substances were developed for ancient wars. In some form or other, these have persisted until now. Toxins from plants, animals, bacteria, cyanobacteria, algae, and fungi have been used by man for fighting and hunting purposes since the earliest times. Some toxins, such as ricin (W), botulinum toxin (X), or saxitoxin (TZ), were formerly proposed as standard fillings for military ammunition. Intensive military research activity was carried out on some toxins like palytoxin, batrachotoxin, and tetrodotoxin (Pitschmann, 2014).

Table 1 List of toxins covered under select agents and toxins list issued by CDC.

Toxin	Source	Type
Abrin	*Abrus precatorius*	Plant
Botulinum neurotoxins	*Clostridium botulinum*	Bacteria
Alpha conotoxins	*Conus imperialis*	Snail
Diacetoxyscirpenol (DAS)	*Fusarium* sp.	Fungus
Ricin	*Ricinus communis*	Plant
Saxitoxin	*Alexandrium* sp. *Gymnodinium* sp. *Pyrodinium* sp.	Dinoflagellates
Staphylococcal enterotoxins (subtypes A, B, C, D, and E)	*Staphylococcus aureus*	Bacteria
T-2 toxin	*Fusarium* sp.	Fungus
Tetrodotoxin	Puffer fish, globe fish, and toad fish	Fish (bacteria)

The use of natural toxins to commit suicide, murder, and to wage war has been known for a long time.

Table 1 provides a select agents and toxins list issued by the Centers for Disease Control and Prevention (CDC) (https://www.selectagents.gov), and includes about nine toxins of different natures.

Most of the biotoxins are very toxic compounds and represent a risk for human health and the viability of organisms. Predation and defense are the two primary uses of these biologically produced molecules. Predators that use toxins include spiders, snakes, jellyfish, wasps, etc. Whereas honeybees, poison dart frogs, deadly nightshade, etc. use toxins for defense. Their high toxicity and relative ease of availability make them potential bio warfare agents (Slater and Greenfield, 2003). The Chemical Weapons Convention (CWC) of 1993 includes toxins as chemical agents, and they are included in its control regime along with other highly toxic chemicals. Protein toxins include ricin, abrin, botulinum, *Clostridium perfringens toxin*, *Corynebacterium diphteriae toxin*, microcystins, *Staphylococcus aureus toxins*, shigatoxin, and tetanus toxin. The Australian Group, a voluntary association of 33 countries, also includes conotoxins, verotoxin, cholera toxin, modeccin, volkensin, and viscumin on its control list (Patocka et al., 2007).

Pathogenic agents are classified by the Centers for Disease Control and Prevention (CDC) into categories A, B, or C based on ease of transmission, severity of morbidity and mortality, and the likelihood of use (http://www.selectagents.gov/ Select AgentsandToxinsList.html). Most of the category A agents are considered especially dangerous due to the potential for airborne transmission or aerosolization. Botulinum neurotoxin is included in this class. Category B agents are regarded as moderately easy to disseminate, may result in moderate morbidity and low mortality.

Staphylococcal enterotoxin B, Clostridium perfringens epsilon toxin, and ricin are included in this class (Clarke, 2005; Berger et al., 2016).

Toxins are nonliving, hence cannot be grown in a culture or identified by simple techniques like sequencing of amino acids, making detection and treatment a more complex challenge. In most cases, poisoning often presents with nonspecific clinical manifestations. Toxins that are airborne tend to cause more severe illness than ingested toxins as they enter deeper parts of the lungs (Yinon, 2002).

This chapter reviews the characteristics and toxicity of natural toxins from dinoflagellates (saxitoxin), bacterial toxins (botulinum neurotoxins and Staphylococcal enterotoxins), fungal toxins (diacetoxyscirpenol and T-2 toxin), snail toxins (alpha conotoxins), fish toxins (tetrodotoxin), and plant toxins (abrin and ricin).

Dinoflagellate toxins
Saxitoxin

Harmful algal blooms (HAB) are natural phenomena caused by the massive growth of phytoplankton that may contain highly toxic chemicals, the so-called marine biotoxins, causing illness and even death to both aquatic organisms and humans. Worldwide, there has been a substantial increase in HABs, commonly called red tide. Ingestion of these algae or contaminated marine products results in massive and dangerous paralytic intoxication (Hernandez et al., 2005; Hundell, 2010). The intensity and occurrence of HAB during the last decades appears to be increasing due to rise in ocean temperature, and growing coastal eutrophication (McCarthy et al., 2015). HAB effect on human health is linked to poisoning after consumption of contaminated seafood, skin contact with contaminated water, and inhalation of aerosolized biotoxins. Saxitoxin, a marine biotoxin contained in bivalve mollusks (mussels, scallops, and clams), is responsible for causing paralytic shellfish poisoning (PSP). In 1957, a PSP toxin was isolated in clams (*Saxidomus giganteus*) living in Alaskan coastal areas and in 1975 the chemical structure was determined.

PSP toxin is a potential neurotoxic agent as it blocks the excitation current in nerve and muscle cells (Schirone et al., 2011). The main producers of PSP toxins are dinoflagellates of the genus *Alexandrium* occurring along the Atlantic and Pacific coast (Bernd and Bernd, 2008) but also in the Mediterranean Sea, where other species such as *Gymnodinium catenatum* can be present (Berti and Milandri, 2014). Saxitoxin is concentrated in shellfish as a result of the continuous filtration of toxic algae produced by several dinoflagellates (including *Alexandrium*, *Gymnodinium*, and *Pyrodinium*) during "red tide" blooms. Predators of bivalve shellfish (scavenging shellfish, lobsters, crabs, and fish) may also be vectors for saxitoxins, thus expanding the potential for human exposure (Halstead and Schantz, 1984). The maximum number of saxitoxin poisoning cases (2124 with 120 deaths) was reported in the Philippines from 1983 to 2002 (Ching et al., 2015). Moderate toxicity by PSP is characterized by incoherent speech, progression of prickly sensation to arms and legs, respiratory difficulty, stiffness of limbs, and backache. During severe exposure cases, symptoms reported include muscular paralysis, pronounced respiratory difficulty, and choking sensation which may lead to death within 2–12 h of exposure (Pierina et al., 2016).

Saxitoxin, including its analogs, is a highly potent neurotoxin produced in both freshwater and marine environments by cyanobacteria and dinoflagellates of the genera *Alexandrium*, *Gymnodinium*, and *Pyrodinium* (Oshima, 1995; Landsberg et al., 2006). The main species responsible for the most abundant production of PSP toxins with saxitoxin as the most toxic analog is *Alexandrium* species. Penetration of marine biotoxins inside the tissue of marine organisms occurs by filter feeding. Saxitoxin is the parent compound of more than 30 naturally occurring derivatives that differ structurally at four positions (Llewellin, 2006). Structurally, saxitoxin is a trialkyl tetrahydropurine (Fig. 1). Saxitoxin is a cynotoxin with a molecular weight of 372 Da and chemical composition $C_{10}H_{17}N_7O_4$ (Schantz et al., 1975). Saxitoxin subtypes are distinguished on the bases of its subsequent side chains, comprising carbamate, *N*-sulfo-carbamoyl, decarbamoyl, and hydroxylated saxitoxins (EFSA, 2009). Saxitoxin can be modified by addition and subtraction of hydroxyl, carbamyl, and/or hydroxysulfate groups, which creates a group of 21 toxins having a wide range of potencies (Plumley, 1997). On the basis of potency, most potent toxins are carbamate toxins (STX, neoSTX, and gonyautoxins 1–4), followed by the decarbamoyl (dc) toxins (i.e., dc-STX, dc-neoSTX, etc.). *N*-sulfocarbamoyl toxins are the least potent (Bricelj and Shumway, 1998).

Saxitoxin is soluble in water and is highly stable in acidic conditions but tends to oxidize when exposed to alkaline environment (WHO, 1984). Saxitoxin is heat stable, which means it is not destroyed by normal food preparation methods (Trevino, 1998). Synthesis of saxitoxin has been possible and published by Akimoto et al. (2013) (Fig. 1).

Saxitoxin is the parent compound of all paralytic shellfish poisons. The primary target of saxitoxin is the voltage gated sodium ion channel in nerve and muscle cells, to which it binds with high affinity, and can result in death via respiratory paralysis (Catterall, 1985). STX is a potent and extremely selective sodium channel blocker. In addition to this, saxitoxin is also known to target the potassium and calcium channels through a mechanism different from the sodium channel. Other toxicity mechanism of STX is the generation of reactive oxygen species (ROS). Fishes, birds, and marine animals are also affected by saxitoxin poisoning. Massive deaths have been observed during red tides. Three main species *Alexandrium* sp.,

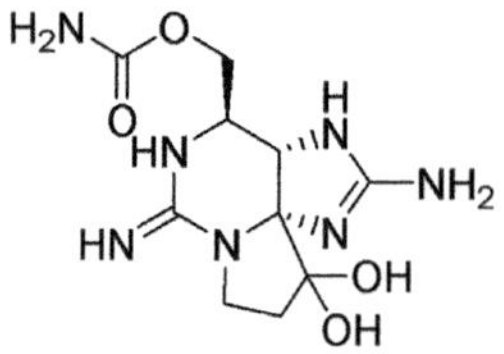

FIG. 1

Chemical structure of saxitoxin (STX).

Pyrodinium sp., and *Gymnodinium* sp. are responsible for contamination of marine water (Felipe et al., 2016).

Consumption of contaminated shellfish leads to poisoning. The first case of PSP was recorded in 1927 near San Francisco, United States, and was caused by *A. catenella*, resulting in 106 human illnesses and six deaths (Wang, 2008). Around 2000 human PSP cases are reported worldwide annually, with 15% fatalities (Dolah, 2000). Saxitoxins can be produced synthetically and are highly potent. For these reasons, saxitoxins are the only cyanotoxins that come under Chemical Weapons Convention Schedule 1. It also comes under the War Weapons List of the German War Weapons Control Act (Merwe, 2015). Symptoms of saxitoxin poisoning include paresthesia and numbness, first around the lips and mouth and then the face and neck, muscular weakness, sensation of lightness and floating, ataxia, impaired motor coordination, drowsiness, incoherence, and progressively decreasing ventilator efficiency. In severe cases, toxicity results in paralysis and death. Usually death occurs within 1–12 h post exposure (Rodrigue et al., 1990). Suffocation with cardiac arrest is the common reason for death.

Acute toxicity of saxitoxin exposure is well known; however, chronic deleterious effects still need to be elucidated. Saxitoxin accumulates in the central nervous system as it crosses the blood brain barrier, but the exact mechanism of transport through the blood brain barrier is still unknown. Saxitoxin's mode of action is to block the depolarization of excitable cells by reversibly binding to the voltage gated sodium channels in the brain, peripheral nervous system, and muscles (Llewellin, 2006; Andrinolo et al., 2002). It is rapidly absorbed through the gastrointestinal tract and excreted in the urine. According to the World Health Organization (WHO) report, LD50 of saxitoxin for intravenous, intraperitoneal, and oral routes of exposures in male and female mice are 3.4, 10, and 263 μg/kg respectively (Wiberg and Stephenson, 1960). Sublethal doses of saxitoxin exposure induces significant changes in neuroactive amino acids, serotonin, dopamine, and its metabolite 3,4-dihydroxyphenylacetic acid (DOPAC) (Cervantes et al., 2009, 2011). Apoptosis in zebrafish brain cells was observed with low doses of saxitoxin and necrosis with high doses (Zhang et al., 2013). Saxiphillin, a saxitoxin binding protein that is present in some PSP containing animals, allows saxitoxin to accumulate to high levels within animal tissues. Puffer fish contain saxiphillins, allowing them to accumulate high levels of saxitoxin in certain organs, primarily the liver (Yotsu et al., 2013).

Saxitoxin concentrations ranging from 0.5 to 64 nM induce indirect genotoxic effects in N2A human cell lines (Perreault et al., 2011). Cells exposed to saxitoxin have shown cytotoxicity and oxidative stress and are reported as an inducer of apoptosis with no effect on caspase 3 activity. It might be possible that it activates some other routes of activation of effecter caspases such as caspase 6 and 7 (Cohen, 1997). Consumption of saxitoxin contaminated water can leads to oxidative stress in both brain and liver, causing deleterious effects on macromolecules such as lipids, and cause increased vulnerability of the hippocampus (Patricia et al., 2014).

Bacterial toxins

Botulinum neurotoxin

Botulinum neurotoxin (BoNT) is derived from the bacteria *Clostridium botulinum* and exposure to it results in a condition known as botulism. It is the most potent toxic substance currently known (Arnon et al., 2001). The first recorded case of botulinum poisoning was recorded in 1735 and the first major outbreak was recorded in Germany in 1793 (Pearce et al., 1997). However, Tchitchkine (1905) first proposed that the soluble factor responsible for botulism was a neurotoxin. Botulinum toxin is a neuro toxin derived from the genus of anaerobic bacteria named Clostridia. Seven antigenic types of botulinum toxin exist, designated from BoNT/A to BoNT/G with a sequence identity of 34%–64%. BoNTs are recognized as the most potent naturally produced neurotoxins known to man, based on the fact that less than 0.2 picomoles of crystalline Type A toxin is the LD50 in the mouse. BoNTs are composed of two peptide chains, a heavy chain (100 KDa) and a light chain (50 KDa) joined by disulfide bonds. Biochemical studies have shown that these extremely potent neurotoxins enzymatically cleave a number of nerve terminal proteins that are critical to normal neurotransmitter release. The light chain is reported to possess a zinc-dependent endopeptidase activity responsible for the cleavage of neuronal proteins. BoNT Types A and E cleave the SNAP25 protein (Blasi et al., 1993; Pearce et al., 1997).

BoNT was used as a biological warfare agent by the Japanese during World War II. Later, other countries, like the United States, USSR, and Iraq, have produced the toxin. In the 1990s, the Japanese cult Aum Shinrikyo attempted to launch several terror attacks with BoNT without any success (Keller et al., 1999; Dembek et al., 2007). Due to the ease of BoNT production, there is a risk of its future use for malicious purposes.

Keeping the past incidents of attempted use of this toxin on military installations and civilian population, botulinum neurotoxin identification and characterization gained renewed importance (Arnon et al., 2001). There is a requirement of developing techniques capable of assaying all proteins in progenitor toxins to identify them at the strain level in a timely manner. Being protein in nature, these toxins can be identified based upon high-performance liquid chromatography (HPLC) followed by mass spectrometry (MS) that can generate the amino acid sequence information essential for progenitor toxin/neurotoxin identification. Using nanoscale HPLC and electrospray ionization techniques, mass spectrometry has the ability to identify proteins at nanomole levels. This would be important for botulinum progenitor toxins because their production and activity are at low levels. However, strain level identification is dependent on factors such as protein variability, database composition, and search engine characteristics (Aebersold, 2003).

Human exposure to BoNT occurs either by disease or during therapeutic exposure for treatment of dystonia. Epidemiological studies have indicated that the human nervous system is resistant to serotypes C and D. Later studies have shown where electrophysiologic techniques were used to monitor toxin effects on neuromuscular transmission in surgically excised human pyramidalis muscles.

Ligand binding studies were carried out to detect and characterize toxin receptors in human nerve membrane preparations, and molecular biologic techniques were used to isolate and sequence a human gene that encodes a substrate for botulinum neurotoxin. In contrast to epidemiologic findings, serotype C also paralyzed human tissues in approximately 65 min. In addition, the human nervous system was found to encode polypeptides of synaptosomal-associated protein and syntaxin1A, which are substrates for botulinum neurotoxin types A and C respectively (Coffield et al., 1997).

Animal testing was the earliest method used to detect botulinum toxin by feeding to or injecting extracts or filtrates from adulterated foods, human tissues, or cultures. Pearce et al. (1994) examined the precision of the mouse lethality assay for measuring botulinum toxin activity. These authors reported that very precise estimates of botulinum toxin activity can be obtained from lethality assays and further indicated that a five dose lethality assay involving 25 mice may be adequate for most laboratory experimentation when only a single estimation of the LD50 is made. Tests like abdominal ptosis assay, mouse hind limb assay, and guinea pig orbicularis oculi assay were also used to assess localized physiological responses as a measure of toxicity (Takahashi et al., 1990; Sugiyama et al., 1975; Pearce et al., 1995; Horn et al., 1993).

Staphylococcal enterotoxins

Staphylococcus aureus is a spherical gram-positive bacterium (ccocus) that produces different virulence factors. They are seen in pairs, short chains, bunches, and grape-like clusters on microscopic observation. They are capable of producing highly heat stable protein enterotoxins and size ranges between 19 and 26 kDa, which are lethal for humans. Seven immunologically different forms of Staphylococcus enterotoxins are known: A, B, C1, C2, C3, D, and E. All these toxin forms cause food poisoning after consumption of food contaminated by Staphylococcus bacteria (Patocka and Streda, 2006). *S. aureus* produces multiple toxins and leukocidins (cytolysins). Among the cytolysins are alpha, beta, gamma, and delta toxins (Lowy, 1998). Alpha toxin is a heptamer pore-forming exotoxin that lyses primarily rabbit erythrocytes but is toxic to human epithelial cells (Gouaux et al., 1994). Gamma toxin is an exotoxin of two-components, which consists of six different combinations of proteins; one of the proteins is leukocidin (Dinges et al., 2000). Delta toxin is a low-molecular-weight exotoxin that forms multimeric structures with the ability to lyse many cell types. Beta toxin is the least known toxin. It functions as sphingomyelinase (SMase) with molecular weight of 35 kDa. It shows a unique property of hot-cold toxin on sheep blood agar plates. Beta toxin is known for this property because it doesn't lyse the sheep RBC's at 37°C, but if the red cells are then placed at 4°C, the cells lyse.

Humans are naturally in contact with *S. aureus* that grows on the skin, mucosal surfaces, or in food items, and can seroconvert to *S. aureus* antigens that include the Staphylococcus enterotoxins. There are three major pathogenic species of Staphylococcus including *S. aureus*, *S. epidermidis*, and *S. saprophyticus*.

Staphylococcal enterotoxin A is the most common toxin associated with staphylococcal food poisoning (Pinchuk et al., 2010). Staphylococcal enterotoxin F is a major component of toxic shock syndrome. The most relevant staphylococcal toxin is Staphylococcal enterotoxin B (SEB), which has been studied as a potential biological weapon. The risk as a biological weapon appears to be as an incapacitating agent rather than a lethal agent.

Staphylococcal enterotoxin B

Staphylococcal enterotoxin B (SEB) is categorized under category B select agent, as it is highly potent enterotoxin and very small quantities are required to exert its toxic effect compared to a synthetic chemical (Bettina and Avanish, 2013). Staphylococcal enterotoxin B is the only one out of more than 25 Staphylococcal enterotoxins that has been characterized to date (Jeffrey et al., 2011). SEB is a 28 kDa protein, consists of 239 amino acids. Structurally and in amino acid content, it shows homology with SEC (Iandolo and Shafer, 1977; Swaminathan et al., 1992). Staphylococcal enterotoxin B is part of a set of exotoxins produced by *S. aureus*, which comprise about 15 antigenically distinct proteins and include the following: SEA, SEB, SEC1, SEC2, SEC3, SED, SEE, SEH, SEG, SEI, SEJ, on the basis of SEK, and the last one discovered recently was identified as SEU (Lefertre et al., 2003). However, SEB is classified as an incapacitating agent because in most cases aerosol exposure does not result in death but in a temporary, though profoundly incapacitating, illness lasting as long as 2 weeks (Ulrich et al., 1997).

Staphylococcus enterotoxins are potent immune activators that lead to severe immune dysfunction. SEB has the ability to function as a superantigen because of its profound effects on the immune system, which result in the secretion of numerous cytokines, chemokines, and growth factors (Bettina and Avanish, 2013; Krakauer et al., 2016a,b). Consumption of contaminated food and water are the main source of SEB poisoning because SEB thrives in unrefrigerated meats, dairy, and bakery products. Initially, it was thought that SEB poisoning is caused by it local interaction with intestinal cells because the toxin stimulates nerve centers in the gut through serotonin (5-hydroxytryptamine or 5 HT) release from intestinal mast cells (Alouf and Muller-Alouf, 2003; Hu et al., 2007). Serotonin binding to its receptors initiates the signaling of emetic reflex center to generate nausea and an emetic response. Enterotoxins exert their effect on the epithelium of the intestinal track, thus they are a common cause of food poisoning. SEB is toxic in minute doses (Gill, 1982), and responsible for a number of extensive pathophysiological changes in humans and mammals. SEB is considered an effective bio-warfare agent because it can be easily aerosolized, is highly soluble in water, and relatively heat resistant. It is also resistant to proteolytic enzymes, including pepsin, trypsin, and papain (Le Loir et al., 2003). Inhalation route also shows toxicity by SEB, but such cases are rarely fatal. The LD50 dose of SEB for oral route and inhalation route is 0.3 and 20 µg/kg (Russmann, 2003). There is no vaccine available to date against SEB toxicity for humans.

Suitability of SEB as a biological warfare agent was studied during the 1960s. Its parameters for use as an aerosolized bioweapon studied at several facilities in the United States and Great Britain (Hale, 2012). Based on those investigations, the effective dose of SEB that would incapacitate 50% of the exposed population was estimated to be 0.0004 µg/kg of body weight (Franz et al., 1997). A number of SEB poisoning cases were reported from different parts of India and lead to death as well (Roy et al., 2015).

The symptoms of SEB poisoning depend on the route of exposure. Leukocytosis is observed in most patients, appearing within 12–24 h of toxin exposure. Vomiting is the hallmark symptom. Inhalation route is the most toxic route of exposure causing hyperpyrexia within 12–13 h of exposure (Tamar et al., 2016). Exposure of SEB by inhalation route results in stimulation of the immune system of rhesus monkeys. Rhesus macaques (*Macaca mulatta*) have been extensively used as an inhalation toxicity model for SEB. Toxicity symptoms include fever, severe respiratory distress, headache, and sometimes nausea and vomiting. The mechanism of toxicity is thought to be massive release of cytokines such as interferon-gamma, interleukin-6, and tumor necrosis factor-alpha (Ulrich et al., 1997). If the toxin is absorbed through the dermis, there is an inflammation of the skin resulting in dermatitis and delayed type hypersensitivity (DTH) (Rusnak et al., 2004). To date, there is no FDA-approved vaccine or therapeutic agent to prevent or treat SEB-intoxication and with its ease of dissemination, SEB remains a serious bioterrorism agent.

Fungal toxins
Trichothecenes (T-2) toxin

Mycotoxins are highly diverse secondary metabolites produced by a wide variety of fungi, which contaminate food and cause a wide variety of diseases in humans and animals. More than 300 species of mycotoxins are known to induce mycotoxicosis effects in mammals (Zain, 2011). The toxic effect of mycotoxins on animal and human health is referred to as mycotoxicosis. T-2 toxin (trichothecene mycotoxin) is a unique mycotoxin. It belongs to a large family of toxins produced by fungi like *Fusarium*, *Trichoderma*, *Myrothecium*, *Stachybotrys*, etc. T-2 toxin is a derivative of a ring system known as tricothecenes. Many field crops such as wheat, maize, barley, and oats, and processed grains, e.g. malt, beer, and bread, are detected with T-2-toxin (Li et al., 2011; Moss, 2002). T-2 is a potent biological warfare agent in that it consists of skin damaging properties several thousand times greater than other chemical warfare agents like sulfur mustard, lewisite, etc. (Bunner et al., 1985; Wannemacher and Wiener, 1997). Agriculturally, trichothecene produced by *Fusarium* species is very important worldwide because of its potential health hazard effects. In terms of toxicity, it affects the gastrointestinal tract, skin, kidney, and liver, with the most sensitive endpoints being neural, reproductive, immunological, and hematological effects (Adhikari et al., 2017). Consumption of T-2 contaminated food results in acute and chronic toxicity, both in humans and animals. The toxicity mechanism of T-2 toxin is oxidative damage and inhibition of protein synthesis. Inactivation of T-2 can be achieved by heating at 900°F for 10 min or 500°F for 30 min (Kachuei et al., 2014).

The trichothecene mycotoxins are low molecular weight (250–500 Da) non-volatile compounds. Structurally, trichothecenes are tetracyclic sesquiterpenoid 12,13-epoxytrichothec-9-ene ring, in which 12,13-epoxy ring is mainly responsible for toxicity. In their chemical structure (Fig. 2) at C-3 position hydroxyl (OH) group is present, at C-4 and C-15 position acetyloxy (–OCOCH$_0$) groups are present and an ester-linked isovaleryl [OCOCH$_2$CH(CH$_3$)$_2$] group is present at the C-8 position (Swanson et al., 1987). On the basis of functional groups, trichothecenes are classified in to four types (A, B, C, and D). Type A consists of T-2 and H-T2 toxin with no carbonyl group present at C-8 position. Type B represents deoxynivalenol and nivalenol toxins with carbonyl group present at the C-8 position. Type C includes crotocin and baccharin. An epoxy ring is present between C-7 and C-8 position. Type D consists of macrocyclic ring between C-4 and C-15; examples are satratoxin and roridin. Trichothecenes are very stable compounds during storage and processing, and are not degraded at high temperatures.

Among other metabolites of trichothecenes, T-2 has received much attention, as it is the most toxic mycotoxin present in wheat, rye, maize, and soybeans (Seeboth et al., 2012). T-2 toxin was first isolated from the mold *F. tricinctum* (*F. sporotrichoides*) (Ueno, 1977; Burmeister et al., 1971). Indirect exposure of mycotoxins to humans can be possible in the form of meat, eggs, and milk as the result of the animal eating contaminated feed (Li et al., 2011; Joffe, 1971).

The acute toxicity, chemical stability, antipersonnel properties, and ease of large-scale production makes T-2 toxin a likely candidate to be used as a weapon for bioterrorism (Seeboth et al., 2012). Several reports show outbreaks of human disease with the presence of trichothecenes in food. T-2 toxin was implicated as the chemical agent of "yellow rain" used against the Laos People Democratic Republic from 1975 through 1981 (Peraica et al., 1999). The reported LD50 of T-2 toxin is approximately 1 mg/kg (Wannemacher and Wiener, 1997). The potential use of the T-2 mycotoxin as a biological weapon was realized during World War II in Orenburg, Russia, when civilians consumed wheat that was unintentionally contaminated with *Fusarium*

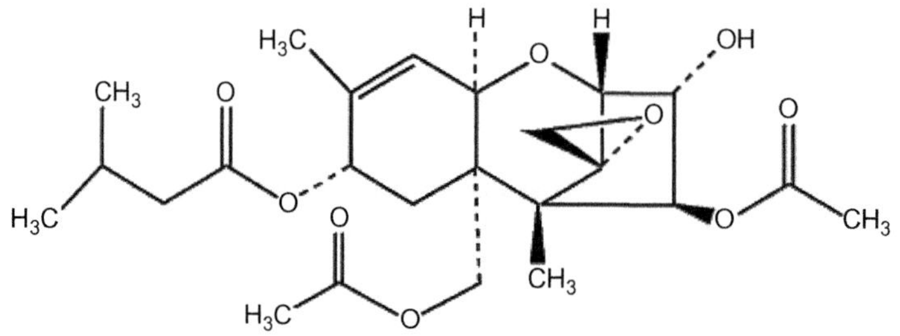

FIG. 2

Structure of T-2 toxin.

fungi. The victims developed diffuse hemorrhage and necrosis of the entire alimentary tract, a disease pattern named alimentary toxic aleukia (ATA). Twenty years after this incident, T-2 mycotoxin was discovered and isolated (Adhikari et al., 2017). The first recognized trichothecene mycotoxicosis was alimentary toxic aleukia in the USSR in 1932; the mortality rate was 60% (Chulze, 2010). In India, at the Kashmir Valley in 1987, an outbreak was reported due to the consumption of breads made up of wheat contaminated with the T-2 toxin (Bhat et al., 1989). Poultry and pigs are adversely affected by consuming feed contaminated with T-2 toxin (Burmeister et al., 1971) leading to loss of animal life.

There are no specific antidotes for T-2 toxin toxicity other than detoxifying with natural substances and replenishing lipids, nutrients, enzymes, amino acids, and probiotics. T-2 mycotoxin is a potent active dermal irritant. Because of its lipophilic property, it is readily absorbed through skin causing systemic toxicity. Other modes of exposure include oral and inhalation routes. As a dermotoxin, it causes necrohemorrhagic dermatitis and blisters; it is alleged to be 400 times more intoxicating than sulphur mustard. Presence of thiol group in the structure of T-2 mycotoxin, makes it a potent protein and DNA synthesis inhibitor. T-2 toxin inhibits the activity of peptidyl transferase, which is an integral part of 60S ribosome, and finally leads to protein synthesis inhibition in the initial phase (Agrawal et al., 2015).

Molecular studies using rodent and human cell lines suggest T-2 toxin also induces apoptosis, programmed cell death, through reactive oxygen species–mediated mitochondrial pathway (Chaudhary et al., 2009; Wu et al., 2011; Agrawal et al., 2012). The amphophilic nature of T-2 helps to enter lipid bilayer and then induce lipid peroxidation by generating free radicals, thereby damaging cellular membranes (Stark, 2005). T-2 induces oxidative stress mediated skin inflammation, activation of myeloperoxidase, MMP and P-38 MAPK activity, infiltration of inflammatory cytokines, and apoptosis of skin epidermal cells resulting in degenerative changes in skin (Moss, 2002). Oxidative stress in the brain and alteration of blood brain barrier permeability after dermal exposure of T-2 toxin was reported (Chaudhary and Rao, 2010; Ravindran et al., 2011).

Clinical symptom occurs within seconds of exposure and plasma concentrations attain peak levels after approx. 30 min. Half-life of T-2 is less than 20 min in plasma. Common symptoms of T-2 exposure includes reduced feed intake, weight loss, skin irritation, itching, diarrhea, bleeding, feed refusal, dyspnea, and vomiting (Gerberick et al., 1984; Cheeke, 1998), hemorrhages and necrosis in the GI tract, reproductive organs and hematopoietic organs, such as the bone marrow and spleen (Pang et al., 1987). Inhalation is one of the most important routes of exposure for this type of agent. Inhalation exposure of T-2 toxin is at least 10 times more toxic than systemic toxicity and 20 times more than dermal (Chaudhary and Rao, 2010). Immune responses were adversely affected by the inhalation route of exposure. The particle of T-2 toxin is too small to be inhaled and deposited in the alveoli, however studies have indicated edema, fibrin deposition, cellular infiltration, and debris present in the alveolar spaces; vascular and alveolar epithelial cell injury has occurred following inhalation exposure of T-2 toxin (Ueno, 1977).

Diacetoxyscirpenol

Diacetoxyscirpenol (DAS) is a toxin of the *Fusarium* species that belongs to the trichothecenes family. *Fusarium* is a species that not only causes loss of quality and yield in plants but also has hazardous effects on both plants and animals (Mehrdad et al., 2011). Trichothecenes are potent protein synthesis inhibitors of eukaryotic cells, which come under the class of mycotoxins. DAS is toxic to fungi, plants, animals, and many animal tissue cultures (Brian et al., 1961; Chi and Mirocha, 1978; Grove and Mortimer, 1969; Reiss, 1973; Weaver et al., 1978). Information on metabolism of trichothecenes is available, which is important for both evaluating and controlling human exposure to trichothecenes in food of animal origin. However, very little information is available about the fate of diacetoxyscirpenol in animals (Ohta et al., 1978). DAS is a secondary metabolite product of fungi of the genus *Fusarium* and may cause toxicosis in farm animals (Hoerr et al., 1981). The potent acute toxicity and chemical stability makes T-2 toxin and DAS potential bioterrorism agents (Stark, 2005).

DAS was discovered first in 1961 as a phytotoxic compound from a culture of *Fusarium equiseti* and *Gibberella intricans*. DAS is similar in structure and potency to T-2 toxin. DAS comes under type A category of trichothecenes (Richardson and Hamilton, 1990). Diacetoxyscirpenol contains the 12,13-epoxytrichothecene group of sesquiterpenes as the core structure. T-2 toxin is metabolized into hydrolyzed, hydroxylated, and de-epoxidation compounds, while DAS is metabolized into hydrolyzed products (Ohta et al., 1978; Bauer et al., 1985). Two *Fusarium* species, namely *Fusarium sporotrichoides* and *F. poae* cultures, are most commonly used for production of T-2 and DAS toxin. During the 1970s, DAS (anguidine) underwent phase I and II clinical trials in humans for development of anticancer drugs, as it inhibits cell cycle progression and cell survival (Jun et al., 2007). However, clinical trials were stopped as it shows minimum antitumor activity and was found to have various side effects such as nausea, vomiting, fever, hypertension, and confusion (Bukowski et al., 1982; Yap et al., 1979).

Absorption of trichothecenes (DAS) can through the gastrointestinal tract, the lungs, and the skin (Madsen, 2001). DAS toxicity information for beagle dogs and rhesus monkeys in the phase I and II clinical studies is available. The study reveals DAS shows similar toxicity in monkeys and dogs on mg/m^2 scale. However, on the basis of mg/kg, dogs are more susceptible for DAS toxicity than monkeys. Toxicity symptoms were the same as other species of type A trichothecenes. General symptoms of toxicity include vomiting, diarrhea, headache, fatigue, dermatitis with focal alopecia, and generalized malaise. Symptoms were reversible and consistency was found in dogs as compared to monkeys (Haschek and Beasley, 2009).

Accidently, people or livestock have been subjected to food poisoning by diacetoxyscirpenol, which contaminates agricultural crops, such as grains, potatoes, peas, and soybeans (Van Egmond et al., 2007). DAS can also cause sea food poisoning because it is also found in marine bacterium that parasitizes red alga. Exposure of trichothecenes to humans occurred in Southeast Asia in the form of a chemical warfare agent named "Yellow rain," which comprises components of T-2, DAS, and DON (deoxynivalenol) toxins. These were detected in blood, urine, and tissue samples of exposed victims (Mirocha et al., 1983; Watson et al., 1984). DAS LD50 has been determined in different animal models. By intraperitoneal route, LD50 was found as 23 mg/kg in mouse, for oral route it is 7.3 mg/kg in rat, and in rabbit, intravenous LD50 is 0.3 mg/kg (Trenholm et al., 1989). DAS effects lymphocyte functioning, however the molecular mechanism is not defined yet. It is cytotoxic to most cell types and tissues in vivo. DAS induced apoptosis can be prevented by overexpressing Bcl-xL, and caspase activation can be suppressed by mitochondrial permeability transition pore inhibitor (CsA) (Jun et al., 2007). DAS also shows toxicity as a teratogen. DAS, when administered intraperitoneally to pregnant mice, resulted in significant fetal body weight reduction and a variety of fetal morphological and skeletal malformations (Mayura et al., 1987).

Snail toxins

Conotoxins

Marine predatory snails use powerful venom to kill their prey. Cone snails belong to the phylum Mollusca, the class Gastropoda, the order Sorbeoconcha, the family Conidae, and the genus Conus (Anderson and Bokor, 2012). The family Conidae comprises many venomous species of cone snails, which produce an array of small disulfide-bridged peptides known as conopeptides or conotoxins. Conopeptides are reported from cone snails from at least 16 super families with different phylogeny and feeding strategies. Cone snails hunt in the darkness of night. After locating prey by smell, they skillfully inject small quantities of venom, which paralyzes prey within seconds. The venom uses a rich cocktail of conopeptides with a size of usually around 5 kDa, which are cleaved from propeptides by specialized venom endoproteases (Milne et al., 2003).

Poisonous species of cone snails include *Conus geographus*, *Conus catus*, *Conus aulicus*, *Conus gloriamaris*, *Conus omaria*, *Conus magus*, *Conus striatus*, *Conus tulipa*, and *Conus textile*. *Conus geographus* is the most lethal to humans. Signs and symptoms of exposure include faintness, ptosis (drooping eyelids), poor coordination, absent gag reflex, areflexia, paresthesias (abnormal sensations such as burning or tingling), urinary retention, diplopia (double vision), blurred vision, speech difficulties, dysphagia (difficulty swallowing), weakness, nausea, generalized numbness, and respiratory arrest (Haddad et al., 2006; Fegan and Andresen, 1997; Rice and Halstead, 1968; Fernandez et al., 2011).

The conotoxins are classified based on gene superfamily, cysteine framework, or by pharmacological effects (Kaas et al., 2012). There are 18 gene superfamilies used by Conoserver, a database on conotoxins maintained by the University of Queensland. Classifying conotoxins by cysteines involves consideration of factors including the number of cysteines, their pattern, and their connectivity within the peptide. *Conus geographus* venom contains α-conotoxin as one of the key components (Gray et al., 1981). Nicotinic receptors are sensitive to activation by nicotine and have ion channels whose activity is induced in microseconds. Knowledge about nicotinic receptors originated through the combination of two natural oddities (Albuquerque et al., 1995). The first was the finding that the electric organ of a fish that produces an electric pulse to stun its prey, such as Torpedo, expresses nicotinic acetylcholine receptors at high densities. The second was the discovery of α-bungarotoxin, a component of krait snake venom that binds muscle-type nicotinic acetylcholine receptors and inhibits their function and promote debilitating paralysis at the neuromuscular junction (Albuquerque et al., 1974). The α-conotoxins are antagonists of nicotinic receptors. Nicotinic receptors serve a variety of functions in the body like contraction of skeletal muscle. Acetylcholine is released by a motor neuron. The acetylcholine then attaches to the nicotinic receptors on the muscle that start a physiological cascade leading to muscle contraction. Physically blocking the nicotinic receptor with a drug or toxin would stop the muscle contraction and cause paralysis. The diaphragm is a muscle located below the lungs and divides the abdomen from the chest cavity. During respiration, the diaphragm is the muscle that causes the lungs to inflate and deflate. Paralysis of the diaphragm muscle results in the cessation of breathing. The diplopia or double vision reported from human exposures probably results from paralysis of the extraocular muscles (Anderson and Bokor, 2012).

Toxic effects of conotoxins are due to muscle paralysis. The paralysis of the diaphragm results in respiratory arrest. The LD50 for α-conotoxins is 10–100 µg/kg in laboratory mice Inhalation of certain α-conotoxins would be expected to produce a clinical presentation similar to the inhalation of botulism toxin (Anderson and Bokor, 2012).

Fish toxin

Tetrodotoxin

Tetrodotoxin (TTX) is one of the most lethal toxins in the marine environment. It is a naturally occurring toxin responsible for human fatalities and intoxication. The name of TTX was established after the Tetraodontidae family of fish. In Japan *fugu*, or puffer fish, is known for its potential for TTX toxicity. More than 20 species of puffer fish have been found to harbor the toxin (Noguchi et al., 2006). Besides puffer fish, other species are also known to harbor TTX include gastropods, crabs, blue-ringed octopuses, ringworms, and land animals such as the atelopid frog of Costa Rica, or newts (Yin et al., 2005; Kim et al., 1975; Mosher et al., 1964). TTX is water-soluble and heat stable, thus, cooking does not affect its toxicity (Saoudi et al., 2007). TTX is especially concentrated in the skin and gonads of fish, amphibians, and reptiles, and in the liver and intestine of mollusks. There is currently no antidote developed to counteract the effects of TTX.

There are 26 naturally occurring analogs of TTX. These analogs are not commercially available; they can be isolated and extracted from living sources of TTX (Vaishali et al., 2014). TTX is a very strong sodium channel inhibitor. The mechanism behind this inhibition is binding of positively charged guanidine group of TTX with the negatively charged carboxylate groups on the side chains in the mouth of the sodium channel (Denac et al., 2000; Hille, 1971; Moran et al., 2003). Degradation of TTX is fastest in heart tissue and slowest in gonads (Wood et al., 2012).

TTX toxicity by pufferfish is well known. TTX and its analogs are widely distributed neurotoxins, found mainly in marine, fresh water, and brackish water species (Noguchi et al., 2006). It is a nonprotein, low molecular weight, neuroparalytic toxin. Arginine is supposed to be the precursor moeity within the organism (Bane et al., 2014). Bacteria are responsible for production of TTX. However, the quantity of TTX produced by these bacteria is very low. Despite low toxin production by bacterial strains in laboratory conditions, even minimal amounts of TTX produced by intestinal microflora of an animal can contribute to its toxicity. TTX-producing bacteria have been isolated from various pufferfish tissues including the skin, intestine, ovaries, and liver (Yu et al., 2011; Chau et al., 2011). The name tetrodotoxin is commonly associated with tetras (four) and odontos (tooth), or the tetraodon puffer fish. The tetraodon puffers are equipped with four large teeth that are nearly fused, forming a beak-like structure used for cracking mollusks and other invertebrates, as well as for scraping corals and general reef grazing.

In 1964, the chemical structure of TTX (Fig. 3) was elucidated by R. B. Woodward, and the racemate was successfully synthesized for the first time in 1972. TTX and its analogs are aminoperhydroxyquinazoline derivatives that exist as dipolar ions with a polarity comparable to saxitoxin. However, TTX consists of only one positively charged guanidinium group and the cation is stabilized by resonance effect. The TTX has a unique heterocyclic structure. Different analogs of TTX have been synthesized by many researchers in the laboratory. Studies have been conducted for analyzing toxicity pattern of different analogs. There are 26 naturally occurring analogs of TTX, which show different toxicity potential based on the number and position of hydroxyl groups present in the structure. It is found that deoxy analogs of TTX are less toxic than TTX, while hydroxyl analogs are more toxic than TTX (Vaishali et al., 2014). In the 1950s, TTX was isolated from bacteria in crystalline form and extracted chromatographically in the 1960s. Four strains of bacteria isolated from red algae and puffer fish were studied for producing TTX and also characterized (Simidu et al., 1990).

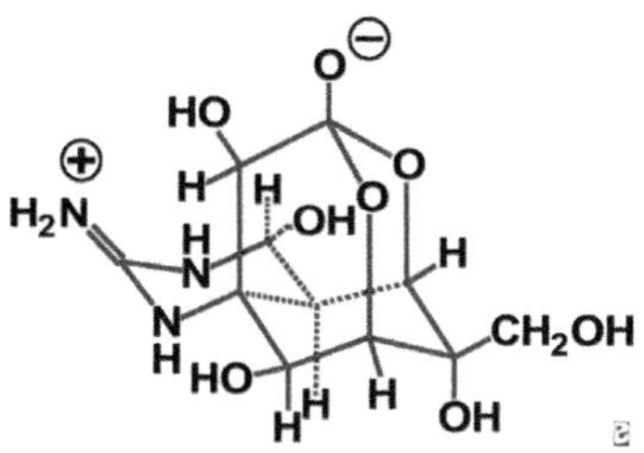

FIG. 3

Chemical structure of tetrodotoxin.

Source: https://emedicine.medscape.com/article/818763-overview#a5.

TTX exists in puffer fish in the forms of a mixture of its analogs (Nakamura and Yasumoto, 1985).

TTX is a hydrophilic heat-stable toxin, produced by bacteria that can be found in certain fish species but also marine gastropods and bivalves. It is about 1200 times more toxic to humans than cyanide and it has no known antidote (Jorge et al., 2015). Generalized symptoms of TTX poisoning are tingling of the tongue and lips, headache, vomiting, muscle weakness, ataxia, and even death due to respiratory and/or heart failure (Noguchi and Ebesu, 2001). Intensity of symptoms is dose dependent. Many studies have been carried out by researchers investigating the LD50 dose of TTX in mouse and rabbit animal models. The LD50 doses of TTX calculated in mice for intraperitoneal (i.p.), subcutaneous (s.c.), and intragastric (i.g.) route of administration were found to be 10.7, 12.5, and 532 µg/kg, respectively (Jorge et al., 2015). TTX is chiefly found in the liver and ovaries of pufferfish, genus, *Fugu.* It is also isolated from the eggs and embryos of the California newt, *Taricha torosa*. TTX is resistant to cooking and is not destroyed by proteases in the gastrointestinal tract. Several studies have suggested that consumption of these puffer fish have caused food intoxications, including deaths, worldwide; viz. Mexico, United States, Hong Kong, Japan, Korea, Taiwan, Malaysia, Bangladesh, and India (Ghosh et al., 2004; Lange, 1990; Loke and Tam, 1997; Nunez et al., 2000; Yang et al., 1996; Yoshikawa et al., 2000).

TTX poisoning results in inhibition of neurotransmittion as it blocks the voltage gated sodium channels and thus effects both action potential generation and impulse conduction, resulting in a blockade of the neuron action potential and in muscle paralysis. Little information is available about the absorption and excretion of TTX and its analogs in humans. TTX induced hepatotoxicity and nephrotoxicity in male wistar rats was studied by Mongi et al. (2011). Raw or boiled tissue extracts of *L. lagocephalus* exhibited hepatotoxic and nephrotoxic effects in rats. TTX also effects the respiratory system. An i.v. injection of TTX in rats leads to respiratory failure, due to paralysis of the respiratory muscles, which were apparently more susceptible to the action of tetrodotoxin than the respiratory and other motor nerves.

Plant toxins
Abrin

Abrin is a protein-based toxin from the plant *Abrus precatorius*, colloquially known as a rosary bead. Abrus is also known as Jequirity, John Crow Bead, Precatory bean, Crab's Eye, and Indian Liquorice. Abrus seeds are also known as Gunja in Sanskrit and Ratti in Hindi. These are found generally in India, and perhaps other parts of tropical Asia. The entire plant is poisonous, but the toxin reaches the greatest concentration in the seeds. Abrin holds considerable potential as a bioweapon for military or terrorist use because of its low cost of isolation, high toxicity, and ease of use either by aerosolization as a dry powder or liquid droplets, or by addition to food and water as a contaminant (Bhaskar et al., 2012). The toxin is lethal in minute doses and causes serious symptoms in even smaller amounts. It is almost identical to a better-known toxin—ricin. Abrin is more lethal than ricin. Toxicologists estimate abrin's lethal dose is between 0.1 and 1μg/kg (Dickers et al., 2003). In comparison, ricin is lethal between 5 and 10μg/kg .

Abrin belongs to the family of type II ribosome inactivating proteins. The molecular mechanism of inhibitory effect on protein synthesis has shown that RIPs act as a RNA N-glycosidase hydrolyzing the CN glycosidic bond of the adenosine residue at position 4 and 324 in rat 28S rRNA (Endo et al., 1987). Abrin exists in two forms, abrin-a and abrin-b. Both forms have two proteins, A chain and B chain. A disulfide bond between Cys247 of the A-chain and Cys8 of the B-chain links the A and B chains. The proteins work together to enter the cell and disrupt its activity. B chain is a galactose specific lectin and grants cell entry by binding to cell membrane, while A chain transports to the ribosome and catalytically inactivates 60S subunit by removing adenine from positions 4 and 324 of 28S rRNA and destroys it. Cells die shortly after ribosome destruction (Chen et al., 1992; Olsnes and Pihl, 1976).

Seeds of *Abrus precatorius* contain a toxic lectin abrin and a nontoxic agglutinin. Both lectins are specific for galactose. The complete protein sequence of A and B chain of abrin has been determined (Chen et al., 1992; Funatsu et al., 1988). An isolation procedure is reported for abrin variants (I, II, III) and two agglutinins (APA-I and APA-II) from *Abrus precatorius* seeds by using lactamyl-sepharose affinity matrix followed by gel filtration and DEAE chromatography (Hegde et al., 1991). There are at least three variants of abrin within *Abrus precatorius* species, which differ in toxicity, binding affinity, and lag period (Hegde and Poddar, 1992). Abrin consists of four isolectins (A to D), which are monovalent compounds with molecular weights ranging from 63,000 to 67,000 Da. Abrus seeds contain a large number of variants, with pI varying over 5.4–8.0. The A chain of abrin I is the least efficient protein-synthesis inhibitor among those of the four toxins studied (Hegde et al., 1993). These variants serve as valuable tools to understand more about toxin associated cellular phenomena. The molecular masses of abrin variants (red and black, black and white seeds) were measured as 61.14, 60.85, and 61.24 kDa, respectively, by the MALDI-TOF/MS. Estimated LD50 values in mice showed that the abrin extracted from white seeds was two to four times more toxic than others. In vivo toxicity studies confirm that white seeds are highly toxic and exhibit high lethal potency (Chaturvedi and Kumar, 2015).

Abrus precatorius beans are known to be among the most toxic plant parts worldwide (Anam, 2001). There are reported deaths both after accidental and intentional poisoning. Fatal incidents have been reported following ingestion of well-chewed seeds of *Abrus precatorius*. The hard coat surrounding the abrus seeds limits the gastrointestinal absorbtion of abrin. The release of abrin from seeds requires proper chewing and grinding of seeds prior to ingestion and nonmasticated seeds pass through the gastrointestinal track without causing toxicity. The high molecular weight of abrin also limits its gastrointestinal absorption. Based on animal studies, elimination of abrin probably occurs by the renal excretion of metabolites (Fodstad et al., 1976).

There have also been reports of people who have ingested abrus seeds and slipped into coma due to demyelinating encephalitis (Sahni et al., 2007). Neuroinflammatory damage mediated by infiltration of inflammatory cells in brain regions after abrin poisoning is also reported in mice (Bhasker et al., 2014). Dose and time dependent effects of abrin on gene expression profile of brain tissue has been reported using mouse microarray data of whole genome indicated the involvement of immunologically important genes influencing neuroinflammation, cell migration, and chemotaxis (Bhaskar et al., 2012). Abrin exposure resulted in rapid immune and inflammatory response in brain. Oxidative stress after abrin intoxication resulted in activation of several signaling pathways, which leads to abrin toxicity (Narayanan et al., 2004). Bhaskar et al. (2008) reported DNA damage after abrin exposure in human leukemic cells. Some investigators have reported that abrin is poorly absorbed from the intestine. However, there have been reports of severe, sublethal toxicity in adults after ingestion of only one-half to two seeds (Hart, 1963). A 37-year-old man was severely poisoned after ingesting half a seed (Gunsolus, 1955). In an another case, cerebral edema, altered sensorium, and seizures developed 4–6 days after ingesting 7–10 crushed abrus seeds (Sahoo et al., 2008). Initial symptoms of abrus seed ingestion include watery diarrhea at first, later nausea, vomiting, abdominal cramps, and chills. The vomiting and diarrhea becomes bloody. Severe dehydration may result, followed by low blood pressure. Other symptoms may include hallucinations, seizures, and blood in the urine. Within days, the person's liver, spleen, and kidneys may stop working. Death could take place within 36–72 h of exposure. Abrin may produce allergic or anaphylactic responses, particularly in atopic patients (Barceloux, 2008). If death has not occurred within 5 days, the person usually recovers, but may suffer long-term organ damage. The toxicity of abrin depends on the route of exposure, inhalation exposure being considered the most hazardous (Audi et al., 2005). There is no specific antidote reported to date. If abrin powder is inhaled, it can cause the death of tissue in the lungs and airways leading to severe inflammation and edema (Griffiths et al., 1995). Abrin can also cause nephrotoxicity, with inflammatory changes in renal tubules, and kidney oxidative stress, leading to kidney degeneration (Sant et al., 2017).

Ricin

Ricin is a toxic glycoprotein found in castor beans from the *Ricinus communis*, in the family Euphorbiaceae (Worbs et al., 2011). Pure ricin is a water-soluble white powder. Accidental exposure to ricin is unlikely, with the exception of ingestion of castor beans. Ricin is a terrorism risk and people could be exposed through air, food, or water (Anderson, 2012b). As a biological weapons agent, ricin has the potential to be inexpensive and easily produced in large quantities, it is lethal, with no known vaccine or treatment, and has the potential to be distributed in aerosolized form. Roxas-Duncan and Smith described > 20 bioterrorist attempts and attacks involving the use of ricin in the period 1990–2011 (Roxas-Duncan and Smith, 2012).

Ricin structure is formed by a disulfide bond linkage between two polypeptide chains (A and B) (Brinkworth, 2010). As a prototype AB toxin, ricin consists of a sugar-binding B chain (~34 kDa) linked via a disulfide bond to the catalytically active A chain (~32 kDa), which acts as an RNA *N*-glycosidase, resulting in a holo-toxin of about 65 kDa (Lappi et al., 1978; Worbs et al., 2011). The A-subunit enters the epithelial cell through receptor-mediated endocytosis after binding of the B sub-unit to glycoproteins on the surface of cells (Brinkworth, 2010). The A-subunit halts peptide elongation (i.e., protein synthesis) and results in cell death (Doan, 2004). One ricin molecule can disable 2000 ribosomes every minute, finally resulting in the death of the cell (Doan, 2004). Because of high concentration of the mannose receptor in reticuloendothelial cell, such as Kupffer cells and macrophages, they are severely susceptible to ricin toxicity (Audi et al., 2005).

Depending on the primary protein sequence and its specific glycosylation and deamidation level, ricin is a heterogeneous molecule with variable molecular weight, protein charge/isoelectric point, and toxicity (Despeyroux et al., 2000; Sehgal et al., 2010, 2011; Bergström et al., 2015). Discrimination of ricin from *R. communis* agglu-tinin (RCA120) is technically challenging. Ricin can be isolated from defatted castor cake after oil has been extracted. Ricin has been purified by lactamyl-sepharose af-finity chromatography followed by carboxy-methyl Sepharose ion exchange column that yield three ricin fractions (I, II, III). Ricin can exist in different isoforms depend-ing on its seed type and plant variety (Despeyroux et al., 2000).

Various isoforms of ricin have been reported and characterized analytically and toxicologically (Helmy and Pieroni, 2000). There are several isoforms of ricin in-cluding ricin D, ricin E, and the closely related *R. communis* agglutinin (RCA), en-coded by a small multigene family of approximately eight members, some of which are nonfunctional (Tregear and Roberts, 1992). It has been reported that there are two toxic ricins: ricin D and ricin E (Lin and Li, 1980). The D form is found in large grain seeds, whereas the small grain seeds contain both D and E forms of ricin toxin (Despeyroux et al., 2000). Several techniques have been devised to detect and char-acterize ricin, among them CE (capillary electrophoresis) for resolving ricin isomers and differentiating ricin D and E forms based on their different pI values (Dong et al., 2001). Three ricin variants were isolated, which give a single band of 64 kDa under nonreducing conditions and two bands of 30–34 kDa region under reducing condi-tions, with differences in their electrophoretic mobility. In disulfide reducing condi-tions, marginally higher molecular mass of ricin I was observed than ricin II and III because of differential glycosylation (Hegde and Poddar, 1992). Ricin exhibits a high degree of microheterogeneity with respect to carbohydrate content (Kimura et al., 1988). The asparagine (Asn) residue of ricin A and B chains were shown to contain mannose, fucose, xylose, and *N*-acetyl galactosamine sugar components. Ricin iso-forms were separated at acidic pH on native gel. Cytotoxicity study in vero cell lines reveals that ricin III is —four to eight times more toxic than the other two forms. LD50 of ricin III ranges from 5 to 20 ng/kg of body weight in mice and having higher lethal potency than the other two forms (Sehgal et al., 2010).

The first recorded isolation of ricin was by the German scientist H. Stillmark in 1888 during his doctoral work. Besides ricin, *R. communis* seeds contain the homolo-gous but less toxic protein *R. communis* agglutinin (RCA120) (Olsnes et al., 1974).

Ricinus agglutinin RCA120 is a heterotetrameric protein with a molecular weight of 120 kDa. Ricinus agglutinin consists of two ricin-like heterodimers linked via a disulfide bond between the two A chains (Sweeney et al., 1997). Ricin and RCA120 show a high sequence homology of 93% and 84% between the A and B chains of ricin and RCA120, respectively (Roberts et al., 1985). Ricin is a potent toxin but a weak hemagglutinin, whereas RCA120 is only a weak toxin but a strong hemagglutinin (Cawley et al., 1978; Saltvedt, 1976). Accidental and intended *R. communis* intoxications in humans and animals have been known for centuries. The toxicity of ricin in vivo is estimated to be 1–20 mg/kg body weight when ingested and 1–10 µg/kg body weight when delivered by inhalation or injection (Worbs et al., 2011).

The murder of the Bulgarian dissident Georgi Markov in London in 1978 with a ricin-containing pellet injected with an umbrella could be considered as an act of biocrime. Injection of a lethal dose of ricin toxin, as much as 500 µg, in Georgi Markov, had caused immediate local pain and weakness, followed by fever, tachycardia, and swollen lymph nodes in the groin after 36 h. On the second day, he suffered from vascular collapse, shock, and leukocytosis; and on the third day, when he passed away, atrioventricular conduction block, anuria, and hematemesis were observed (Crompton and Gall, 1980; Papaloucas et al., 2008). There have been many incidents involving illegal possession of ricin in various forms meant for inflicting damage to human life.

Different mechanisms of ricin toxicity include direct membrane damage, apoptosis pathway, and promoting the release of cytokines. Ricin is toxic in many routes of administration: ingestion, injection, or even inhalation (Audi et al., 2005). However, due to possibly partial enzymatic degradation of ricin in the digestive tract, oral ingestion is less toxic than the others. Symptoms of toxicity depends upon route of administration, ingestion of ricin would produce vomiting and diarrhea (Franz and Jaax, 1997). Severe dehydration and low blood pressure may follow. Other signs can include hallucinations, seizures, and blood in the urine.

Intramuscular exposure of ricin induces severe localized pain, necrosis of muscle and regional lymph nodes, gastrointestinal hemorrhage, liver necrosis, diffuse nephritis, and diffuse splenitis (Schep et al., 2009). Inhalation of ricin results in respiratory distress with airway and pulmonary lesions. Pulmonary inflammatory biomarkers such as total protein and the number of broncho-alveolar fluid inflammatory cells increases within one-half day. Aerosolized ricin absorbed through the pulmonary tract induces systemic inflammation secondary to cytokine and chemokine release arthralgias, and fever (Pincus et al., 2011). The clinical manifestations appear within 2–24 h and the victim may die within 36–72 h after exposure (Audi et al., 2005). Fluid losses during ricin toxicity may lead to electrolyte imbalances, dehydration, hypotension, and circulatory collapse (Koch and Caplan, 1942). Laboratory abnormalities may include leukocytosis, elevated transaminases and creatinine kinase, hyperbilirubinemia, renal insufficiency, and anemia (Wedin et al., 1986). Ricin causes nuclear DNA damage (Brigotti et al., 2002), activates cell stress responses, and induces programmed cell death via both the intrinsic and extrinsic pathways of apoptosis (Rao et al., 2005; Tesh, 2012; Walsh et al., 2013). Dermal exposure is usually insignificant because of poor absorption through the skin, unless enhanced with a strong solvent like DMSO. Exposure to ricin via direct contact of skin or mucous membranes is not typical and may lead to erythema and pain (Darling and Woods, 2004; Poli et al., 2007; Bradberry et al., 2003).

No medicine has been specifically approved for ricin poisoning. The progressive nature of the toxin's effects requires hospitalization and continual supportive care. Nonetheless, as a possible BW agent, ricin is considered a real threat and therefore research continues for development of prophylactic and therapeutic therapy for it.

Conclusion

Natural toxins are a group of extremely potent toxic compounds. Toxins are placed in between chemical and biological agents as they can be produced by living organisms as well as chemically synthesized. As compared to chemical or biological agents, toxins are a possible favorable choice due to, for example, their ease of production and transport. Progress in biotechnology and aerobiology techniques make them most potent bioterrorism agents, which may cause massive destruction. With effective and specific medical counter measures not clearly available against most of the toxin agents, they pose severe risks to armed forces as well as civilian populations. There is a need for augmenting the evaluation and deployment of existing technologies. Developing state of the art technologies is needed in detection, identification, decontamination, and drug development against natural toxins of warfare importance.

References

Adhikari, M., Negi, B., Kaushik, N., Adhikari, A., Abdulaziz, A., Al-Khedhairy, A.A., Kaushik, N.K., Choi, E.H., 2017. T-2 mycotoxin: toxicological effects and decontamination strategies. Oncotarget 8 (20), 33933–33952.

Aebersold, R., 2003. A mass spectrometric journey into protein and proteome research. J. Am. Soc Mass Spectrom. 14, 685–695.

Agrawal, M., Yadav, P., Lomash, V., Bhaskar, A.S.B., Rao, P.V.L., 2012. T-2 toxin induced skin inflammation and cutaneous injury in mice. Toxicology 302, 255–265.

Agrawal, M., Bhaskar, A.S.B., Rao, P.V.L., 2015. Involvement of mitogen-activated protein kinase pathway in T-2 toxin-induced cell cycle alteration and apoptosis in human neuroblastoma cells. Mol. Neurobiol. 51 (3), 1379–1394.

Akimoto, T., Masuda, A., Yotsu-Yamashita, M., et al., 2013. Synthesis of saxitoxin derivatives bearing guanidine and urea groups at C13 and evaluation of their inhibitory activity on voltage-gated sodium channels. Org. Biomol. Chem. 11 (38), 6642–6649.

Albuquerque, E.X., Barnard, E.A., Porter, C.W., Warnick, J.E., 1974. The density of acetylcholine receptors and their sensitivity in the postsynaptic membrane of muscle endplates. Proc. Natl. Acad. Sci. U. S. A. 71, 2818–2822.

Albuquerque, E.X., Pereira, E.F.R., Castro, N.G., Alkondon, M., Reinhardt, S., Schroder, H., Maelicke, A., 1995. Nicotinic receptor function in the mammalian central nervous system. Ann. N. Y. Acad. Sci. 757, 48–72.

Alouf, J.E., Muller-Alouf, H., 2003. Staphylococcal and streptococcal superantigens: molecular, biological, and clinical aspects. Int. J. Med. Microbiol. 292, 429–440.

Anam, E.M., 2001. Anti-inflammatory activity of compounds isolated from the aerial parts of Abrus precatorius (Fabaceae). Phytomedicine 8 (1), 24–27.

Anderson, P.D., 2012a. Emergency management of chemical weapons injuries. J. Pharm Pract. 25, 61–68.

Anderson, P.D., 2012b. Bioterrorism: toxins as weapons. J. Pharm. Pract. 25 (2), 121–129.

Anderson, P.D., Bokor, G., 2012. Conotoxins: potential weapons from the sea. Bioterr. Biodef. 3, 3. https://doi.org/10.4172/2157-2526.1000120.

Andrinolo, D., Iglesias, V., Garcia, C., Lagos, N., 2002. Toxicokinetics and toxicodynamics of gonyautoxins after an oral toxin dose in cats. Toxicon 40 (6), 699–709.

Arnon, S.S., Schechter, R., Inglesby, T.V., Henderson, D.A., Bartlett, J.G., Ascher, M.S., Eitzen, E., Fine, A.D., Hauer, J., Layton, M., Lillibridge, S., Osterholm, M.T., O'Toole, T., Parker, G., Perl, T.M., Russell, P.K., Swerdlow, D.L., Tonat, K., Working Group on Civilian Biodefense, 2001. Botulinum toxin as a biological weapon: medical and public health management. JAMA 285, 1059–1070.

Audi, J., Belson, M., Patel, M., Schier, J., Osterloh, J., 2005. Ricin poisoning: a comprehensive review. JAMA 294, 2342–2351.

Bane, V., Lehane, M., Dikshit, M., O'Riordan, A., Furey, A., 2014. Tetrodotoxin: chemistry, toxicity, source, distribution and detection. Toxins (Basel) 6, 693–755.

Barceloux, D.G., 2008. Jequirity bean and abrin. In: Medical Toxicology of Natural Substances. John Wiley & Sons, New Jersey, USA, pp. 729–732. Chapter 115.

Bauer, J., Bollwahn, W., Gareis, M., Gedek, B., Heinritzi, K., 1985. Kinetic profiles of diacetoxyscirpenol and two of its metabolites in blood serum of pigs. Appl. Environ. Microbiol. 49, 842–845.

Berger, T., Eisenkraft, A., Bar-Haim, E., Kassirer, M., Aran, A.A., Fogel, I., 2016. Toxins as biological weapons for terror—characteristics, challenges and medical countermeasures: a mini-review. Disaster Mil. Med. 2, 1–7.

Bergström, T., Fredriksson, S.A., Nilsson, C., Åstot, C., 2015. Deamidation in ricin studied by capillary zone electrophoresis- and liquid chromatography-mass spectrometry. J. Chromatogr. B Anal. Technol. Biomed. Life Sci. 974, 109–117.

Bernd, C., Bernd, L., 2008. Determination of marine biotoxins relevant for regulations: from the mouse bioassay to coupled LC-MS methods. Anal. Bioanal. Chem. 391, 117–134.

Berti, M., Milandri, A., 2014. Le biotossine marine. In: Schirone, M., Visciano, P. (Eds.), Igiene Degli Alimenti. Edagricole, Bologna, pp. 163–198.

Bettina, C.F., Avanish, K.V., 2013. Bacterial toxins staphylococcal enterotoxin B. Microbiol. Spectr. 1 (2), https://doi.org/10.1128/microbiolspec.AID-0002-2012.

Bhaskar, A.S.B., Deb, U., Kumar, O., Rao, P.V.L., 2008. Abrin induced oxidative stress mediated DNA damage in human leukemic cells and its reversal by N-acetylcysteine. Toxicol. In Vitro 22, 1902–1908.

Bhaskar, A.S.B., Gupta, N., Rao, P.V.L., 2012. Transcriptomic profile of host response in mouse brain after exposure to plant toxin abrin. Toxicology 299 (1), 33–43.

Bhasker, A.S.B., Sant, B., Yadav, P., Agrawal, M., Rao, P.V.L., 2014. Plant toxin abrin induced oxidative stress mediated neurodegenerative changes in mice. Neurotoxicology 44, 194–203.

Bhat, R.V., Ramakrishna, Y., Beedu, S.R., Munshi, K.L., 1989. Outbreak of trichothecene mycotoxicosis associated with consumption of mould-damaged wheat products in Kashmir valley, India. Lancet 7, 35–37.

Blasi, J., Chapman, E.R., Link, E., Binz, T., Yamasaki, S., De Camilli, P., Sudhof, T.C., Niemann, H., Jahn, R., 1993. Botulinum neurotoxin A selectively cleaves the synaptic protein SNAP25. Nature 365, 160–163.

Bradberry, S.M., Dickers, K.J., Rice, P., Griffiths, G.D., Vale, J.A., 2003. Ricin poisoning. Toxicol. Rev. 22, 65–70.

Brian, P.W., Dawkins, A.W., Grove, J.F., Hemming, H.G., Lowe, D., Norris, G.L.F., 1961. Phytotoxic compounds produced by F. equiseti. J. Exp. Bot 12, 1–12.

Bricelj, V., Shumway, S., 1998. Paralytic shellfish toxins in bivalve molluscs: occurrence, transfer kinetics, and biotransformation. Rev. Fish. Sci. 6, 315–383.

Brigotti, M., Alfieri, R., Sestili, P., Bonelli, M., Petronini, P.G., Guidarelli, A., et al., 2002. Damage to nuclear DNA induced by Shiga toxin and ricin in human endothelial cells. FASEB J. 16, 365–372.

Brinkworth, C.S., 2010. Identification of ricin in crude and purified extracts from castor beans using on-target tryptic digestion and MALDI mass spectrometry. Anal. Chem. 82, 5246–5252.

Bukowski, R., Vaughn, C., Bottomley, R., Chen, T., 1982. Phase II study of anguidine in gastrointestinal malignancies: a Southwest Oncology Group study. Cancer Treat Rep. 66, 381–383.

Bunner, D.L., Neufeld, H.A., Brennecke, L.H., Campbell, Y.G., Dinterman, R.E., Pelosi, J.G., 1985. Clinical and Hematological Effects of T-2 Toxin in Rats. United States Army Medical Research Institute of Infectious Diseases, Fort Detrick. DTIC ADA 158874.

Burmeister, H.R., Ellis, J.J., Yates, S.G., 1971. Correlation of biological to chromatographic data for two mycotoxins elaborated by Fusarium. Appl. Microbiol. 21, 673–675.

Catterall, W.A., 1985. The voltage sensitive sodium channel: a receptor for multiple neurotoxins. In: Anderson, D.M., White, A.W., Baden, D.G. (Eds.), Toxic Dinoflagellates. Elsevier Science Publishing Co., Inc., New York, pp. 329–342.

Cawley, D.B., Hedblom, M.L., Houston, L.L., 1978. Homology between ricin and *Ricinus communis* agglutinin: amino terminal sequence analysis and protein synthesis inhibition studies. Arch. Biochem. Biophys. 190, 744–755.

Cervantes, C.R., Durán, R., Faro, L.F., Alfonso, P.M., 2009. Effects of systemic administration of saxitoxin on serotonin levels in some discrete rat brain regions. Med. Chem. 5, 336–342.

Cervantes, C.R., Faro, L.F., Durán, B.R., Alfonso, P.M., 2011. Alterations of 3,4-dihydroxyphenylethylamine and its metabolite 3,4-dihydroxyphenylacetic produced in rat brain tissues after systemic administration of saxitoxin. Neurochem. Int. 59, 643–647.

Chaturvedi, K., Kumar, O., 2015. Purification and characterization of abrin variants from three different varieties of Abrus precatorius seeds. Planta Med. 81, PC16.

Chau, R., Kalaitzis, J.A., Neilan, B.A., 2011. On the origins and biosynthesis of tetrodotoxin. Aquat. Toxicol. 104, 61–72.

Chaudhary, M., Rao, P.V.L., 2010. Brain oxidative stress after dermal and subcutaneous exposure of T-2 toxin in mice. Food Chem. Toxicol. 48, 3436–3442.

Chaudhary, M., Jayaraj, R., Bhaskar, A.S., Lakshmana Rao, P.V., 2009. Oxidative stress induction by T-2 toxin causes DNA damage and triggers apoptosis via caspase. Toxicology 262, 153–161.

Cheeke, P.R., 1998. Mycotoxins in cereal grains and supplements. In: Cheeke, P.R. (Ed.), Natural Toxicants in Feeds, Forages, and Poisonous Plants. Interstate Publishers, Inc., Danville, IL, pp. 87–136.

Chen, Y.L., Chow, L.P., Tsugita, A., Lin, J.Y., 1992. The complete primary structure of abrin a B chain. FEBS Lett. 309, 115–118.

Chi, M.S., Mirocha, C.J., 1978. Necrotic oral lesions in chickens fed diacetoxyscirpenol, T-2 toxin, and crotocin. Poult. Sci 57, 807–808.

Ching, P.K., Ramos, R.A., De los Reyes, V.C., Sucaldito, M.M., Tayag, E., 2015. Lethal paralytic shellfish poisoning from consumption of green musselbroth, Western Samar, Philippines, August 2013. Western Pac. Surveill. Response J. 6, 22–26.

Chulze, S.N., 2010. Strategies to reduce mycotoxin levels in maize during storage: a review. Food Addit. Contam. Part A 27, 651–657.

Clarke, S.C., 2005. Bacteria as potential tools in bioterrorism, with an emphasis on bacterial toxins. Br. J. Biomed. Sci. 62, 40–46.

Coffield, J.A., Bakry, N., Zhang, R.D., Carlson, J., Gomella, L.G., Simpson, L.L., 1997. *In vitro* characterization of botulinum toxin types A, C and D action on human tissues: combined electrophysiologic, pharmacologic and molecular biologic approaches. J. Pharmacol. Exp. Ther. 280, 1489–1498.

Cohen, G.M., 1997. Caspases: the executioners of apoptosis. Biochem. J. 326, 1–16.

Crompton, R., Gall, D., 1980. Georgi Markov—death in a pellet. Med. Leg. J. 48, 51–62.

Darling, R.G., Woods, J.B., 2004. Medical Management of Biological Casualties Handbook, fifth ed. US Army Medical Research Institute of Infectious Diseases, Fort Detrick, MD80–91.

Dembek, Z.F., Smith, L.A., Rusnak, J.M., 2007. Botulinum toxin. In: Dembek, Z.F. (Ed.), Medical Aspects of Biological Warfare. Office of the Surgeon General, US Army Medical Department Center and School; Borden Institute, Walter Reed Army Medical Center, Washington, DC, pp. 337–353.

Denac, H., Mevissen, M., Scholtysik, G., 2000. Structure, function and pharmacology of voltage-gated sodium channels. Naunyn Schmiedebergs Arch. Pharmacol. 362, 453–479.

Despeyroux, D., Walker, N., Pearce, M., Fisher, M., McDonnell, M., Bailey, S.C., Griffiths, G.D., Watts, P., 2000. Characterization of ricin heterogeneity by electrospray mass spectrometry, capillary electrophoresis, and resonant mirror. Anal. Biochem. 279, 23–36.

Dickers, K.J., Bradberry, S.M., Rice, P., Griffiths, G.D., Vale, J.A., 2003. Abrin poisioning. Toxicol. Rev. 22, 137–142.

Dinges, M.M., Orwin, P.M., Schlievert, P.M., 2000. Exotoxins of Staphylococcus aureus. Clin. Microbiol. Rev. 13, 16–34.

Doan, L.G., 2004. Ricin: mechanism of toxicity, clinical manifestations, and vaccine development. A review. J. Toxicol. Clin. Toxicol. 42, 201–208.

Dolah, F.M.V., 2000. Marine algal toxins: origins, health effects, and their increased occurrence. Environ. Health Perspect. 108, 133–141.

Dong, H.N., Eun, J.P., Myung, S.K., Cheong, K.C., Byung, H.W., Hye, S.L., Kang, C.L., 2001. Characterization of two ricin isoforms by sodium dodecyl sulfate-capillary gel electrophoresis and capillary isoelectric focusing. Bull. Kor. Chem. Soc. 32, 12.

EFSA, 2009. Scientific opinion of the panel on contaminants in the food chain on are quest from the European Commission on marine biotoxins in shellfish—Saxitoxin group. EFSA J. 1019, 1–76.

Endo, Y., Mitsui, K., Tsurungi, K., 1987. The mechanism of action of ricin and related lectins on eukaryotic ribosomes. The site and characteristic of the modification in 28S ribosomal RNA caused by toxins. J. Biol. Chem. 262, 5908–5912.

Fegan, D., Andresen, D., 1997. Conus geographus envenomation. Lancet 349, 1672.

Felipe, D., Patricia, B.R., Juliane, M.S., Daniela, M.B., Joao, S.Y., 2016. Behavioral alterations induced by repeated saxitoxin exposure in drinking water. J. Venom. Anim. Toxins Incl. Trop. Dis. 22, 18.

Fernandez, I., Valladolid, G., Varon, J., Sternbach, G., 2011. Encounters with venomous sea-life. J. Emerg. Med. 40, 103–112.

Fodstad, O., Olsnes, S., Pihl, A., 1976. Toxicity, distribution and elimination of the cancer-ostatic lectins abrin and ricin after parenteral injection into mice. Br. J. Cancer 34, 418–425.

Franz, D.R., Jaax, N.R., 1997. Ricin toxin. In: Zajtchuk, R., Bellamy, R.F. (Eds.), Textbook of Military Medicine—Medical Aspects of Chemical and Biological Warfare. Borden Institute, Washington, DC, pp. 631–642.

Franz, D.R., Parrott, C.D., Takafuji, E.T., 1997. The U.S. biological warfare and biological defense programs. In: Sidell, F.R., Takafuji, E.T., Franz, D.R. (Eds.), Textbook of Military Medicine. Part I. Warfare, Weaponry and the Casualty. vol. 3. U.S. Government Printing Office, Washington, DC, pp. 425–436.

Funatsu, G., Taguchi, Y., Kamenosono, M., Yanaka, M., 1988. The complete amino acid sequence of the A-chain of abrin-a, a toxic protein from the seeds of Abrus precatorius. Agric. Biol. Chem. 52, 1095–1097.

Gerberick, G.F., Sorenson, W.G., Lewis, D.M., 1984. The effects of T-2 toxin on alveolar macrophage function in vitro. Environ. Res 33, 246–260.

Ghosh, S., Hazra, A.K., Banerjee, S., Mukherjee, B., 2004. The seasonal toxicological profile of four puffer fish species collected along Bengal coast, India. Indian J. Mar. Sci. 33, 276–280.

Gill, D.M., 1982. Bacterial toxins: a table of lethal amounts. Microbiol. Rev. 46, 86–94.

Gouaux, J.E., Braha, O., Hobaugh, M.R., Song, L., Cheley, S., Shustak, C., Bayley, H., 1994. Subunit stoichiometry of staphylococcal α-hemolysin in crystals and on membranes: a heptameric transmembrane pore. Proc. Natl. Acad. Sci. U. S. A. 91, 12828–12831.

Gray, W.R., Luque, A., Olivera, B.M., Barrett, J., Cruz, L.J., 1981. Peptide toxins from Conus geographus venom. J. Biol. Chem. 256, 4734–4740.

Griffiths, G.D., Lindsay, C.D., Allenby, A.C., Bailey, S.C., Scawin, J.W., Rice, P., Upshall, D.G., 1995. Protection against inhalation toxicity of ricin and abrin by immunization. Hum. Exp. Toxicol. 14, 155–164.

Grove, J.F., Mortimer, P.H., 1969. The cytotoxicity of some transformation products of diacetoxyscirpenol. Biochem. Pharmacol. 18, 1473–1478.

Gunsolus, J.M., 1955. Toxicity of jequirity beans. J. Am. Med. Assoc. 157, 779.

Haddad 2nd, V., de Paula Neto, J.B., Cobo, V.J., 2006. Venomous mollusks: the risks of human accidents by conus snails (gastropoda: conidae) in Brazil. Rev. Soc. Bras. Med. Trop. 39, 498–500.

Hale, M.L., 2012. Staphylococcal Enterotoxins, Staphylococcal Enterotoxin B and Bioterrorism. In: Morse, Stephen Dr (Eds.), Bioterrorism. InTech. ISBN: 978-953-51-0205-02.

Halstead, B.W., Schantz, E., 1984. Paralytic Shellfish Poisoning. WHO, Geneva, Switzerland. World Health Organization Offset Publication No. 79.

Hart, M., 1963. Jequirity bean poisoning. N. Engl. J. Med. 268, 885–886.

Haschek, W.M., Beasley, V.R., 2009. Trichothecene mycotoxins. In: Gupta, R.C. (Ed.), Handbook of Toxicology of Chemical Warfare Agents. Academic Press, London, pp. 353–369. Chapter 26.

Hegde, R., Poddar, S.K., 1992. Studies on the variants of the protein toxins ricin and abrin. Eur. J. Biochem. 204, 155–164.

Hegde, R., Maiti, T.K., Podder, S.K., 1991. Purification and characterization of three toxins and two agglutinins from Abrus precatorius seed by using lactamyl-Sepharose affinity chromatography. Anal. Biochem. 194, 101–109.

Hegde, R., Karande, A.A., Podder, S.K., 1993. The variants of the protein toxins abrin and ricin A useful guide to understanding the processing events in the toxin transport. Eur. J. Biochem. 215 (41), 1–419.

Helmy, M., Pieroni, G., 2000. RCA60: purification and characterization of ricin D isoforms from Ricinus sanguineus. J. Plant Physiol. 156, 477–482.

Hernandez, C., Ulloa, J., Vergara, J., Espejo, R., Cabello, F., 2005. Infecciones por Vibrio para-haemolyticus e intoxicaciones por algas: problemas emergentes de salud pública en Chile. Rev. Med. Chile 133, 1081–1088.

Hille, B., 1971. The permeability of the sodium channel to organic cations in myelinated nerve. J. Gen. Physiol. 58, 599–619.

Hoerr, F.J., Carlton, W.W., Yagen, B., 1981. Mycotoxicosis caused by a single dose of T-2 toxin or diacetoxyscirpenol in broiler chickens. Vet. Pathol. 18, 652–664.

Horn, A.K., Porter, J.D., Evinger, C., 1993. Botulinum toxin paralysis of the orbicularis oculi muscle. Types and time course of alterations in muscle structure, physiology and lid kine-matics. Exp. Brain Res. 96, 39–53.

Hu, D.L., Zhu, G., Mori, F., Omoe, M., Wakabayashi, K., Kaneko, S., Shinagawa, K., Nakane, A., 2007. Staphylococcal enterotoxin induces emisis through increasing serotonin release in intestine and it is downregulated by cannabinoid receptor 1. Cell. Microbiol. 9, 2267–2277.

Hundell, H., 2010. The state of U.S. freshwater harmful algal blooms assessment, policy and legislation. Toxicon 55, 1024–1034.

Iandolo, J.J., Shafer, W.M., 1977. Regulation of staphylococcal enterotoxin B. Infect. Immun. 16 (2), 610–616.

Jeffrey, W.F., Bradley, S., Thibaut, P., Philippe, T., 2011. Antibodies for biodefense. mAbs 3, 517–527.

Joffe, A.Z., 1971. Alimentary toxic aleukia. In: Kadis, S., Ciegler, A., Ajl, S.J. (Eds.), Microbiol Toxins. Algal and Fungal Toxins, vol 7. Academic Press, New York, NY, pp. 139–189.

Jorge, L., Laura, P.R., Lucía, B., Juan, M.V., Ana, G.C., 2015. Tetrodotoxin, an extremely potent marine neurotoxin: distribution, toxicity, origin and therapeutical uses. Mar. Drugs 13, 6384–6406.

Jun, D.Y., Kim, J.S., Park, H.S., Song, W.S., Bae, Y.S., Kim, Y.H., 2007. Cytotoxicity of di-acetoxyscirpenol is associated with apoptosis by activation of caspase-8 and interruption of cell cycle progression by down-regulation of cdk4 and cyclin B1 in human Jurkat T cells. Toxicol. Appl. Pharmacol. 222, 190–201.

Kaas, Q., Yu, R., Jin, A.H., Dutertre, S., Craik, D.J., 2012. ConoServer: updated content, knowl-edge, and discovery tools in the conopeptide database. Nucleic Acids Res. 40, 325–330.

Kachuei, R., Rezaie, S., Yadegari, M.H., Safaie, N., Allameh, A.A., Aref-poor, M.A., Fooladi, A.A.I., Riazipour, M., Abadi, H.M.M., 2014. Determination of T-2 Mycotoxin in Fusarium strains by HPLC with fluorescence detector. J. Appl. Biotech. Rep. 1, 38–43.

Keller, J.E., Neale, E.A., Oyler, G., Adler, M., 1999. Persistence of botulinum neurotoxin ac-tion in cultured spinal cord cells. FEBS Lett. 456, 137–142.

Kim, Y.H., Brown, G.B., Mosher, H.S., Fuhrman, F.A., 1975. Tetrodotoxin: occurrence in Atelopid frogs of Costa Rica. Science 189, 151–152.

Kimura, Y., Hase, S., Kobayashi, Y., Kyogoku, Y., Ikenaka, K., Funatsu, G., 1988. Structures of sugar chains of ricin D. J. Biochem. 103, 944–949.

Koch, L.A., Caplan, J., 1942. Castor bean poisoning. Am. J. Dis. Child. 64 (3), 485–486.

Krakauer, T., Pradhan, K., Stiles, B.G., 2016a. Staphylococcal superantigens spark hostmedi-ated danger signals. Front. Immunol. 7, 23.

Krakauer, T., Pradhan, K., Stiles, B.G., 2016b. Staphylococcal superantigens spark host-mediated danger signals. Front. Immunol. 7, 23. https://doi.org/10.3389/fimmu.2016.00023.

Landsberg, J.H., Hall, S., Johannessen, J.N., White, K.D., Conrad, S.M., Abbott, J.P., et al., 2006. Saxitoxin puffer fish poisoning in the United States, with the first report of Pyrodinium bahamense as the putative toxin source. Environ. Health Perspect. 114, 1502–1507.

Lange, W.R., 1990. Puffer fish poisoning. Am. Fam. Physician 42, 1029–1033.

Lappi, D.A., Kapmeyer, W., Beglau, J.M., Kaplan, N.O., 1978. The disulfide bond connecting the chains of ricin. Proc. Natl. Acad. Sci. U. S. A. 75, 1096–1100.

Le Loir, Y., Baron, F., Gautier, M., 2003. Staphylococcal aureus and food poisoning. Genet. Mol. Res. 2, 630–676.

Lefertre, C., Perelle, S., Dilasser, F., Fach, P., 2003. Identification of a new putative enterotoxin SEU encoded by the egc cluster of staphylococcus aureus. J. Appl. Microbiol. 95, 38–43.

Li, Y., Wang, Z., Beier, R.C., et al., 2011. T-2 toxin, a trichothecene mycotoxin: review of toxicity, metabolism, and analytical methods. J. Agric. Food Chem. 59, 3441–3453.

Lin, T.T., Li, S.L., 1980. Purification and physicochemical properties of ricins and agglutinins from Ricinus communis. Eur. J. Biochem. 105, 453–459.

Llewellin, L.E., 2006. Saxitoxin, a toxic marine natural product that targets a multitude of receptors. Nat. Prod. Rep. 23, 200–222.

Loke, Y.K., Tam, M.H., 1997. A unique case of tetrodotoxin poisoning. Med. J. Malays. 52, 172–174.

Lowy, F.D., 1998. Staphylococcus aureus infections. N. Engl. J. Med. 339, 520–532.

Madsen, J.M., 2001. Toxins as weapons of mass destruction. A comparison and contrast with biological-warfare and chemical warfare agents. Clin. Lab. Med. 21, 593–605.

Mayura, K., Smith, E.E., Clement, B.A., Harvey, R.B., Kubena, L.F., Phillips, T.D., 1987. Developmental toxicity of diacetoxyscirpenol in the mouse. Toxicology 45, 245–255.

McCarthy, M., Bane, V., García-Altares, M., Van Pelt, F.N.A.M., Furey, A., O'Halloran, J., 2015. Assessment of emerging biotoxins (pinnatoxinGand spirolides) at Europe's first marine reserve: LoughHyne. Toxicon 108, 202–209.

Mehrdad, S., Rudolf, M., Roberto, C., Gerlinde, W., Chiara, D.A., Rainer, S., Rudolf, K., Gerhard, A., Franz, B., 2011. Isolation and characterization of a new less-toxic derivative of the fusarium mycotoxin diacetoxyscirpenol after thermal treatment. J. Agric. Food Chem. 59, 9709–9714.

Merwe, D.V., 2015. Cyanobacterial (Blue green algae) toxins. In: Gupta, R.C. (Ed.), Handbook of Toxicology of Chemical Warfare Agents. Second ed. Academic Press, London, pp. 421–429. Chapter 31.

Milne, T.J., Abbenante, G., Tyndall, J.D., Halliday, J., Lewis, R.J., 2003. Isolation and characterization of a cone snail protease with homology to CRISP proteins of the pathogenesis-related protein superfamily. J. Biol. Chem. 278, 31105–31110.

Mirocha, C.J., Pawlosky, R.A., Chatterjee, K., Watson, S., Hayes, A.W., 1983. Analysis for Fusarium toxins in various samples implicated in biological warfare in Southeast Asia. J. Assoc. Off. Anal. Chem 66, 1485–1499.

Mongi, S., Mahfoud, M., Amel, B., Abdelwaheb, A., Wassim, K., Kamel, J., Abdelfattah, E.F., 2011. Extracted tetrodotoxin from puffer fish Lagocephalus lagocephalus induced hepatotoxicity and nephrotoxicity to Wistar rats. Afr. J. Biotechnol. 10, 8140–8145.

Moran, O., Picollo, A., Conti, F., 2003. Tonic and phasic guanidinium toxin-block of skeletal muscle Na channels expressed in mammalian cells. Biophys. J. 84, 2999–3006.

Mosher, H.S., Fuhrman, F.A., Buchwald, H.D., Fischer, H.G., 1964. Tarichatoxin tetrodotoxin: a potent neurotoxin. Science 144, 1100–1110.

Moss, M.O., 2002. Mycotoxin review—2. Fusarium. Mycologist 16, 158–161.

Nakamura, M., Yasumoto, T., 1985. Tetrodotoxin derivatives in puffer fish. Toxicon 23, 271–276.

Narayanan, S., Surolia, A., Karande, A.A., 2004. Ribosome inactivating protein and apoptosis: abrin causes cell death via mitochondrial pathway in Jurkat cells. Biochem. J. 377, 233–240.

Noguchi, T., Ebesu, J.S.M., 2001. Puffer poisoning: epidemiology and treatment. J. Toxicol. Toxin Rev. 20, 1–10.

Noguchi, T., Arakawa, O., Takatani, T., 2006. TTX accumulation in pufferfish—review. Comp. Biochem. Physiol. D 1, 145–152.

Nunez, E.J., Yotsu-Yamashita, M., Sierra-Beltran, A.P., Yasumoto, T., Ochoa, J.L., 2000. Toxicities and distribution of tetrodotoxin ion the tissue of puffer fish found in the coast of Baja California Peninsula, Mexico. Toxicon 38, 729–734.

Ohta, M., Matsumoto, H., Ishii, K., Ueno, Y., 1978. Metabolism of trichothecene mycotoxins. II. Substrate specificity of microsomal deacetylation of trichothecenes. J. Biochem. 84, 697–706.

Olsnes, S., Pihl, A., 1976. Kinetics of binding of the toxic lectins abrin and ricin to surface receptors of human cells. J. Biol. Chem. 251, 3977–3984.

Olsnes, S., Saltvedt, E., Pihl, A., 1974. Isolation and comparison of galactose-binding lectins from Abrus precatorius and Ricinus communis. J. Biol. Chem. 249, 803–810.

Oshima, Y., 1995. Chemical and enzymatic transformation of paralytic shellfish toxins in marine organisms. In: Lassus, P., Arzul, G., Erard-Le Denn, E., Gentien, P., Marcaillou-Le baut, C. (Eds.), Harmful Marine Algal Blooms. Lavoisiers Science Publishers, Paris, France, pp. 475–480.

Pang, V.F., Lambert, R.J., Felsburg, P.J., Beasley, V.R., Buck, W.B., Hascheck, W.M., 1987. Experimental T-2 toxicosis in swine following inhalation exposure: effects on pulmonary and systemic immunity, and morphologic changes. Toxicol. Pathol. 15, 308–319.

Papaloucas, M., Papaloucas, C., Stergioulas, A., 2008. Ricin and the assassination of Georgi Markov. Pak. J. Biol. Sci. 11, 2370–2371.

Patocka, J., Streda, L., 2006. Protein biotoxins of military significance. Acta Med. (Hradec Kralove) 49, 3–11.

Patocka, J., Hon, Z., Streda, L., Kuca, K., Jun, D., 2007. Biohazards of protein biotoxins. Def. Sci. J. 57, 825–837.

Patricia, B.P., Felipe, D., Juliane, M.S., Jose, M.M., Joao, S.Y., 2014. Oxidative stress in rats induced by consumption of saxitoxin contaminated drink water. Harmful Algae 37, 68–74.

Pearce, L.B., Borodic, G.E., First, E.R., MacCallum, R.D., 1994. Measurement of botulinum toxin: assessment of the lethality assay. Toxicol. Appl. Pharmacol. 128, 69–77.

Pearce, L.B., First, E.R., Borodic, G.E., 1995. Botulinum toxin-death versus localized denervation. J. R. Soc. Med. 88, 239–240.

Pearce, L.B., First, E.R., MacCallum, R.D., Gupta, A., 1997. Pharmacologic characterization of Botulinum toxin for basic science and medicine. Toxicon 35, 1373–1412.

Peraica, M., Radic, B., Lucic, A., Pavlovic, M., 1999. Toxic effects of mycotoxins in humans. Bull. World Health Organ. 77, 754–766.

Perreault, F., Matias, M.S., Melegari, S.P., Creppy, E.E., Popovic, R., Matias, W.G., 2011. Investigation of animal and algal bioassays for reliable saxitoxin ecotoxicity and cytotoxicity risk evaluation. Ecotoxicol. Environ. Safety 74, 1021–1026.

Pierina, V., Maria, S., Miriam, B., Anna, M., Rosanna, T., Giovanna, S., 2016. Marine biotoxins: occurrence, toxicity, regulator, limits and reference methods. Front. Microbiol. 7, 1051.

Pinchuk, I.V., Beswick, E.J., Reyes, V.E., 2010. Staphylococcal enterotoxins. Toxins 2, 2177–2197.

Pincus, S.H., Smallshaw, J.E., Song, K., Berry, J., Vitetta, E.S., 2011. Passive and active vaccination strategies to prevent ricin poisoning. Toxins 3, 1163–1184.

Pitschmann, V., 2014. Overall view of chemical and biochemical weapons. Toxins 6, 1761–1784.

Plumley, F., 1997. Marine algal toxins: biochemistry, genetics, and molecular biology. Limnol. Oceanogr. 42, 1252–1264.

Poli, M.A., Roy, C., Huebner, K.D., et al., 2007. Ricin. In: Dembek, Z.F. (Ed.), Medical Aspects of Biological Warfare. Office of the Surgeon General, Washington, DC, pp. 323–335.

Rao, P.V.L., Jayaraj, R., Bhaskar, A.S.B., Kumar, O., Bhattacharya, R., Saxena, P., Dash, P.K., Vijayaraghavan, R., 2005. Mechanism of ricin-induced apoptosis in human cervical cancer cells. Biochem. Pharmacol. 69, 855–865.

Ravindran, J., Agrawal, M., Gupta, N., Rao, P.V.L., 2011. Alteration of blood brain barrier permeability by T-2 toxin: role of MMP-9 and inflammatory cytokines. Toxicology 280, 44–52.

Reiss, J., 1973. Influence of the mycotoxins patulin and diacetoxyscirpenol on fungi. J. Gen. Appl. Microbiol. 19, 415–420.

Rice, R.D., Halstead, B.W., 1968. Report of fatal cone shell sting by Conus geographus Linnaeus. Toxicon 5, 223–224.

Richardson, K.E., Hamilton, P.B., 1990. Comparative toxicity of scirpentriol and its acetylated derivatives. Poult. Sci. 69, 397–402.

Roberts, L.M., Lamb, F.I., Pappin, D.J., Lord, J.M., 1985. The primary sequence of *Ricinus communis* agglutinin. Comparison with ricin. J. Biol. Chem. 260, 15682–15686.

Rodrigue, D.C., Etzel, R.A., Hall, S., De Porras, E., Velasquez, O.H., Tauxe, R.V., Kilbourne, E.M., Blake, P.A., 1990. Lethal paralytic shellfish poisoning in Guatemala. Am. J. Trop. Med. Hyg. 42, 267–271.

Roxas-Duncan, V.I., Smith, L.A., 2012. Ricin perspective in bioterrorism. In: Morse, S.A. (Ed.), Bioterrorism. Rijeka, Croatia, InTech, pp. 133–158.

Roy, P., Sahni, A.K., Kumar, A., 2015. A fatal case of staphylococcal toxic shock syndrome. Med. J. Armed Forces India 71, S107–S110.

Rusnak, J.M., Kortepeter, M., Ulrich, R., Poli, M., Boudreau, E., 2004. Laboratory exposures to Staphylococcal enterotoxin B. Emerg. Infect. Dis. 10, 1544–1549.

Russmann, H., 2003. Toxine—Biogene Gifte und potenzielle Kampfstoffe. 46. Springer-Verlag, Heidelberg989–996.

Sahni, V., Agarwal, N.P., Sikdar, S., 2007. Acute demyelinating encephalitis after jequirty pea ingestion (Abrus precatorious). Clin. Toxicol. 45, 77–79.

Sahoo, R., Hamide, A., Amalnath, S., Narayan, B.S., 2008. Acute demyelinating encephalitis due to abrus precatorious poisioning—complete recovery after steroid therapy. Clin. Toxicol. 46, 1071–1079.

Saltvedt, E., 1976. Structure and toxicity of pure ricinus agglutinin. Biochim. Biophys. Acta 451, 536–548.

Sant, B., Rao, P.V.L., Nagar, D.P., Pant, S.C., Bhasker, A.S.B., 2017. Evaluation of abrin induced nephrotoxicity by using novel renal injury markers. Toxicon 131, 20–28.

Saoudi, M., Rabeh, F.B., Jammoussi, K., Abdelmouleh, A., Belbahri, L., Feki, A.E., 2007. Biochemical and physiological responses in Wistar rat after administration of puffer fish (Lagocephalus lagocephalus) flesh. J. Food Agric. Environ. 5, 107–111.

Schantz, E.J., et al., 1975. The structure of saxitoxin. J. Am. Chem. Soc. 97, 1238.

Schep, L.J., Temple, W.A., Butt, G.A., Beasley, M.D., 2009. Ricin as a weapon of mass terror—separating fact from fiction. Environ. Int. 35, 1267–1271.

Schirone, M., Berti, M., Zitti, G., Ferri, N., Tofalo, R., Suzzi, G., et al., 2011. Monitoring of marine biotoxins in *Mytilus galloprovincialis* of central Adriatic Sea. Ital. J. Food Sci. 23, 431–435.

Seeboth, J., Solinhac, R., Oswald, I.P., Guzylack-Piriou, L., 2012. The fungal T-2 toxin alters the activation of primary macrophages induced by TLR-agonists resulting in a decrease of the inflammatory response in the pig. Vet. Res. 43, 35.

Sehgal, P., Khan, M., Kumar, O., Vijayaraghavan, R., 2010. Purification, characterization and toxicity profile of ricin isoforms from castor beans. Food Chem. Toxicol. 48, 3171–3176.

Sehgal, P., Kumar, O., Kameswararao, M., Ravindran, J., Khan, M., Sharma, S., Vijayaraghavan, R., Prasad, G.B.K.S., 2011. Differential toxicity profile of ricin isoforms correlates with their glycosylation levels. Toxicology 282, 56–67.

Simidu, U., Kita-Tsukamoto, K., Yasumoto, T., Yotsu, M., 1990. Taxonomy of four marine bacterial strains that produce tetrodotoxin. Int. J. Syst. Bacteriol. 40, 331–336.

Slater, L.N., Greenfield, R.A., 2003. Biological toxins as potential agents of bioterrorism. J. Okla State Med. Assoc. 96, 73–76.

Stark, A.A., 2005. Threat assessment of mycotoxins as weapons: molecular mechanisms of acute toxicity. J. Food Prot. 68, 1285–1293.

Sugiyama, H., Brenner, S.L., Dasgupta, B.R., 1975. Detection of Clostridium botulinum toxin by local paralysis elicited with intramuscular challenge. Appl. Microbiol. 30, 420–423.

Swaminathan, S., Furey, W., Pletcher, J., Sax, M., 1992. Crystal structure of staphylococcal enterotoxin B, a superantigen. Nature 359, 801–806.

Swanson, S.P., Nicoletti, J., Rood, H.D., Buck, W.B., Cote, L.M., Yoshizawa, T., 1987. Metabolism of three trichothecene mycotoxins, T-2 toxin, diacetoxyscirpenol and deoxynivalenol by bovine rumen microorganisms. J. Chromatogr. 414, 335–342.

Sweeney, E.C., Tonevitsky, A.G., Temiakov, D.E., Agapov, I.I., Saward, S., Palmer, R.A., 1997. Preliminary crystallographic characterization of ricin agglutinin. Proteins Struct. Funct. Bioinf. 28, 586–589.

Takahashi, M., Kameyama, S., Sakaguchi, G., 1990. Assay in mice for low levels of Clostridium botulinum toxin. Int. J. Food Microbiol. 11, 271–277.

Tamar, B., Arik, E., Erez, B.H., Michael, K., Adi, A.A., Itay, F., 2016. Toxins as biological weapons for terror-characteristics, challenges and medical countermeasures: a mini review. Disaster Mil. Med. 2, 7.

Tchitchkine, A., 1905. Essai d'immunisation par la voie gastro-intestinale contre la toxine botulique. Ann. Inst. Pasteur xix, 335.

Tesh, V.L., 2012. The induction of apoptosis by Shiga toxins and ricin. Curr. Top. Microbiol. Immunol. 357, 137–178.

Tregear, J.W., Roberts, L.M., 1992. The lectin gene family of Ricinus communis: cloning of a functional ricin gene and three lectin pseudogenes. Plant Mol. Biol. 18, 515–525.

Trenholm, H.L., Friend, D.W., Hamilton, R.M.G., Prelusky, D.B., Foster, B.C., 1989. Lethal toxicity and non specific effects. In: Beasley, V.R. (Ed.), Trichothecene Mycotoxicosis: Pathophysiologic Effects. CRC Press, Boca Raton, FL, pp. 107–141.

Trevino, S., 1998. Fish and shellfish poisoning. Clin. Lab. Sci. J. Am. Soc. Med. Technol. 11, 309–314.

Ueno, Y., 1977. Trichothecenes: overview address. In: Rodricks, J.V., Hesseltine, D.W., Mehlman, M.A. (Eds.), Mycotoxins in Human and Animal Health. Pathotox Publishers, Park Forest South, IL, pp. 189–207.

Ulrich, R.G., Sidell, S., Taylor, T.J., Wilhelmsen, C.L., Franz, D.R., 1997. Staphylococcal enterotoxin B and related pyrogenic toxins. In: Textbook of Military Medicine. Part I. Warfare, Weaponry and the Casualty, vol. 3. US Government Printing Office, Washington, DC, pp. 621–631.

Vaishali, B., Mary, L., Madhurima, D., Alan, O.R., Ambrose, F., 2014. Tetrodotoxin: chemistry, toxicity, source, distribution and detection. Toxins 6, 693–755.

Van Egmond, H.P., Schothorst, R.C., Jonker, M.A., 2007. Regulations relating to mycotoxins in food: perspectives in a global and European context. Anal. Bioanal. Chem. 389, 147–157.

Walsh, M.J., Dodd, J.E., Hautbergue, G.M., 2013. Ribosome-inactivating proteins. Potent poisons and molecular tools. Virulence 4, 774–784.

Wang, D.Z., 2008. Neurotoxins from marine dinoflagallates: a brief review. Mar. Drugs 6, 349–371.

Wannemacher, J.R., Wiener, S.L., 1997. Chapter 34: Trichothecene Mycotoxins. In: Sidell, F.R., Takafuji, E.T., Franz, D.R. (Eds.), Medical Aspects of Chemical and Biological Warfare. Textbook of Military Medicine Series. Office of The Surgeon General, Department of the Army, United States of America.

Watson, S.A., Mirocha, C.J., Hayes, A.W., 1984. Analysis for trichothecenes in samples from southeast Asia associated with "yellow rain". Fundam. Appl. Toxicol. 4, 700–717.

Weaver, G.A., Kurtz, H.J., Mirocha, C.J., Bates, F.Y., Behrens, J.C., 1978. Acute toxicity of the mycotoxin diacetoxyscirpenol in swine. Can. Vet. J. 19, 267–271.

Wedin, G.P., Neal, J.S., Everson, G.W., Krenzelok, E.P., 1986. Castor bean poisoning. Am. J. Emerg. Med. 4, 259–261.

Wiberg, G.S., Stephenson, N.R., 1960. Toxicologic studies on paralytic shellfish poison. Toxicol. Appl. Pharmacol. 2, 607–615.

Woodward, R.B., 1964. The structure of tetrodotoxin. Pure Appl. Chem. 9, 49–74.

Wood, S., Casas, M., Taylor, D., McNabb, P., Salvitti, L., Ogilvie, S., Cary, S.C., 2012. Depuration of tetrodotoxin and changes in bacterial communities in Pleurobranchea maculata adults and egg masses maintained in captivity. J. Chem. Ecol. 38, 1342–1350.

Worbs, S., Köhler, K., Pauly, D., Avondet, M.A., Schaer, M., Dorner, M.B., Dorner, B.G., 2011. *Ricinus communis* intoxications in human and veterinary medicine—a summary of real cases. Toxins 3, 1332–1372.

World Health Organization (WHO), 1984. Environmental health criteria for aquatic (Marine and Freshwater) biotoxins. www.inchem.org/documents/ehc/ehc37.htm. ISBN: 92 4 154097 4.

Wu, J., Jing, L., Yuan, H., Peng, S.Q., 2011. T-2 toxin induces apoptosis in ovarian granulosa cells of rats through reactive oxygen species-mediated mitochondrial pathway. Toxicol. Lett. 202, 168–177.

Yang, C.C., Liao, S.C., Deng, J.F., 1996. Tetrodotoxin poisoning in Taiwan: an analysis of poison centre data. Vet. Hum. Toxicol. 38, 282–286.

Yap, H.Y., Murphy, W.K., DiStefano, A., Blumenschein, G.R., Bodey, G.P., 1979. Phase II study of anguidine in advanced breast cancer. Cancer Treat Rep. 63, 789–791.

Yin, H.L., Lin, H.S., Huang, C.C., et al., 2005. Tetrodotoxication with nassauris glans: a possibility of tetrodotoxin spreading in marine products near Pratas Island. Am. J. Trop. Med. Hyg. 73, 985–990.

Yinon, A., 2002. Introduction to toxins. In: Brener, B., Catz, L., Rubinstok, A., et al. (Eds.), The Biology Book: Medical Aspects and Responses [Hebrew]. SAREL Logistics Solutions& Products for Advanced Medicine, Netanya, pp. 109–112.

Yoshikawa, J.S.M., Hokama, Y., Noguchi, T., 2000. Tetrodotoxin. In: Hui, Y.H., Kitts, D., Stanfield, P.G. (Eds.), Food Borne Disease Handbook: Sea Food and Environmental Toxins. Marcel Deckker, New York, pp. 253–285.

Yotsu, Y.M., Okoshi, N., Watanabe, K., Araki, N., Yamaki, H., Shoji, Y., Terakawa, T., 2013. Localization of pufferfish saxitoxin and tetrodotoxin binding protein (PSTBP) in the tissues of the pufferfish, Takifugu pardalis, analyzed by immunohistochemical staining. Toxicon 72, 23–28.

Yu, V.C., Yu, P.H., Ho, K.C., Lee, F.W., 2011. Isolation and identification of a new tetrodotoxin-producing bacterial species, Raoultella terrigena, from Hong Kong marine puffer fish Takifugu niphobles. Mar. Drugs 9, 2384–2396.

Zain, M.E., 2011. Impact of mycotoxins on humans and animals. J. Saudi Chem. Soc. 15, 129–144.

Zhang, D., Hu, C., Wang, G., Li, D., Li, G., Liu, Y., 2013. Zebrafish neurotoxicity from aphantoxins-cyanobacterial paralytic shellfish poisons (PSPs) from Aphanizomenon flos-aquae DC-1. Environ. Toxicol. 28 (5), 239–254.

Further Reading

Mirocha, C.J., Pathre, S.W., Schauerhamer, B., Christensen, C.M., 1976. Natural occurrence of Fusarium toxins in feedstuff. Appl. Environ. Microbiol. 32, 553–556.

Viral agents including threat from emerging viral infections

4

Archna Panghal, S.J.S. Flora

National Institute of Pharmaceutical Education and Research-Raebareli, Lucknow, India

Introduction

Microbes are ubiquitous in nature and exist everywhere, whether it is soil, air, water, or human and animal bodies. They play a crucial role in the environment, human health, animal husbandry, etc. The beneficial role of microbes in human welfare has been well known since antiquity. Their beneficial effects are such that without them, human life cannot be imagined. Many microbes are even helpful in normal physiological processes such as digestion, absorption, etc. Microbes and livestock have been symbiotic to each other, but at the same time they are also pathogenic to mankind. The symbiotic and pathogenic nature of microbes has always been in a balanced state. According to a previously published study, "viruses are not just pathogens but they are symbiotic partners playing an important role in the health of hosts" (Roossinck, 2015). The dual nature of microbes, i.e., being beneficial and harmful at the same time, fascinated researchers and potentiated the research in the field of microbiology. However, in recent years, most research has been inclined towards the pathogenic nature of microbes and, therefore, in the current era their virulent behavior overweighs their symbiotic nature. Currently, the microbiological field has made so much advancement that microbes are practiced for the purpose of mass destruction, which is also known as a bioattack or bioterrorism (Mayor, 2019). Viruses have been used as biothreat agents since the 16th century, but recent advancements in the field of virology have made viruses more devastating for populations, and these pose serious problems. Terrorists and even the military organizations of various nations are using viruses as potential bioweapons in order to create havoc in large populations (Bronze et al., 2002).

Viruses have been used as bioweapons since the era when just limited information was known regarding their virulent nature, dissemination, and transmission. The first ever use of viruses as biothreat agents dates back to the 16th century, when variola-contaminated clothing was gifted to the natives of South America by a Spanish explorer, leading to a smallpox epidemic. Later on, smallpox was used as a bioweapon against Native Americans during the French-Indo war. At the same time, a severe outbreak of smallpox happened in the Ohio River valley during 1763–64,

Handbook on Biological Warfare Preparedness. https://doi.org/10.1016/B978-0-12-812026-2.00004-9

due to variola-contaminated blankets, which were gifted to Native Americans by British army representatives (Barras and Greub, 2014); although it is not certain whether this outbreak occurred due to gifted blankets or the virus was already present in the environment of valley.

Besides practicing biothreat agents against mankind, viruses also have been part of bioweapon stores of various countries such as the Soviet Union and United States. Their efficacy has been checked from time to time by various countries. In the 1970s, the Soviet Union released variola virus in the atmosphere in order to check its efficacy. It was found to be so efficacious that a lady 15 km far from the place of release of virus was infected (Shoham and Wolfson, 2004).

The intentional use of viruses is not just limited against human populations. They have also been targeted against animal livestock. In 1997, calcivirus was deliberately introduced in the south island of New Zealand with the purpose of killing feral rabbits (Carus, 2001). Therefore, the use of a bioweapon agent has the potential to destroy completely the life of the exposed ecosystem. Various viral agents that have been used as biological warfare agents are presented in Table 1.

Table 1 List of viral agents outbreaks from the 16th century to the contemporary period.

Year	Incident	Outcome
1532	Franciso Pizarro gifted variola-contaminated clothing to the natives of South America	Smallpox epidemic in Incan population, which unexpectedly killed an Incan emperor
1763–64	Variola contaminated blankets were distributed to Native Americans by British Commander	Severe outbreak of smallpox in tribes of the Ohio River Valley
1775	British distributed slaves infected with smallpox	Serious outbreak of smallpox among continental American forces
1959	A freeze dried vial was accidently dropped by Soviet medical personnel	20 Laboratory staff were infected with Venezuelan equine encephalitis virus
1970	Soviet Union released 400 g of the variola virus on Voz-rozhdeniye Island	Variola virus transmitted over 15 km area around island, resulting in many deaths including laboratory technician
1992	The Aum Shinrikyo group sent a team to Zaire with the purpose of getting knowledge and samples of the Ebola virus	Failed in getting samples
1997	Calicivirus was illicitly introduced in the South Island by some anonymous farmers of New Zealand	Serious outbreak of hemorrhagic disease in feral rabbits
2000	Klingerman Foundation was mailing envelopes containing "Klingerman virus"	According to the mail alert, this virus killed 23 people. However, it was later found to be a hoax

With the advancements in the field of virology, other viruses such as Ebola virus, Marburg virus, Hanta virus, West Nile virus, Dengue virus, Rift valley fever virus, Nipah virus, and Venezuelan equine encephalitis virus are also appealing biothreat agents for terrorists. They possess several unique characteristics due to which they have been in high demand as bioweapons. Ease of production, cost-effectiveness, difficulty in detection, high dissemination and transmission rate, along with high morbidity and mortality rate are a few of the characteristics that make them different from conventional means of war. They do not produce immediate effects and thus provide a chance to terrorists to escape from the government or secret agencies. Beside this, the dual behavior of viruses, i.e., living cell inside living system and nonliving outside the body, is a special feature of viruses, which make them unique among other biowarfare agents (Morse and Meyer, 2017). Various viral biological warfare agents along with their category, mode of dissemination, and vectors involved in their transmission are given in Table 2.

Usually, the signs and source of biological warfare are unidentifiable and epidemiological observations provide valuable findings regarding the source of a bioattack (Treadwell et al., 2003). Sudden onset of disease due to unidentifiable reason or the pathogen whose history is lacking, unexpected spread of disease to wide geographical area, high morbidity and mortality rate in exposed population without any known reason, serious respiratory inability because of aerosolized bioweapons are some of the signs that are indicative of bioattack.

The devastating effects of viral warfare agents urge the early and accurate detection of these pathogens so that their pathogenic effects can be controlled. An ideal detecting system should be highly sensitive, specific, cheap, easy to handle, portable, and capable of detecting a wide range of pathogens simultaneously (Thavaselvam and Vijayaraghavan, 2010). Several detecting systems, including culture based assays, immunoassays, nucleic acid based assays, biosensors, biodetectors, flow cytometry, etc., have been adopted for the purpose of detecting biothreat agents. Immunoassays exploit antigen-antibody reaction while nucleic acid based assay follows the principle of gene amplification and detect pathogens even when only a minute quantity of genetic material is present. Biosensors reflect a signal in terms of fluorescence, electrochemical, and optical signals if biological warfare agent is present.

Biological warfare agents as a means of terrorism

Warfare agents are the chemical, nuclear, and biological agents, used by military services, which have potential to cause catastrophic medical disasters that could overwhelm the healthcare system. Their brutal effects are not just limited to life, but may also cause damage to nonliving things including buildings, monuments, and the environment. They are weapons of mass destruction, which are powerful tools for terrorists to create havoc in a population. However biological warfare (BW, or germ warfare) is the deliberate employment of living organisms (bacteria, viruses,

Table 2 List of viral biological warfare agents along with their category, transmission mode, incubation period, and detection techniques.

Virus	Category	Dissemination mode	Vector	Incubation period (days)	Diagnosis techniques
Variola virus or Smallpox virus	A	Aerosols	None	10–12	Microscopic examination, ELISA, PCR
Influenza virus	C	Aerosols	None	1–4	Rapid molecular assay, PCR
Marbrug virus	A	Aerosols and droplet nuclei	Nonhuman mammals, especially bats	2–21	ELISA, PCR of blood specimens
Ebola virus	A	Aerosols and droplet nuclei	Infected fruit bats or nonhuman primates (apes and monkeys)	2–21	ELISA, serological tests
Yellow fever virus	C	Aerosols	Aedes or Haemagogus species mosquitoes	3–6	Blood test along with PCR, ELISA
Venezuelan equine encephalitis and western equine encephalitis	B	Aerosols	Aedes, Coquillettidia, and Culex species of mosquito	4–10	Serum and CSF testing, ELISA, immunoassays
Tick-borne encephalitis virus	C	Aerosols	Ticks	7–14	Blood and CSF testing for IgM antibody detection
Hanta virus	C	Aerosols	Rodent excreta or rodent bites	7–28	PCR, serologic, immunoassays
Nipah virus	C	Infected secretions	Bats and pigs	5–14	RT-PCR, ELISA

or fungi) or their toxic products with the desire to cause death, disability, or damage in livestock including both flora and fauna. In the current era, biological warfare are given more preference over conventional methods of war-fighting and their increasing use has raised an alarm, alerting the agencies throughout the globe to develop and revise the current strategies in order to combat the challenges posed by them (Jansen et al., 2014).

They have several unique characteristics due to which they are considered superior to the conventional weapons. They do not produce immediate effects and start to develop symptoms in population only after the incubation period of causative organism and are capable of self-replication (Thavaselvam and Vijayaraghavan, 2010). They are appealing for the terrorist due to difficulty in their detection, cost-effectiveness, ease in handling and use of simpler techniques for their production. Due to these properties they are termed a "Poor Man's Nuclear Arsenal" (Beeching et al., 2002; Lawrence and Dennis, 2001).

The effects of biological warfare agents primarily depend on their ability of multiplication in the host. The BW is classified on the basis of source organism (bacteria, viruses, fungi, insects) used to develop terror among people. The alleged use of these agents is a serious problem and their use in bioterrorist attack is increasing (Wheelis et al., 2006).

BWA can be spread through air, water, or food and they could be modified to improve their capability to cause disease and to make them resistant to drugs. Centre for Disease Control and Prevention (CDC) has classified them into three categories depending on the severity of disease that they cause and on their ability to spread (Centers for Disease Control and Prevention (CDC), 2015a):

Category A. Pathogens that would create the greatest risk during a bioterrorism attack because of high spreadability and transmission from person to person. They cause serious health issues among the exposed population.
Category B. Pathogens that would produce moderate risk, resulting in moderate morbidity and mortality.
Category C. Includes emerging pathogens that could be the cause of mass destruction in the future because of their availability, ease in production, and potential for high morbidity and mortality rate.

Viruses as biological warfare agents

Viruses are small, highly infectious agents of dual nature, as they are in the phase between living and nonliving. They are a nonliving entity outside a cell and are capable of replication only inside a living cell. They can cause harm to all types of living forms ranging from animals and humans to plants. Viruses mainly consist of proteins and nucleic acids (DNA and RNA) and are fast multiplying as well as spreadable. The aforementioned characteristics make viruses unique among all the bioweapons and attract the attention of terrorists. Developments in the field of viral culture during the second half of the 20th century assisted the large-scale production of viral agents for aerosol dissemination (Morse and Meyer, 2017). They are capable of person to person transmission via aerosolized form. Despite advancements in the detection technology, viruses also have developed their own mechanisms to escape from detection and even remain unnoticed from the defense system of the body, which ultimately reflects the healthcare system of the exposed population (Lucas et al., 2001). Even nowadays, targeting infectious viral agents is a prime interest of terrorists for mass destruction due to easy availability, high transmission rate, economical use, and ability to cause high morbidity and mortality to livestock. Terrorists can handle viral agents safely without any risk of being infected themselves, and can acquire them more easily than other BWA (Keremidis et al., 2013a,b).

Viral bioweapons have received considerable attention in the past few decades, but their use to terrorize people started long ago. The potential use of these agents is a risk to common people due to serious societal impact and lack of full-fledged effective tools for their detection and identification (Cenciarelli et al., 2013). These agents mostly enter the body through the respiratory tract by means of aerosols and are readily inhaled. Besides this, other routes may be mucosal surfaces (nose/mouth), gastrointestinal tract (by means of contaminated food and water), and injection.

Historical perspectives

The utilization of viral WA is an emerging threat for modern civilization, although practicing it as weapons started hundreds of years ago (Christopher et al., 1999). The earliest use of smallpox was found to be in the 16th century, when Francisco Pizarro, a Spanish explorer, gifted variola-contaminated clothing to the natives of South America, leading to a widespread smallpox epidemic (Morse and Meyer, 2017). Later on, during the French-Indo War (1745–67), Sir Jeffrey Amherst, the commander of the British army, granted approval for the use of smallpox as a biological weapon against Native Americans hostile to British forces. Caption Ecuyer, one of Amherst's assistants, obtained two variola-contaminated blankets and a scarf from a smallpox hospital, which were gifted by William Trent and other British army representatives to the Native Americans as a "symbol of good will." Consequently, a severe outbreak of smallpox happened in several tribes of the Ohio River Valley, during 1763–64 (Hopkins, 1983). Although apologists questioned whether the epidemic was due to the distributed blankets or the virus was already present in the valley. During 1957–65, smallpox by means of formites, measles, and influenza infections were spread into the Native Americans of the Amazonian basin by crooked agents of the Brazilian Indian Protective Service (Wheelis, 2004). In 1775, at the time of the American Revolutionary War, the British made efforts to cause an outbreak of smallpox among the continental forces by inoculating the persons fleeing Boston with variola virus. The evidence supports that the British distributed slaves infected with smallpox to the rebellious opposition with the intention of spreading the disease (Hopkins, 1983).

Viral agents have been a part of the biological weapon store of both the Soviet Union and the United States (Leitenberg, 2001). Venezuelan equine encephalitis virus (VEE) was kept in stock by both countries as a devastating pathogen, while variola major and Marburg viruses were stockpiled by the Soviet Union as fatal pathogens. In 1959, a freeze-dried vial of VEE was accidently dropped by Soviet medical personnel, resulting in infection of 20 laboratory staff (Croddy Eric et al., 2002). The efficacy of BWA has been tested from time to time by various countries. In the 1970s, the Soviet Union released 400 g of the variola virus in the atmosphere of Voz-rozhdeniye Island in the Aral Sea, with the aim of testing the efficacy. The virus was found to be so efficacious in terms of transmission and infection that a laboratory technician collecting plankton samples even at a distance of 15 km from the island became infected. She transmitted the infection to several people, resulting in their deaths (Shoham and Wolfson, 2004; Enserink, 2002).

With the advancements in the virological field, viruses other than variola major came in picture for the purpose of having societal impact and challenge the health-care system. There were severe outbreaks of swine fever in 1971 and dengue fever in 1980 that created havoc in Cuba. Later on, the Cuban government accused the US Central Intelligence Agency (CIA) for these outbreaks; however, subsequent inquiry failed to collect any solid proof regarding involvement of the CIA in these outbreaks (Zilinskas, 1999; Leitenberg, 2001). The Aum Shinrikyo, a religious group, was found to be planning an Ebola viral attack. To achieve the goal, this group sent a team of 16 doctors and nurses to Zaire with the purpose of getting knowledge and the samples of the virus (Olson, 1999). In 1997, a virus (calicivirus) responsible for rabbit hemorrhagic disease was intentionally and illicitly introduced in the South Island by some anonymous farmers of New Zealand with the intention of killing feral rabbits. Some farmers mixed the homogenized organs of the infected rabbits with carrot and oat sand in order to further propagate the disease in the rabbits (Carus, 2001).

Beside the generalized use of viruses as bioweapons, they are also considered to be involved in suspected incidents or hoaxes. An article of 1999 suggested the enquiry of the CIA against Iraq for the accountability of Iraq for West Nile fever outbreak in the New York City (Preston, 1999). This investigation was in response to the story written by an Iraqi traitor, claiming that Saddam Hussein intended to use West Nile virus strain SV1417 to create havoc. Later on, evidence revealed that this strain of West Nile virus was already prevalent in the Mediterranean region since 1998. A "fictional virus" was found to be involved in one of the largest bioterrorism hoaxes, in 2000. Email messages circulating on the internet suggested that an organization called the Klingerman Foundation was mailing envelopes containing sponges packed in plastic infected with an imaginary virus called "Klingerman virus." According to that mail alert, this fictional virus infected and killed 23 people. This hoax appeared to be inspired by the September 11, 2001, terrorist attacks on the United States, followed by anthrax attacks (Keremidis et al., 2013a,b).

Viral biological warfare agents

Several viral agents such as Orthopox viruses (Variola virus), Filoviruses (Marburg virus and Ebola virus), Influenza virus, Flaviviruses (Dengue, Japanese encephalitis virus, Tick-borne encephalitis), Rift valley fever, Venezuelan equine encephalitis virus, and Nipah virus, are the most likely to be used for bioterrorism purposes. A brief description of the chief representatives of viral agents used as bioweapons is as follows.

Variola virus

The variola virus is one of the devastating infectious viral agents causing serious threat even after its eradication Evidences suggest that variola virus was the first virus to be exploited for the purpose of bioterrorism. It is the largest animal virus, which is brick to ovoid shape morphologically. Structurally, it contains double stranded DNA, surrounded by double lipoprotein membrane. This virus exists in two strains: variola major, responsible for severe smallpox and variola minor, causing mild manifestations. Both these strains are nondistinguishable based on immunologic methods and differ just in their clinical manifestations. The common mode of transmission includes direct contact, droplet nuclei, aerosols expelled from the infected persons, and contaminated clothing. The clinical symptoms, ranging from headache to high fever, make for a high fatality rate. The high virulence and infective nature, long incubation period (12 days), and high fatality rate make this a most attractive bioweapon for terrorists. Furthermore, high susceptibility of human population to variola virus due to nonimmunization makes these viruses a serious threat (Klietmann and Ruoff, 2001). Despite these facts, due to the high biohazards associated with its use, the WHO decided to destroy virus in its stores for research purposes (Breman and Henderson, 1998).

Case study: In 1763, a British colonel decided to distribute variola-contaminated blankets to the Native Americans aggressive against the British. This act of distribution resulted in a severe outbreak of smallpox in the Ohio River Valley and affected the lives of Native Americans (Smart, 1997).

Influenza virus

The influenza virus is a RNA virus, which has an envelope containing hemagglutinin and neuraminidase glycoproteins. It is a pleomorphic virus, which has elliptical to filamentous shape and is extremely small (80–120 nm) in diameter. Influenza virus has been classified in four types (type A–D), out of which, types A–C are infective to humans, while type D is noninfective to humans but has potential to cause severe damage (Longo, 2012). The symptoms of influenza appear after an incubation period of 1–4 days and often are nondistinguishable from that of common cold and pneumonia including fever, cough, runny nose, and sneezing. It can be transmitted in three ways: direct transmission from person to person (sneezing), airborne route (aerosols exhaled by infected person), and contact with contaminated surfaces (door knobs, handshake, etc.) (Hall, 2007). The scope of genetic alterations in influenza virus to increase its virulence, dissemination ability via aerosol, along with easy accessibility makes it an attractive bioweapon (Madjid et al., 2003). Recently, scientists reconstructed the virus responsible for the 1918 pandemic, which raised various questions regarding reason behind necessity to reconstruct such an "extinct" virus, assumed to be responsible for the deaths of millions people across the world. Later on, a US scientist suggested that reconstruction of this virus would help in understanding the reason behind its unusual virulence and prepare against the situation if any pandemic happen due to this virus in future (Taubenberger et al., 2012).

Case study: 1918 influenza pandemic caused by H1N1 virus is one of the most severe pandemics to date. It spread throughout the globe in 1918–19 following World War I and infected 500 million people, approximately one third of the total population at that time (Johnson and Mueller, 2002).

Filoviruses

Filoviruses, including the Marburg and Ebola virus, are among the terrible pathogens that cause devastating effects to populations. They are enveloped, thread-like viruses, having RNA as a genetic material. Contact with infected person or nonhuman primates, body fluids, breast feeding, and reuse of unsterilized needles or syringes are the possible ways of transmission of these viruses (Salvaggio and Baddley, 2004). They can cause hemorrhagic fever, characterized by acute onset of fever, followed by maculopapular rashes and episodic bleeding. The diagnosis of these viruses may be hazardous for healthcare personnel, as there are chances of transmission of the viruses from tissues or fluids of infected persons. Its dissemination through airborne routes (aerosol or small-droplets nuclei), difficulty in diagnosis, and fatality rates higher than 80% suggests its potential to be used as a BWA. In the same context, the Ebola virus and Marburg virus have been weaponized by the Soviet Union for the purpose of bioattack (Borio et al., 2002; Davis, 1999).

Case study 1: In 2004, the worst outbreak of MARV occurred in Angola, resulting in 227 deaths out of 252 cases (90% fatality) (Ndayimirije and Kindhauser, 2005). In this outbreak, the virus was spread through funeral rituals and injections.

Case study 2: The most terrible and largest Ebola epidemic occurred in West Africa (Guinea, Liberia, Sierra Leone) during 2014–16. A total of 28,610 cases of Ebola were reported, out of which 11,308 persons died (39% fatality) due to poor measures to control the spread of the virus and weak surveillance systems (Bell et al., 2016). However, there is no evidence suggesting this outbreak was a result of bioattack by any country.

Flaviviruses

The flavivirus family includes Dengue virus, Japanese encephalitis virus, tick-borne encephalitis virus, and yellow fever virus. Most of these viruses have similar structure and range between 40 and 60 nm in diameter. They all have single-stranded RNA enveloped by capsid. Transmission of these viruses from their natural reservoirs occurs via arthropods like ticks and mosquitoes. The severity of symptoms caused by viruses of this group ranges from mild to severe and even fatal if neglected. Clinically, the most common symptoms include fever, flu-like condition, and bleeding. High mortality and morbidity rate, easy access in nature, transmission via aerosols, and nonavailability of even a single vaccine are a few characteristics that put them in the category of potential threats as bioweapons (Schwind, 2016; Robenshtok et al., 2002). Reports suggest that North Korea weaponized yellow fever virus (Center for Nonproliferation Studies, 2000).

Case study 1: Yellow fever posed troubles for the American army in Cuba at the time of the Spanish-American war. During that war, more soldiers died because of disease than the battle causalities (CDC).

Case study 2: Dengue virus affected 40,000 people in New Orleans in 1873 (Bemiss, 1880).

Hantaviruses

Hantaviruses belong to the Bunyavirideae family and include Sin Nombre virus, Seoul virus, Puumala virus, and Hantaan virus. They cause either fever with renal syndrome, or pulmonary infections whose pathophysiological hallmark is vascular dysfunction. The transmission of these viruses has been found to be closely related with rodents' habitats and, therefore, rodents such as mice and rats are considered as vectors for the virus (McCaughey and Hart, 2000; Chandy and Mathai, 2017). Their easy replication in the laboratory, worldwide existence, low immunity in people against viruses, availability of several strains, and nonavailability of vaccines to counter effects are a few of the characteristics that make them potential bioweapons. The most likely mode of attack of these viruses is aerosolized formulations. The serious concern for the use of Hantaviruses as bioweapon is raised after the incidence of Old World Hantavirus infection in the United States (Bronze et al., 2002).

Case study: During the Korean war (1950–53), a serious outbreak of Korean hemorrhagic fever occurred among soldiers of America and Korea. This outbreak affected 3000 soldiers with 10% mortality rate (Lee, 1989).

Nipah virus

The Nipah virus is a zoonotic paramyxovirus that is primarily transmitted by infected secretions. This virus is highly infectious to pigs. The symptoms in humans, such as fever, dizziness, and diminished consciousness, developed after the incubation period of 2 weeks. The symptoms can progress to complex neurological manifestations, encephalitis, and coma in 50% of infected persons, and results in a high mortality rate. It has been put in category C of the biological weapon classification, indicating that currently it is not considered as a source of bioterrorism but in the near future it may emerge as a serious threat. Lack of effective treatment strategies and high morbidity and mortality rate due to the Nipah virus render it a potential bioweapon (Chua et al., 2000; Bronze et al., 2002).

Identification of signs of viral attack

A wide array of signs is developed by bioweapons. Although most of these signs are difficult to identify, broad epidemiological observations prove helpful in identifying the source of bioattacks (Dorner et al., 2016; Song et al., 2005). Early identification of a bioattack helps to protect society from the devastating effects of bioweapons. The signs common to most bioattacks are as follows:

1. Sudden onset of an epidemic resulting in high mortality and morbidity.
2. Spread of an epidemic to a wide area either by direct contact or by various indirect means such as by aerosols, food, or water sources.
3. Epidemic occurs mainly due to an unidentifiable pathogen whose earlier epidemiological or clinical history is unavailable.
4. Sudden development of clinical presentations such as flu-like symptoms due to influenza virus.
5. Unexpected, unidentifiable symptoms without any known cause.
6. Unexpected illness or death of persons due to unexplainable reasons.
7. The most severe symptoms include respiratory symptoms along with lung disorders.
8. Infection of the brain, causing encephalitis (e.g., rabies and West Nile virus induced encephalitis) unusually affects large population.
9. Skin infections in form of warts, blemishes (e.g., smallpox virus induced blisters) may present in wide epidemiological area.
10. Recurrence of the symptoms in a population simultaneously.

Detection techniques for viral biological warfare agents

The devastating effects of viral warfare agents on a wide population demand the early and accurate detection of these pathogens so that the outbreak of disease can be limited to a small population and the risk from these bioweapons can be mitigated (Thavaselvam and Vijayaraghavan, 2010). The biological properties of an ideal detecting system should be as follows: (1) highly sensitive and specific along with high accuracy, (2) cheap, (3) capable of detecting a wide range of pathogens simultaneously, (4) portable, (5) easy to handle, and (6) able to discriminate pathogens from other biological or nonbiological threats (pollen grains, dust, etc.).

Although various innovative and sophisticated detecting systems have been developed, none of them meet all the aforementioned properties of an ideal detecting tool (Ludovici et al., 2015). Numerous efforts have been made to develop perfect tools for detection of bioweapons. Some of these detection systems include (1) culture-based assays, (2) immunological assays, (3) nucleic acid-based detection, (4) DNA sequencing techniques, (5) biosensors, and (6) biological detectors.

Bioculture-based detection is considered as the gold standard method. Viral agents are grown in selective culture medium, which only allow the growth of target agent and don't allow the nontarget cell to grow. Furthermore, an array of tests such as morphological study, staining, and biochemical parameters is used to detect biothreat agent. This is an efficient method to generate valuable information, but it is time consuming and laborious (Lim et al., 2005). Detection of bioweapons by means of immunoassays is based on the antigen-antibody interactions. Antibodies bind to a specific antigen (viral agent) resulting in formation of an identifiable complex. ELISA based immunoassay has been used for detection of Ebola and Marburg viruses (Saijo et al., 2001). Immunoassays are highly sensitive, economical, simple, and reliable assays for detection of biothreat agents. Nucleic acid-based assays are able to generate multiple copies of target nucleic acid even from very minute genetic material. The only limitation of this detection system is that it is not able to detect proteins; therefore, toxins cannot be detected by this assay. Polymerase chain reaction (PCR) is implicated to detect specific DNA sequences of organisms in nucleic acid based detection systems. Filoviruses, Hanta viruses, and Arena viruses have been detected on the basis of PCR-based assays (Trombley et al., 2010). Loop mediated isothermal amplification is a variant technique of PCR that has been used for detection of Ebola virus (Kurosaki et al., 2016), Marburg virus (Kurosaki et al., 2010). Next generation sequencing (NGS) is a recent advancement in the field of DNA sequencing, which represents a superior detection system over conventional DNA sequencing. This technique provides high throughput at low cost and in minimum time and is capable for detection of viral pathogens in biological and environmental samples (Buermans and den Dunnen, 2014). Use of biosensors is another recent advancement, which has revolutionized the field of detection of biothreat agents. In this method of detection, an analyte (bioweapon) interacts with a biological component to produce a detectable signal in terms of luminescence, optical, or electrochemical signal, which is read by a transducer. Detection by biosensors may impart a crucial role for detection of viral agents and this method is superior to conventional methods of detection in terms of sensitivity and selectivity.

Impact of viral bioterrorism

The tremendous advancements in the field of microbiology over the course of time have made viral biothreat agents highly dangerous to human health, agriculture, and national security. The terrible effects of previous viral biothreat agents such as smallpox attack or Ebola virus attack provides proof regarding the highly devastating impact of these viral agents on human healthcare systems and the economic status of a country (Pimentel and Pimentel, 2006). Viral pathogens used as bioweapons may cause moderate to high morbidity and mortality in the targeted population. Using viruses as biological weapons disturbs the social, economic, political, and healthcare status of a nation, and creates terror among the public. In the worst situation, viral bioterrorism has a potential even to eradicate human civilization and push the economy of a country on to the back foot in such a way that it may take several years to regain its previous status. Their attack is so powerful that people once affected by such an attack may live the rest of their lives in fear. Ecological relationships between man, lower animals, and vectors responsible for transmission may be altered by massive attacks of such powerful bioweapons. Their large-scale use may result in everlasting and unpredictable changes to the social environment and can cause mass destruction and pose a serious threat to mankind in the near future (Harigel, 2001; Morea et al., 2018).

Future perspectives

Although the art of using viral agents as a weapon is an old one but modern technologies such as gene editing by Clustered Regularly Interspersed Short Palindromic Repeats (CRISPER), techniques set off alarms for the near future. With the advent of CRISPER techniques, new doors for a highly disastrous bioattack are opened. The nucleic acid of viruses, even including those which are not threats currently, can be genetically engineered, resulting in development of more serious biothreat viral agents (DiEuliis and Giordano, 2018). Biotechnological techniques also safeguard the future of viral agents as bioweapons as they can be exploited for the development of novel viral agents. Such types of techniques ensure a bright future for bioweapons and, therefore, they can be termed as "weapons of future generations." They have potential of completely replacing the conventional weapons of war-fighting in the near future.

Technology has "dual use"; that is, beside harmful misuses it also has potential for beneficial impacts. With the advancements in technology, highly sensitive techniques such as PCR have been developed, which enable the rapid and accurate diagnosis of bioweapons. Early diagnosis is a key criterion for development of valuable therapeutic strategies in order to handle such types of disasters and protect people from the devastating effects of biological weapons (Thavaselvam and Flora, 2014; Casadevall, 2012).

Conclusion

Although the concept of using viruses for the purpose of terrorizing people is not a new one, advancements in the field of virology have revolutionized the widespread use of these agents as bioweapons. Ease in production, cost-effectiveness, dual nature, and high dissemination and transmission rates are a few of the characteristics of viruses that attract the attention of terrorists. Furthermore, they are highly susceptible to mutations and can be easily manipulated in the laboratory to introduce desired characteristics. Intentional and indiscriminate use of viral biothreat agents leads to long-term devastating effects on the environment, healthcare, and economic status of the exposed population or nation. Highly damaging impacts of these bioweapons demands intensive research for the development of specific, accurate, and sensitive techniques so that they can be diagnosed at an early stage and necessary steps for their mitigation can be taken at an early stage. Furthermore, technology for virus identification, genotyping, surveillance, and therapeutics, along with training and education programs should be strengthened to combat viral outbreaks.

References

Barras, V., Greub, G., 2014. History of biological warfare and bioterrorism. Clin. Microbiol. Infect. 20 (6), 497–502.

Beeching, N.J., Dance, D.A., Miller, A.R., Spencer, R.C., 2002. Biological warfare and bioterrorism. BMJ 324 (7333), 336–339.

Bell, B.P., Damon, I.K., Jernigan, D.B., et al., 2016. Overview, Control Strategies, and Lessons Learned in the CDC Response to the 2014–2016 Ebola Epidemic. MMWR Morb. Mortal. Wkly Rep. 65 (3), 4–11.

Bemiss, S.M., 1880. Dengue. New Orleans Med. Surg. J. 8, 501–512.

Borio, L., Inglesby, T., Peters, C.J., Schmaljohn, A.L., Hughes, J.M., Jarhling, P.B., et al., 2002. Hemorrhagic fever viruses as biological weapons. JAMA 287 (18), 2391–2405.

Breman, J.G., Henderson, D.A., 1998. Poxvirus dilemmas-monkeypox, smallpox, and biologic terrorism. N. Engl. J. Med. 339, 556–559.

Bronze, M.S., Huycke, M.M., Machado, L.J., Voskuhl, G.W., Greenfield, R.A., 2002. Viral agents as biological weapons and agents of bioterrorism. Am. J. Med. Sci. 323 (6), 316–325.

Buermans, H.P., den Dunnen, J.T., 2014. Next generation sequencing technology: advances and applications. Biochim. Biophys. Acta 1842 (10), 1932–1941.

Carus, W.S., 2001. Bioterrorism and Biocrimes. The Illicit Use of Biological Agents Since 1900. Fredonia Books, Amsterdam, The Netherlands.

Casadevall, A., 2012. The future of biological warfare. Microb. Biotechnol. 5 (5), 584–587.

Cenciarelli, O., Rea, S., Carestia, M., et al., 2013. Bio-weapons and bioterrorism: a review of history and biological agents. Defence S&T Tech. Bull. 6, 111–129.

Center for Nonproliferation Studies, 2000. Chemical and Biological Weapons: Possession and Programs Past and Present. Middlebury Institute of International Studies.

Centers for Disease Control and Prevention (CDC), 2015a. Emergency Preparedness and Response. Bioterrorism Overview. http://emergency.cdc.gov/bioterrorism/overview.asp.

Chandy, S., Mathai, D., 2017. Globally emerging hantaviruses: an overview. Indian J. Med. Microbiol. 35, 165–175.

Christopher, G.W., Cieslak, T.J., Pavlin, J.A., Eitzen, E.M., 1999. Biological warfare: a historical perspective. In: Lederberg, J. (Ed.), Biological Weapons. Limiting the Threat. The MIT Press, Cambridge, MA, pp. 17–35.

Chua, K.B., Bellini, W.J., Rota, P.A., et al., 2000. Nipah virus: a recently emergent deadly paramyxovirus. Science 288, 1432–1450.

Croddy Eric, C., Hart, C., Perez-Armendariz, J., 2002. Chemical and Biological Warfare. Springer, 30–31.

Davis, C.J., 1999. Nuclear blindness: an overview of the biological weapons programs of the former Soviet Union and Iraq. Emerg. Infect. Dis. 5 (4), 509–512.

DiEuliis, D., Giordano, J., 2018. Gene editing using CRISPER/Cas9: implications for dual-use and biosecurity. Protein Cell 9 (3), 239–240.

Dorner, B.G., Zeleny, R., Harju, K., Hennekinne, J.-A., Vanninen, P., Schimmel, H., et al., 2016. Biological toxins of potential bioterrorism risk: current status of detection and identification technology. TrAC Trends Anal. Chem. 85, 89–102.

Enserink, M., 2002. Did bioweapons test cause deadly smallpox outbreak? Science 296, 2116–2117.

Hall, C.B., 2007. The spread of influenza and other respiratory viruses: complexities and conjectures. Clin. Infect. Dis. 45 (3), 353–359.

Harigel, G.G., 2001. Chemical and Biological Weapons: Use in Warfare, Impact on Society and Environment. Carnegie Endowment for International Peace.

Hopkins, D.R., 1983. Princes and Peasants: Smallpox in History. University of Chicago Press, Chicago, IL.

Jansen, H.J., Breeveld, F.J., Stijnis, C., et al., 2014. Biological warfare, bioterrorism, and biocrime. Clin. Microbiol. Infect. 20 (6), 488–496.

Johnson, N.P., Mueller, J., 2002. Updating the accounts: global mortality of the 1918–1920 "Spanish" influenza pandemic. Bull. Hist. Med. 76, 105–115.

Keremidis, H., Appel, B., Menrath, A., et al., 2013a. Historical perspective on agroterrorism: lessons learned from 1945 to 2012. Biosecur. Bioterror. 11 (Suppl. 1), S17–S24.

Keremidis, H., Appel, B., Menrath, A., et al., 2013b. Historical perspective on agro terrorism: lessons learned from 1945 to 2012. Biosecur. Bioterror. 11 (Suppl. 1), S17–S24.

Klietmann, W.F., Ruoff, K.L., 2001. Bioterrorism: implications for the clinical microbiologist. Clin. Microbiol. Rev. 14, 364–381.

Kurosaki, Y., Grolla, A., Fukuma, A., Feldmann, H., Yasuda, J., 2010. Development and evaluation of a simple assay for Marburg virus detection using a reverse transcription loop-mediated isothermal amplification method. J. Clin. Microbiol. 48 (7), 2330–2336.

Kurosaki, Y., Magassouba, N., Oloniniyi, O.K., Cherif, M.S., Sakabe, S., Takada, A., et al., 2016. Development and evaluation of reverse transcription-loop-mediated isothermal amplification (RT-LAMP) assay coupled with a portable device for rapid diagnosis of Ebola virus disease in Guinea. PLoS Negl. Trop. Dis. 10 (2), e0004472.

Lawrence, C.M., Dennis, L.K., 2001. Basic considerations in infectious diseases. In: Braunwald, E., Fauci, A.S., Kasper, D.L., et al. (Eds.), Harrison's Principles of Internal Medicine. 15th ed.. vol. 1. McGraw-Hill Professional, pp. 763–764.

Lee, H.W., 1989. Hemorrhagic fever with renal syndrome in Korea. Rev. Infect. Dis. 11 (Suppl. 4), S864–S876.

Leitenberg, M., 2001. Biological weapons in the twentieth century: a review and analysis. Crit. Rev. Microbiol. 27, 267–320.

Lim, D.V., Simpson, J.M., Kearns, E.A., Kramer, M.F., 2005. Current and developing technologies for monitoring agents of bioterrorism and biowarfare. Clin. Microbiol. Rev. 18 (4), 583–607.

Longo, D.L., 2012. Chapter 187: Influenza. In: Harrison's Principles of Internal Medicine. 18th ed.. McGraw-Hill, New York.

Lucas, M., Karrer, U., Lucas, A.D., et al., 2001. Viral escape mechanisms—escapology taught by viruses. Int. J. Exp. Pathol. 82 (5), 269–286.

Ludovici, G.M., Gabbarini, V., Cenciarelli, O., Malizia, A., Tamburrini, A., Pietropaoli, S., Carestia, M., Gelfusa, M., Sassolini, A., Di Giovanni, D., Palombi, L., 2015. A review of techniques for the detection of biological warfare agents. Defence S&T Tech. Bull. 8 (1), 17–26.

Madjid, M., Lillibridge, S., Mirhaji, P., et al., 2003. Influenza as a bioweapon. J. R. Soc. Med. 96 (7), 345–346.

Mayor, A., 2019. Chemical and biological warfare in antiquity. In: History of Toxicology and Environment Health In Toxicology in Antiquity. second ed.. pp. 243–255.

McCaughey, C., Hart, C.A., 2000. Hantaviruses. J. Med. Microbiol. 49, 587–599.

Morea, D., Poggi, L.A., Tranquilli, V., 2018. Economic impact of biological incidents: a literature review. In: Enhancing CBRNE Safety & Security: Proceedings of the SICC 2017 Conference. Springer, pp. 291–297.

Morse, S.A., Meyer, R.F., 2017. Viruses and Bioterrorism. Centers for Disease Control and Prevention.

Ndayimirije, N., Kindhauser, M.K., 2005. Marburg hemorrhagic fever in Angola—fighting fear and a lethal pathogen. N. Engl. J. Med. 352 (21), 2155–2157.

Olson, K.B., 1999. Aum Shinrikyo: once and future threat? Emerg. Infect. Dis. 5, 513–516.

Pimentel, D., Pimentel, M., 2006. Bioweapon impacts on public health and the environment. William Mary Environ. Law Policy Rev. 30 (3), 625–656.

Preston, R., 1999. West Nile mystery. The New Yorker 90–91. October 18, 1999.

Robenshtok, E., Laster, M., Katz, L., et al., 2002. Viral hemorrhagic fever as a biological weapon. Harefuah 141, 96–99.

Roossinck, M.J., 2015. Move over, bacteria! Viruses make their mark as mutualistic microbial symbionts. J. Virol. 89 (13), 6532–6535.

Saijo, M., Niikura, M., Morikawa, S., Ksiazek, T.G., Meyer, R.F., Peters, C.J., et al., 2001. Enzyme-linked immunosorbent assays for detection of antibodies to Ebola and Marburg viruses using recombinant nucleoproteins. J. Clin. Microbiol. 39 (1), 1–7.

Salvaggio, M.R., Baddley, J.W., 2004. Other viral bioweapons: Ebola and Marburg hemorrhagic fever. Dermatol. Clin. 22 (3), 291–302.

Schwind, V., 2016. Viral hemorrhagic fever attack: Flaviviruses. In: Ciottone's Disaster Medicine. second ed. Elsevier, pp. 763–765.

Shoham, D., Wolfson, Z., 2004. The Russian biological weapons program: vanished or disappeared? Crit. Rev. Microbiol. 30, 241–261.

Smart, J.K., 1997. History of chemical and biological warfare: an American perspective. In: Sidell, F.R., Takafuji, E.T., Franz, D.R. (Eds.), Medical Aspects of Chemical and Biological Warfare. Borden Institute, Walter Reed Army Medical Center, Washington, DC.

Song, L., Ahn, S., Walt, D.R., 2005. Detecting biological warfare agents. Emerg. Infect. Dis. 11 (10), 1629.

Taubenberger, J.K., Baltimore, D., Doherty, P.C., et al., 2012. Reconstruction of the 1918 influenza virus: unexpected rewards from the past. MBio 3 (5). e00201-12.

Thavaselvam, D., Flora, S.J.S., 2014. Chemical and biological warfare agents. In: Biomarkers in Toxicology. Elsevier, pp. 521–538.

Thavaselvam, D., Vijayaraghavan, R., 2010. Biological warfare agents. J. Pharm. Bioallied Sci. 2 (3), 179.

Treadwell, T.A., Koo, D., Kuker, K., Khan, A.S., 2003. Epidemiologic clues to bioterrorism. Public Health Rep. 118 (2), 92.

Trombley, A.R., Wachter, L., Garrison, J., Buckley-Beason, V.A., Jahrling, J., Hensley, L.E., et al., 2010. Comprehensive panel of real-time TaqMan polymerase chain reaction assays for detection and absolute quantification of filoviruses, arena viruses, and New World hantaviruses. Am. J. Trop. Med. Hyg. 82 (5), 954–960.

Wheelis, M., 2004. A short history of biological warfare and weapons. In: Chevrier, M.I., Chomiczewski, K., Garrigue, H., Granasztoi, G., Dando, M.R., Pearson, G.S. (Eds.), The Implementation of Legally Binding Measures to Strengthen the Biological and Toxin Weapons Convention. NATO Science Series II: Mathematics, Physics and Chemistryvol. 150. Springer, Dordrecht, pp. 15–31.

Wheelis, M., Rozsa, L., Dando, M., 2006. Deadly Cultures: Biological Weapons Since 1945. Harvard University Press. pp. 284–293, 301–303.

Zilinskas, R.A., 1999. Cuban allegations of biological warfare by the United States: assessing the evidence. Crit. Rev. Microbiol. 25, 173–227.

Further Reading

Alibek, K., Handelman, S., 1999. Biohazard: The Chilling True Story of the Largest Covert Biological Weapons Program in the World—Told From Inside by the Man Who Ran It. Random House, New York.

Cenciarelli, O., Pietropaoli, S., Gabbarini, V., et al., 2014. Use of non-pathogenic biological agents as biological warfare simulants for the development of a stand-off detection system. J. Microb. Biochem. Technol. 6, 375–380.

Centers for Disease Control and Prevention (CDC), 2015b. History Timeline Transcript. https://www.cdc.gov/travel-training/local/HistoryEpidemiologyandVaccination/HistoryTimelineTranscript.pdf.

Advance detection technologies for select biothreat agents

5

M.M. Parida, Paban Kumar Dash, Jyoti Shukla

Division of Virology, Defence Research and Development Establishment (DRDE), Defence Research and Development Organization, Ministry of Defence, Gwalior, India

Introduction

Biological warfare agents are a unique class of weapons that pose dangers to all biodiversity; the future threat is directly linked to the technological advancement in modern biotechnology (Prockop, 2006; Atlas, 2002). The armed forces and the civilians are under constant threat from a variety of microorganisms that can be used as weapons of mass destruction (Christophe et al., 1997). Advanced biological warfare agents will pose the greatest challenge to the development of appropriate medical countermeasures. Meeting this challenge will require an effective biodefense strategy in terms of a robust biodefense program that delivers the diagnostic technologies, medicines, and vaccines needed to counter the range of advanced bioweapons of the 21st century.

Biological weapons are characterized by low visibility, high potency, substantial accessibility, and relatively easy delivery (Klietmann and Ruoff, 2001). Bioterrorism is the intentional use of microorganisms—bacteria, viruses, fungi, and toxins—to produce disease and death in humans, livestock, and/or crops (Eneh, 2012). The bio agents could be carried by winds, insects, or birds, none of which respect national borders. It thus becomes very difficult to detect the biological agent or to determine whether the victim has been deliberately infected (Noah et al., 2002; Riedel, 2004). The potential spectrum of bioterrorism ranges from hoaxes and use of non-mass casualty agents by individuals or small groups to state-sponsored terrorism that employs classic biological warfare (BW) agents and can produce large-scale outbreaks and mass casualties (Centers for Disease Control and Prevention, 1999). Such scenarios would present serious challenges for patient treatment and for prophylaxis of exposed persons. Further, environmental contamination could pose continuing threats. Harmfulness or hazardous (threat) of biological agents like Nipah, Ebola, Anthrax, Plague, etc., has increased significantly with the involvement of non-state actors who would exploit not only for mass casualties, but also for a variety of other purposes from the strategic to the tactical, leading to huge economic loss and general social disruption (Szinicz, 2005; Eitzen, 2001). Public awareness of the growing

Handbook on Biological Warfare Preparedness. https://doi.org/10.1016/B978-0-12-812026-2.00005-0

threat of bioterrorism is gaining momentum all over the world. Thus there is an increased demand for overall preparedness to address the challenges pertaining to the diagnosis, treatment, and prophylaxis of new and re-emerging maliciously incited infectious diseases (Snowden, 2008; Grundmann, 2014). Anti-bioterrorism measures depend on the rapid biomonitoring of the situation as a part of pre-warning for immediate implementation of proper control measures (Parnell et al., 2010).

Bio-detection technologies

The threat of attack on military and civilian targets employing chemical and biological weapons is a growing national concern. Technologies for detection of these materials in the natural environment are being developed worldwide. While several technologies show great promise as broadband detectors, there is no silver bullet that detects all biological agents at the requisite levels of sensitivity and specificity.

Capability for detecting and identifying multiple biological warfare agents quickly and accurately is required to protect both troops on the battlefields and civilians confronted with terrorist attacks. The main focus of biodefense is therefore to develop fast, sensitive, automatic technologies for the detection and identification of biological warfare agents with a high degree of selectivity, sensitivity, and specificity (Lim et al., 2005). The detection technologies therefore focus on a variety of technologies including surface properties, genomic signatures, proteomics, etc. Ideally, a platform should be portable, easy to use, and capable of detecting multiple agents simultaneously. Platforms that integrate sample processing will have the benefit of reduced complexity for the operator. The sample processing method should be applicable for all sample types and all target analytes. In addition, instrumental techniques are widely used to detect toxins that are not amenable for DNA-based assays. The systems currently available for sensing biological analytes rely primarily on two technologies: reporter molecules that attach to antibodies and give off fluorescent signals, and the molecular PCR technology that amplifies suspect DNA. Because two steps are required to identify biological weapons, the procedure is both labor- and time-intensive (Huang et al., 2012; Martin et al., 2003).

Rapid, early, and accurate detection is the cornerstone in preventing loss of life and further spread of a disease leading to an epidemic in a biothreat scenario. From a preparedness perspective, early detection and response are crucial to minimize the potential consequences (Bravata et al., 2004). Systems to detect bioterrorism agents in clinical and environmental samples and to diagnose bioterrorism-related illnesses are essential components of responses to both hoaxes and actual bioterrorism events. The ability to identify rapidly the introduction of a bioterrorist agent into the civilian population will require highly sensitive, specific, inexpensive, high-throughput, and easy-to-use diagnostic tools (Peruski and Peruski, 2003; Rotz and Hughes, 2004). Ideally, these tests could also evaluate the possible spectrum of antimicrobial resistance and be connected to a central database. Centralized confirmatory testing also should be expanded to include routine evaluations of positive samples for genetic profiling, and bioengineered properties. The theoretical ability to design and develop such assays exist, e.g., microchip-based platforms, which could contain thousands of microbial signature profiles that are either nucleic acid or protein based, providing standards for validation and comparison of potential products (Lillehoj et al., 2010; Ewalt et al., 2001). The scope of the chapter is to give an updated comprehensive review about technological developments happening in the field of biothreat agent detection.

Culture

The culture of a microbial agent has long been considered a gold standard of diagnosis. However, owing to several limitations including cost, time, expertise, and containment facility, alternate culture independent methods are now being explored. These methods offer better sensitivity and are capable of simultaneous detection of multiple agents, even novel pathogens. Further, the rapidity of the culture independent assay is critical for decision making, particularly in a biothreat scenario (Doggett et al., 2016; Hong et al., 2013).

Immunological assays

Enzyme linked immunosorbent assay (ELISA)-based systems are generally widely used for diagnosis of single microbial infection. Recent advancements using the Luminex xMAP technology offer multiplex capability (Reslova et al., 2017). The MagPix assay system based on ELISA principles is centered on paramagnetic microsphere technology, which can transition ELISAs to a more sensitive and consistent system with the added capability of multiplexing. It employs color-coded magnetic microspheres with antibodies covalently coupled to the beads as the solid support for an ELISA-like sandwich immunoassay. A charged coupled device (CCD) camera is used for detecting the fluorescence from each microsphere excited by light-emitting diodes and facilitates measuring the median fluorescence intensity (MFI) of each sample. The MFI is then used as the basis for sample analysis (Yan et al., 2017; Andreotti et al., 2003).

Immunochromatographic test (ICT)

The assay system is based on the lateral flow/flow-through principle, employing the colloidal gold as the indicator. The ICT system has revolutionized the field of immunodiagnosis by offering an easy-to-perform test at the patient's bedside, providing results in 5–10 min.

Lateral flow rapid strip test

Lateral flow tests are also called immunochromatographic tests (ICT). They have been a popular platform for rapid tests since their introduction in the late 1980s.

ICTs are used for the specific qualitative or semi-quantitative detection of many analytes including antigens, antibodies, and even the products of nucleic acid amplification tests. One or several analytes can be tested for simultaneously on the same strip. Urine, saliva, serum, plasma, or whole blood can be used as specimens. Extracts of patient exudates or fluids have also been used successfully. Lateral flow immunoassay (LFIA) are easy to perform and fast, but they are not too sensitive and give more false-positive results. However, they may be useful for rapid initial screening of samples for the presence of biological agents, although, as a matter of principle, any positive result must be confirmed by other tests, such as PCR. Lateral flow devices have been developed by many companies for large number of biological agents as *Bacillus anthracis*, *Francisella tularensis*, *Yersinia pestis*, *Clostridium botulinum*, and several toxins, such as ricin and staphylococcal enterotoxin B (Cox et al., 2015; Li et al., 2015; Gessler et al., 2007).

Flow through spot test

The test principle involves a flow of fluid containing the analyte through a porous membrane and into an absorbent pad. A second layer, or sub-membrane, inhibits the immediate back-flow of fluids, which can obscure results. These tests can be used to detect both antibodies and antigens. To detect antibodies and antigens, the corresponding analyte is bound or immobilized as a dot or line on the membrane. This reagent "captures" the analyte as it flows through the membrane. To perform the test, a sample is applied to the membrane and allowed to wick through by capillary action. Thereafter, sequentially, there is a wash step, addition of the signal reagent, and a second wash to clear the membrane. Recent flow-through tests have successfully used colored latex particles or colloidal gold. This is a very rapid test procedure (3–5 min). However, these are not "walk-away" tests as the lateral-flow test is. Test sensitivity is good for serological assays, but for solid-phase tests, detection of antigens is often less sensitive than lateral-flow or enzyme immunoassays (EIA) methods (Fan, 1991).

In response to the 2001 anthrax cases, considerable interest was generated in the handheld antibody-based detection tests such as the Sensitive Membrane Antigen Rapid Test (SMART) and the Antibody-based Lateral Flow Economical Recognition Ticket (ALERT) (Bravata et al., 2004). Such systems use antibodies to recognize specific targets on the toxins, antigens, or cells of interest. Limitations of these tests include nonspecific binding of the antibodies, which may lead to false-positive results, and degradation of the antibodies over time, which may lead to false-negative results. Additionally, these tests are limited by the availability of antibodies.

Molecular assays

Molecular detection methods rely on the unique nucleic acid (DNA/RNA) signature of a biological agent. These methods also tend to be more sensitive than antibody-based detection methods, with real-time PCR assays being able to detect 10 or fewer microorganisms (Drosten et al., 2002). The major limitation of PCR is the inability to distinguish live and dead agents and multiplexing which is limited to 4–6 targets at the current time for real-time PCR. Much higher levels of multiplexing are possible with endpoint PCR methods using the Luminex system, but sensitivity, quantitative dynamic range, and specificity are reduced. A number of assay formats based on isothermal and non-isothermal are available and currently widely used for the gene amplification.

Polymerase chain reaction (PCR)

PCR remains the most popular and widely used technology. The major advantage of the nucleic acid-based system is its specificity due to the uniqueness of genome in living systems. The careful designing of primer and probes enables specific detection of organisms. It also offers the highest sensitivity due to exponential amplification of the genomic signature. During PCR, a short piece of genome (DNA/RNA) of the bio-threat agent is amplified, resulting in millions of copies of DNA within a short time (Towner et al., 2004; Mourya et al., 2012). A number of commercial versions of PCR systems are now available which consist of a disposable assay cartridge containing consumable reagents, an instrument that integrates the thermal components required for gene amplification, and the optical components required to quantify the amplified products (generally tagged with a fluorescent dye). Further, positive and negative controls are included as part of the assay to rule out reaction failure and validate both the assay and instrument performance. Sample preparation remains one of the important critical areas for the realization of on-site nucleic acid amplification systems. The results of PCR vary widely depending on the presence of inhibitors in the sample and are more crucial from complex environmental matrices, which require sample processing. Different modifications to the conventional PCR have been made that enable the simultaneous detection of multiple threat agents. The multiplex PCR substantially reduces cost and time (Nazarenko et al., 2002). The general contamination of source DNA in toxin samples is now exploited to identify the toxins.

Real-time RT-PCR

Compared to conventional end-point PCR, real-time (RT) PCR is quantitative in nature. It can simultaneously detect and quantify the nucleic acid in any sample. RT-PCR measures changes in fluorescence intensity which is proportional to the increase of the amplicon. RT-PCR are primarily of two types: specific and non-specific. In non-specific RT-PCR, a universal DNA interacting dye like SYBR Green is used that emits fluorescence when bound to DNA. Further, a melting curve analysis is performed following amplification that provides specificity based on the length and composition of amplicon. This format is simpler to perform and inexpensive due to requirement of only primers. The specific detection relies on the use of unique target genome-specific fluorogenic labeled probes. The increase in fluorescence indicates the hybridization of probes to the target DNA, leading to physical separation of the fluorophore from the quencher (Liu et al., 2012). In contrast to non-specific SYBR Green format, probe-based assays provide multiplex capability through use of different fluorescent dyes. TaqMan Probes have been successfully employed for detection of multiple biothreat agents including *Bacillus anthracis*, *Yersinia pestis*, *Coxiella burneti*, Cat A biothreat viral agents including smallpox, Ebola, and other hemorrhagic viruses (Buzard et al., 2012).

Isothermal gene amplification assays

In contrast to RT-PCR, isothermal assays provide the advantage of a fast turnaround time. Isothermal loop-amplification (LAMP) is a novel powerful gene amplification technique that is widely adapted to early detection and identification of numerous microbial agents including biothreat pathogens. The isothermal nature makes this assay simple and rapid, and the whole amplification process can be completed within 1h. It employs a set of six specially designed primers spanning eight distinct sequences of a target gene, making the assay extremely specific. The gene amplification products can be detected by agarose gel electrophoresis as well as by real-time monitoring in an inexpensive turbidimeter. This assay is amenable to field application, as the amplified product can be visualized by the naked eye either as turbidity or in the form of a color change when a fluorescent dsDNA intercalating dye like SYBR Green I (Parida et al., 2011; Kurosaki et al., 2007).

Helicase-dependent amplification (HDA) is an isothermal amplification technology that closely mimics the PCR. However, it works in isothermal conditions due to the use of helicase capable of unwinding a DNA duplex. Recently, a novel isothermal real-time detection method (HDA-TaqMan) that combines the advantages of both HDA and a TaqMan assay was reported for detection of biothreat organisms: *Vibrio cholerae* and *Bacillus anthracis*. In this technique, the reactions of DNA unwinding, primer annealing, polymerization, probe hybridization, and subsequent hydrolysis by the polymerase are coordinated and synchronized to perform at a single temperature (Barreda-García et al., 2016).

Nucleic acid sequence-based amplification (NASBA) is another popular isothermal gene amplification technique used to detect viable organisms using mRNA as a template in both clinical and environmental matrix. In this method, the primer binds to the RNA target sequence, and reverse transcriptase produces a technique, a primer binds to mRNA and cDNA strand is generated. The parental RNA is then degraded by RNase. A second primer binds to cDNA, which is reverse transcribed and a double-stranded cDNA is synthesized. Finally, T7 RNA polymerase enzyme is used to synthesize RNA transcripts during the amplification process. This method has been applied to detect a number of pathogens including viruses, bacteria, fungi, and protozoa (Khaled et al., 2005; Birgit et al., 2002).

Understanding the dynamics of emerging and reemerging infections is critical to efforts to reduce the morbidity and mortality of such infections, to establish policy related to preparedness for infectious threats and for decisions on where to deploy limited resources against infection. Therefore there is a need for the creation of a genome data bank for selected viral agents and circulating organisms in the country causing natural epidemics for reference and mapping. This will help to distinguish natural epidemics versus intentional biological warfare (BW) attack as well as to pinpoint the source and origin of the organism. Furthermore, this will also help to update the primer data bank and strategies for vaccine designing.

Next-generation sequencing (NGS) technology

NGS technology has the potential to revolutionize the field of molecular detection technology. This technology enables the sequencing of complete genome content of a clinical or environmental sample, creating a metagenome. This metagenome is not only capable of identification of known biothreat agent, but also can identify a hitherto unknown agent in a very short duration of time. NGS fills a vital role in characterization of disease outbreaks by whole-genome sequencing of isolates and in the identification of infectious agents when other diagnostics fail due to involvement of rare and novel pathogens. With suitable sample extraction technology that is compatible with a molecular detection platform, NGS can play a very important role in the rapid detection of agents from complex environmental matrix. A trade-off between yield and purity in the sample processing step can lead to efficient use of NGS technology. This technology has led to identification of a large number of novel species of non-culturable microbial and viral agents. The major limitation of NGS includes the high cost of instruments and complex interpretation of data (Karlsson et al., 2013; Gilchrist et al., 2015).

The focus of the nucleic acid sequence technology is the development of a biochip that contains an array of engineered molecules that react with the genome of biological warfare agents. The biochip is embedded in a platform that is portable, automated, and allows for direct sampling of the environment. A biochip platform to identify the anthrax bacteria is in the testing stages, and additional biochips for identifying other harmful bacteria and viruses are in development.

Biomonitoring

The most important step of the biodefence strategies is the rapid detection and identification of the causative agents. Detection is the unspecific demonstration of increased concentration of microorganisms in a particular environment, whereas identification is the species determination of the detected microorganism. The attack with BW agents is difficult to detect owing to the inherent intrinsic properties of the organisms. In cases of suspected use of BW agents, rapid detection and identification of the infectious agent are critical for early implementation of specific countermeasures. Therefore, the detection systems for BW agents should have the properties of rapidity, reliability, sensitivity, and specificity so as to diagnose quickly the correct etiological agent from complex environmental samples before the spreading of illness on a large scale (Lim et al., 2005).

The test systems that are suitable for the detection and identification of the agent in the laboratory cannot be directly applied for BW agents considering the nature of the environmental samples (air, water, soil, and foodstuffs). The environmental samples have highly complex structures and are therefore difficult to analyze. In addition, the concentration of the organism may be very low compared to the sensitivity or detection limit of the existing test systems. Therefore, it is essential to concentrate the microorganisms from the environmental samples in order to achieve the detectable concentrations prior to analysis.

Aerosol detection technologies

Real-time detection and measurement of biological agents in the environment are daunting. A myriad of microorganisms are present in the environment and each organism has its own signature. Most detection schemes are specific for a particular biological agent. Detection technologies are categorized by their requirement to come in direct physical contact with the biological agent. Depending on the need, detection system architecture and sensors involved will be different. For an early warning of a biological event, a "stand-off" detection system may be sufficient. For early warning systems, the sensitivity of the detection system is not important. The presence of live biological agents needs to be determined.

Air sampling of microorganisms is governed by the same principles of collection as other particulates; however, the viability of the organisms complicates their collection. The main objective is to keep the viability of the organism in a viable state so that subsequent identification steps become easier. The project envisages development of simpler and appropriate system for air sampling, analysis, and detection of biological agents in the aerosol/environment (Mary et al., 2003).

Sensor technologies

DNA array-based sensors: The DNA array class of detectors relies on comparing the DNA taken from microorganisms in a sample with the known DNA of known biowarfare agents. Researchers have in recent years been sequencing the DNA of an array of potential agents, using common gene sequencing methods (ESpehar-Délèze et al., 2015).

Protein array-based sensors: Multiplex protein liquid arrays will also be developed to detect more than one agent simultaneously. The system uses a liquid suspension array of sets of beads, each internally dyed with different ratios of two spectrally distinct fluorophores to assign it a unique spectral address. Each set of beads can be conjugated with different capture molecule. The conjugated beads can then be mixed and incubated with sample in a micro plate well to react with specific analytes. Captured molecules can be enzyme substrates, DNA, antigens, or antibodies. The bead array system can be used to detect several agents simultaneously (Birgit and Ehricht, 2006).

Immunological sensors: Although antibody technology has progressed steadily over the last five years with antibody biosensors becoming significantly smaller, such technology is still not truly man-portable and available in our country. It is proposed to develop antibody-based biosensors for environmental aerosol monitoring. One of the most sensitive designs relies on antibodies adsorbed onto the surface of colloidal gold particles, which gives a visual indication of the result (Karlsson et al., 1991).

Tissue-based biosensors: A drawback with the DNA and antibody tests is that they require prior knowledge of the bioagents. Many toxins, for example, trigger reactions in living cells. These reactions can be measured and differentiated. The tissue-based biosensor will be constructed with immobilized live cells which are seeded into a cartridge, and after exposure the response is measured. This is a completely new area of work and attempts will be made to initiate research in this new area (Wijesuriya, 1993).

MIP-based sensors: Molecular imprinting is a technique for the fabrication of biomimetic polymeric recognition sites or plastic antibodies/receptors which is attracting rapidly increasing interest. By this technology, recognition matrices can be prepared which possess high substrate selectivity and specificity. In the development of this technology, several applications have been foreseen in which imprinted materials may be exchanged for natural recognition elements. Thus MIPs have been used as antibody/receptor binding mimics in immunoassay-type analyses, as enzyme mimics catalytic applications and as recognition matrices in biosensors. Sensitive detection systems will be developed using MIPs against selected toxins (Selvolini and Giovanna, 2017).

Nanomaterials biosensors

Nanomaterials are efficient in addressing many of the limitations of existing sensors including speed, cost, mobility, and the stringent requirements of sample processing. Their small size and disposability make them excellent candidates for field-based sensors. As a substrate they provide high surface area on a platform that can be dispersed in an analytic sample and often provide feedback in less than a minute. Magnetic nanomaterials help in concentrating an analyte from a complex matrix and can also provide feedback even in opaque solutions. Quantum-confined semiconducting nanomaterials possess photophysical properties that can be exploited for tagging analytes and participating in energy transfer, while their physical properties make them more durable than dyes and suited to a non-laboratory environment. The diversity of nanomaterials in conjugation with different assay formats is useful to create superior sensors (Clare et al., 2016).

Instrumental technologies
Mass spectrometry

The unique cellular fatty acid profile can be used effectively to identify the bio organism. Fatty acid analysis is more objective and less prone to human error.

Fatty acid analysis can identify to the strain level versus the species level for most DNA-based methods. It is proposed to develop initially GC-FID generated library comprising profiles of cellular fatty acids followed by generation of a library of fingerprint programs of available organisms. The data generated will be used to identify the potential toxins and bioagents in the environmental samples. The findings can be used as preliminary analytical data. The instrumental techniques to be used for generating fingerprints of toxins and bioagents include GC-MS, LC-MS, and MALDI-TOF (Boyer et al., 2015).

Mass spectrometry techniques have been widely reported following the development of soft ionization techniques (MALDI and ESI), and by the continuous development of MS technologies (high resolution, accurate mass HR/AM instruments, novel analyzers, hybrid configurations). Multiplexed toxin detection, discovery of new markers, and identification of untargeted novel molecular targets are successfully achieved. A proof of concept study has been reported for successful post-exposure recovery of biological agent in a simulated biothreat scenario using tandem mass spectrometry (Wang et al., 2014; Alam et al., 2012).

Toxins represent one of the most dreaded forms of bioterrorism agents and have been successfully employed for bioterrorism/biocrime events. Due to the proteinaceous nature of toxins, gene amplification assays are of limited utility. The detection of toxins relies mainly on immunological assays. However, proteomics approach based on MS/MS is sensitive, rapid, and allows absolute quantitation and multiplexing capabilities. Targeted LC-MS/MSA based assay was reported for specific detection and quantification of multiple toxins, namely ricin, *Clostridium perfringens* epsilon toxin (ETX), *Staphylococcus aureus* enterotoxins (SEA, SEB and SED), shigatoxins from Shigella dysenteriae, and entero-hemorragic *Escherichia coli* strains (STX1 and STX2) in complex food matrices. However, instrument cost and service contracts remain high, limiting developing countries use to large diagnostic laboratories.

PLEX-ID is a powerful technology that combines the power of both PCR and mass spectrometry (Murillo et al., 2013). PLEX-ID allows for rapid identification and genotyping of microorganisms including, bacteria, viruses, fungi, etc. Here initially either monoplex or multiplex PCR is carried out on the nucleic acid extracted from the sample. The amplicon is subjected to electrospray ionization and time-of-flight mass spectrometry that allows very accurate determination of molecular size and weight of both strands. Unique identification of the sample is possible through comparison with a reference database. There are a number of detection panels available now in this technology (respiratory virus, biothreat, broad bacteria, broad viruses, food-borne, multi-drug resistance, etc). The advantages of such technology include extremely high multiplexing capability (up to thousands of agents) and significant throughput. These features make PLEX-ID an excellent device in the case of analyzing samples of unknown origin.

Raman chemical imaging

Chemical imaging combines molecular spectroscopy and digital imaging, and has been demonstrated to be a powerful tool for the rapid molecular analysis of biological threat agents in complex matrices. Chemical imaging microscopy provides molecular composition and structural information, without the use of dyes or stains, at sub-micron spatial resolution, in a non-contact, non-invasive, reagent-less detection mode. Optical techniques for specimen interrogation include Raman scattering and fluorescence emission (Kathryn et al., 2007; Gregory et al., 2012).

Biodetectors

Biodetectors are analytical devices that combine the precision and selectivity of biological systems with the processing power of microelectronics. They act as powerful analytical tools in medicine, environmental diagnostics, and food industries, as well as forensic analysis and counterterrorism. Biodetectors usually consist of a biological recognition system, typically enzymes or binding proteins immobilized on a surface acting as a physicochemical transducer. One typical example of a biodetector is the immunosensor, which uses antibodies as the biorecognition system. In addition to enzymes and antibodies, the recognition systems can consist of nucleic acids, whole bacteria, and single-cell organisms, and even tissues of higher organisms. Specific interactions between the target molecule or analyte and the complementary biorecognition layer produce a detectable physico-chemical change, which can then be measured by the detector. The detection system can take many forms depending upon the parameters being measured. Electrochemical, optical, mass, or thermal changes are the most common parameters providing both qualitative or quantitative data. Electrochemical biosensors are promising platforms that could achieve rapid highly sensitive and selective on-site detection of such agents (Qian and Bau, 2004; Berchebru et al., 2014).

Commercially available biodetectors

A large number of commercial firms have developed a large number of detection systems for on-site detection of biothreat agents. These are based on either molecular or immunological detection platforms. These systems are highly suitable for handling by first responders. The immunological systems are based on immunochromatographic tests, which are rapid and result can be interpreted both on site as well as can be transmitted for offsite interpretation. The molecular detection platforms are based on both isothermal and real-time PCR. They provide high accuracy, sensitivity, and multiplexing capability to the detection platform. A list of advanced technologies based on molecular and immunological detection platform has been provided in Tables 1 and 2.

Table 1 List of PCR-based technologies for detection of biothreat agents

Sl. no.	Product name	Manufacturer	Principle	Run time	Sample preparation	Automatic result display	Freeze dried	No. of agents
1	BioFire Defense, LLC: FilmArray	BioFire Diagnostics, Inc.; Biomerieux, France	Multiplex PCR	60 min	Minimal	Yes	Yes	16 agents (27 targets)
2	Bio-Seeq PLUS	BioFire Diagnostics, Inc.; Biomerieux, France	LATE-PCR	60 min	Minimal	Yes	Yes	4 agents (5 targets)
3	RAZOR EX	BioFire Defense, LLC, USA	RT-PCR	30 min	Minimal	Yes	Yes	10-agent assay
4	one3	Biomeme, USA	PCR	60 min	Minimal	No	Yes	1-agent assay
5	POCKIT	GeneReach Biotechnology Corporation, Taiwan	PCR	60 min	Moderate	Yes	Yes	1-agent assays
6	POCKIT Micro Nucleic Acid Analyzer	GeneReach Biotechnology Corporation, Taiwan	Convection PCR	30 min	Moderate	Yes	Yes	1-agent assays
7	T-COR 8	Tetracore, Inc. Rockville, USA	Multiplexed PCR	20–45 min	Minimal	Yes	Yes	2–3 agent assays
8	FilmArray	BioFire, Salt Lake City, UT	Nested multiplex PCR	60 min	Minimal	Yes	Yes	16 biothreats

LCCD, lowest concentration consistently detected; LATE-PCR, linear after the exponential polymerase chain reaction technology.

List of agents	Instrument cost	Assay cost	Detection limit	Wt	LCCD	Analyzing capacity	Shelf life of assay
1,2,3,4,5,6,7,8,9, 11,12,13,14,15	$39,500	$1110/6 pack	Different for different agents (in pdf)	20 lb	1000	Single sample for 27 targets	6 months
1,5,8,10,2,17, 3,18,9,16	$35,000	$46,199.18	100 organisms	6.6 lb	20,000	6 agents simultaneously	–
1,5,8,10,2, 17,3,18,9,16	$38,500	$768/64 reactions	100 cfu/mL	10 lb.	1000	single sample	6 months
1,9,19,16,20,3, 2,5,8,13,11,7, 21,6,22,12,23,24	$4950	$760 field kit	Different for different agents (in pdf)	1 lb.	100	Single sample	Five years
1,17,5 (5.1,5.2),9,25, 26,27	$8000	$380/48 reactions	Different for different agents (in pdf)	4.6 lb	2000	8 samples	24 months
1,17,5,5.1,5.2,9	$900	$380/48 reactions	Different for different agents (in pdf)	0.84 lb	10	8 samples	24 months
19,9,1,11, 16,6, 20,5,3,2	$28,500	$768/64 reactions	1–100 PFU per mL	10 lb.	2000	4 samples	12 months
28,29,30,31,17	$39,500	$3870	1.0E+04 CFU/mL	20 lbs	675	Single sample at a time	12 months

Table 2 List of immunoassays for detection of biothreat agents

S.no.	Product name	Manufacturer	Run time	Sample preparation	Automatic result display	No. of agents
1	BADD	AdVnt Biotechnologies	~15 min	Minimal	Yes	6
2	Pro Strips	AdVnt Biotechnologies, LLC	~15 min	Minimal	Yes	5
3	RAID Multi-Test Strips	Alexeter Technologies, LLC	~15 min	Minimal	Yes	8
4	NIDS assays and optical reader	ANP Technologies, Inc.	~15 min	Minimal	Yes	11
5	IMASS assays	BBI Detection, LLC.	~15 min	No	Yes	7
6	ENVI Assay System and optional reader	Environics, Inc.	~15 min	Minimal	Yes	4
7	PR2 1800	Meso Scale Defense™	15–60 min	Moderate	Yes	16
8	Smart II CANARY Zephyr	PathSensors, Inc.	~15 min	Minimal	Yes	6
9	RAPTOR: Automated, Multianalyte Bioassay Detection System		~15 min	Minimal	Yes	10
10	RAMP assays and optical reader	Response Biomedical Corp.	~20 min	Minimal	Yes	4
11	BioThreat Alert assays and optical reader	Tetracore, Inc.	~15 min	Minimal	Yes	9

Note: Bacillus anthracis-1, Francisella tularensis-2, Yersinia pestis-3, Clostridium botulinum-4, Brucella species-5 Brucella melitensis (5.1), Brucella abortus (5.2), Burkholderia mallei/pseudomallei-6, Ebola virus-7.1, MarBurg virus-7.2, Lassa virus-7.3, Coxiella burnetii-8, Ricin toxin-9, E. coli 157:H7-10, Variola virus-11, Rickettsia prowazekii-12, Venezuelan equine encephalitis virus-13, Eastern equine encephalitis virus-14, Western equine encephalitis virus-15, Plague-16, Salmonella-17, Small Pox-18, Abrin toxin (Abrus precatorius)-19, Staphylococcal Enterotoxin B (SEB)-20, Epsilon Toxin (Clostridium perfringens)-21, Chlamydophila psittaci-22, Vibrio cholerae-23, Cryptosporidium parvum-24, Dengue virus-25, Middle east respiratory syndrome coronavirus-26, Rift valley fever virus-27, Rotavirus-28, Campylobacter-29, Clostridium difficile-30, Norovirus-31, Shigatoxin-32, Viral Encephalitis-33, Glander-34, Malaria-35, Protozoan infection-36, Q-fever-37, T2 toxin-38, Saxitoxin-39.

List of agents	Assay shelf life	Assay cost	Analyzing capacity	Certified
1,4,16,9,20,2	24 months from date of manufacture	$257/10 pack	Single	DHS
1,4,16,9,20	24 months from date of manufacture	$735/10 pack	Multiple	DHS
1,16,2,18,4,9,20,5	18 months from date of manufacture	$995/10 pack	Multiple	–
1,16,37,2,18,4,9,20,5,33,23	24 months from date of manufacture	Complete kit—$9000	Multiple	–
1,16,2,9,20,5,34	12 months	$1270/10 pack	Multiple	ISO 9001:2008-certified
1,4,9,20	12–24 months	$400–$650/10 pack	Single	ISO 9001:2008-certified
1,16,37,2,18,4,9,19,20,32, 38,39, 7.1,7.2,7.3,35	12 months	$1–$4/assay		–
1,16,37,4,9,20	12 months	$575/25 pack	Single	–
1,16,2,18,4,9,20,5,23,36		$2000/10 pack	4-agent assay	–
1,18,4,9	12 months	$675/25 pack	Single	–
1,16,2,4,9,19,20,5	2 years	$605/25 pack	Single	–

Conclusion

The detection of biological agents is a challenging task, particularly in the outdoor environment. The developments of technologies with rapid and sensitive detection capabilities have become crucial in the present scenario of emerging bioterrorism. Researchers around the world have taken a number of different avenues in their search for a biosensor. Most of the studies report bench-top studies for proof of concept. Very few describe robust analytical results from the field. Successful implementation of a national biodefense strategy will require integration of a variety of independent efforts across the government agencies, bioscience research, and medical/public health communities.

Commercial molecular detection platforms

BioFire Defense (FilmArray)
[BioFire Diagnostics/
Biomerieux, France]

Bio-Seeq PLUS
[BioFire Diagnostics/ Biomerieux, France]

RAZOR EX
[BioFire Defense, LLC, USA]

T-COR 8
[Tetracore, USA]

POCKIT
[GeneReach Biotech Corp, Taiwan]

POCKIT Micro Nucleic Acid Analyzer
[GeneReach Biotech Corp, Taiwan]

Commercial immunological detection platforms

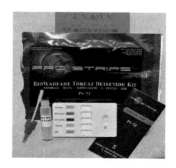

Pro Strips
[AdVnt Biotechnologies,LLC]

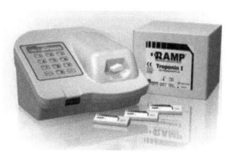

RAMP
[Response Biomedical Corp.]

BioThreat Alert
[Tetracore, Inc.]

Envi Assay
[Environics, Inc.]

RAPTOR 4 channel
[Research International, Inc.]

NIDS
[ANP Technologies Inc.]

References

Alam, S.I., Kumar, B., Kamboj, D.V., 2012. Multiplex detection of protein toxins using MALDI-TOF-TOF tandem mass spectrometry: application in unambiguous toxin detection from bioaerosol. Anal. Chem. (23), 10500–105017.

Andreotti, P.E., Ludwig, G.V., Peruski, A.H., Tuite, J.J., Morse, S.S., Peruski, L.F., 2003. Immunoassay of infectious agents. Biotechnology 35, 850–859.

Atlas, R.M., 2002. Bioterrorism: from threat to reality. Annu. Rev. Microbiol. 56, 85–167.

Barreda-García, S., Miranda-Castro, R.N., Lobo-Castañón, M.J., 2016. Comparison of isother-mal helicase-dependent amplification and PCR for the detection of *Mycobacterium tuberculosis* by an electrochemical genomagnetic assay. Anal. Bioanal. Chem. 408, 8603–8610.

Berchebru, L., Rameil, P., Gaudin, J.C., Gausson, S., Larigauderie, G., Pujol, C., Morel, Y., Ramisse, V., 2014. Normalization of test and evaluation of biothreat detection systems: overcoming microbial air content fluctuations by using a standardized reagent bacterial mixture. J. Microbiol. Methods 105, 141–149.

Birgit, H., Ehricht, R., 2006. A simple and rapid protein array based method for the simultaneous detection of biowarfare agents. Proteomics 6, 2972–2981.

Birgit, D., Aarle, P., Sillekens, P., 2002. Characteristics and applications of nucleic acid sequence-based amplification (NASBA). Mol. Biotechnol. 20, 163–179.

Boyer, A.E., Gallegos-Candela, M., Quinn, C.P., Woolfitt, A.R., Brumlow, J.O., Isbell, K., Hoffmaster, A.R., 2015. High-sensitivity MALDI-TOF MS quantification of *Anthrax* lethal toxin for diagnostics and evaluation of medical countermeasures. Anal. Bioanal. Chem. 10, 2847–2858.

Bravata, D.M., Sundaram, V., McDonald, K.M., Smith, W.M., Szeto, H., Schleinitz, M.D., Owens, D.K., 2004. Evaluating detection and diagnostic decision support systems for bioterrorism response. Emerg. Infect. Dis. (1), 100–108.

Buzard, G.S., Baker, D., Wolcott, M.J., Norwood, D.A., Dauphin, L.A., 2012. Multi-platform comparison of ten commercial master mixes for probe-based real-time polymerase chain reaction detection of bioterrorism threat agents for surge preparedness. Forensic Sci. Int. 223, 292–297.

Centers for Disease Control and Prevention, 1999. Bioterrorism alleging use of anthrax and interim guidelines for management United States. MMWR 48, 69–74.

Christophe, G., Cieslak, T., Pavlin, J., Eitzen, E., 1997. Biological warfare: a historical perspective. JAMA 278, 77–412.

Clare, E.R., Carl, B., James, B., 2016. Nanomaterial based sensors for detection of biological threat agents. Mater. Today 19, 464–477.

Cox, C.R., Jensen, K.R., Mondesire, R.R., Voorhees, K.J., 2015. Rapid detection of *Bacillus anthracis* by γ phage amplification and lateral flow immunochromatography. J. Microbiol. Methods 118, 6–51.

Doggett, N.A., Mukundan, H., Lefkowitz, E.J., Slezak, T.R., Chain, P.S., Morse, S., Anderson, K., Hodge, D.R., Pillai, S., 2016. Culture independent diagnostics for health security. Health Secur. (3), 122–142.

Drosten, C., Göttig, S., Schilling, S., Asper, M., Panning, M., Schmitz, H., Günther, S., 2002. Rapid detection and quantification of RNA of *Ebola* and viruses, *Lassa* virus, *Crimean-Congo hemorrhagic fever* virus, *Rift Valley fever* virus, *dengue* virus, and *yellow fever* virus by real-time reverse transcription-PCR. J. Clin. Microbiol. (7), 2323–2330.

Eitzen, E.M., 2001. Reducing the bioweapons threat: international collaboration efforts. Public Health Rep. 116, 17–18.

Eneh, O.C., 2012. Biological weapons agents for life and environmental destruction. Res. J. Environ. Toxicol. 6, 65–87.

ESpehar-Délèze, A.M., Gransee, R., Martinez-Montequin, S., Bejarano, N., Dulay, D., 2015. Electrochemiluminescence DNA sensor array for multiplex detection of biowarfare agents. Anal. Bioanal. Chem. 407, 6657–6667.

Ewalt, K.L., Haigis, R.W., Rooney, R., Ackley, D., Krihak, M., 2001. Detection of biological toxins on an active electronic microchip. Anal. Biochem. 289, 162–172.

Fan, E., 1991. Immunochromatographic assay and method of using same. International Patent: WO 91/12336.

Gessler, F., Wieder, S., Avondet, M.A., Böhnel, H., 2007. Evaluation of lateral flow assays for the detection of *botulinum* neurotoxin type A and their application in laboratory diagnosis of botulism. Diagn. Microbiol. Infect. Dis. 57, 243–249.

Gilchrist, C.A., Turner, S.D., Riley, M.F., Petri Jr., W.A., Hewlett, E.L., 2015. Whole-genome sequencing in outbreak analysis. Clin. Microbiol. Rev. 283, 541–563.

Gregory, M., Borland, L., Brickhouse, M., 2012. Raman spectroscopy for homeland security applications. Int. Spectrosc. 21, 2121–2134.

Grundmann, O., 2014. The current state of bioterrorist attack surveillance and preparedness in the US. Risk Manag. Health Care Policy 9, 177–187.

Hong, W.S., Young, E.W., Tepp, W.H., Johnson, E.A., Beebe, D.J., 2013. A microscale neuron and Schwann cell coculture model for increasing detection sensitivity of *botulinum* neurotoxin type a. Toxicol. Sci. (1), 64–72.

Huang, Y., Wei, H., Shi, Y.Z., Raoul, H., Yuan, Z., 2012. Rapid detection of filoviruses by real-time TaqMan polymerase chain reaction assays. Virol. Sin. 27, 273–812.

Karlsson, R., Michaelsson, A., Mattsson, L., 1991. Kinetic analysis of monoclonal antibody-antigen interactions with a new biosensor based analytical system. Immunol. Methods 145, 229–240.

Karlsson, O.E., Hansen, T., Knutsson, R., Löfström, C., Granberg, F., Berg, M., 2013. Metagenomic detection methods in biopreparedness outbreak scenarios. Biosecur. Bioterror. 11 (Suppl), 46–57.

Kathryn, S.K., Ted, H., April, A., Shea, F., Victor, K., Matthew, P., Nelson, N., Patrick, J.T., 2007. Raman chemical imaging spectroscopy reagentless detection and identification of pathogens: signature development and evaluation. Anal. Chem. 79, 2658–2673.

Khaled, H., Abd, M.E., Galil, A., Sokkary, E., Kheira, S.M., Andre, M., Yates, M.V., Chen, W., Mulchandani, A., 2005. Real-time nucleic acid sequence-based amplification assay for detection of *hepatitis* A virus. Appl. Environ. Microbiol. 71, 7113–7116.

Klietmann, W.F., Ruoff, K.L., 2001. Bioterrorism: implications for the clinical microbiologist. Clin. Microbiol. Rev. 14, 364–381.

Kurosaki, Y., Takada, A., Ebihara, H., Grolla, A., Kamo, N., Feldmann, K., Yasuda, J., Kurosaki, Y., 2007. Rapid and simple detection of Ebola virus by reverse transcription-loop-mediated isothermal amplification. J. Virol. Methods 141, 78–83.

Li, C., Zhang, P., Wang, X., Liu, X., Zhao, Y., Sun, C., Wang, C., Yang, R., Zhou, L., Zhonghua, Y., Fang, Y., Xue, Z.Z., 2015. Development and comparative evaluation of up-converting phosphor technology based lateral flow assay for rapid detection of *Yersinia pestis, Bacillus anthracis* spore and *Brucella* spp. PLoS ONE 49, 3–8.

Lillehoj, P.B., Weib, F., Ho, C., 2010. A self-pumping lab-on-a-chip for rapid detection of *botulinum* toxin. Lab Chip 10, 2265–2270.

Lim, D.V., Simpson, J.M., Kearns, E.A., Kramer, M.F., 2005. Current and developing technologies for monitoring agents of bioterrorism and biowarfare. Clin. Microbiol. 18, 583.

Liu, Y., ZX, S., Ma, Y.K., Wang, H.T., Wang, Z.Y., Shao, D.H., Wang, J.C., Liu, X.H., 2012. Development of SYBR green I real-time RT-PCR for the detection of *Ebola* virus. Bing Du Xue Bao 28, 567–571.

Martin, K., Steinberg, T.H., Cooley, L.A., Gee, K.R., Beechem, J.M., et al., 2003. Quantitative analysis of protein phosphorylation status and protein kinase activity on microarrays using a novel fluorescent phosphorylation sensor dye. Proteomics 3, 1244–1255.

Mary, T., Don, M.B., Masquelier, B., Hindson, J., Makarewicz, A., Keith, B., Thomas, M., Richard, G., Langlois, K., Wing, T., Colston, B.W., 2003. Autonomous detection of aerosolized *Bacillus anthracis* and *Yersinia pestis*. Anal. Chem. 75, 5293–5299.

Mourya, D.T., Yadav, P.D., Mehla, R., Barde, P.V., Yergolkar, P.N., Kumar, S.R., 2012. Diagnosis of Kyasanur forest disease by nested RT-PCR, real-time RT-PCR and IgM capture ELISA. J. Virol. Methods (2), 49–54.

Murillo, L., Hardick, J., Jeng, K., Gaydos, C.A., 2013. Evaluation of the Pan Influenza detection kit utilizing the PLEX-ID and influenza samples from the 2011 respiratory season. J. Virol. Methods 193, 173–176.

Nazarenko, I., Lowe, B., Darfler, M., Ikonomi, P., Schuster, D., Rashtchian, A., 2002. Multiplex quantitative PCR using self-quenched primers labelled with a single fluorophore. Nucleic Acids Res. 9, 1–7.

Noah, D.L., Huebner, K.D., Darling, R.G., Waeckerle, J.F., 2002. The history and threat of biological warfare and terrorism. Emerg. Med. Clin. North Am. 20, 255–271.

Parida, M.M., Shukla, J., Sharma, S., Ranghia, S., Ravi, V., Mani, R.M., Thomas, S., Khare, A., Rai, R.K., Mishra, B., Rao, P.V.L., Vijayaraghavan, R., 2011. Development and evaluation of reverse transcription loop-mediated isothermal amplification assay for rapid and real-time detection of the swine-origin influenza a H1N1 virus. J. Mol. Diagn. 13, 100–107.

Parnell, G.S., Smith, C.M., Moxley, F.I., 2010. Intelligent adversary risk analysis: a bioterrorism risk management model. Risk Anal. 30, 32–48.

Peruski, L., Peruski, A.H., 2003. Rapid diagnostic assays in the genomic biology era: detection and identification of infectious disease and biological weapon agents. BioTechniques (4), 840–846.

Prockop, L.D., 2006. Weapons of mass destruction: overview of the CBRNEs (chemical, biological, radiological, nuclear, and explosives). J. Neurol. Sci. 249, 4–50.

Qian, S., Bau, H.H., 2004. Analysis of lateral flow biodetectors: competitive format. Anal. Biochem. 62, 211–224.

Reslova, N., Michna, V., Kasny, M., Mikel, P., Kralik, P., 2017. xMAP technology: applications in detection of pathogens. Front. Microbiol. 8, 55–62.

Riedel, S.M., 2004. Biological warfare and bioterrorism: a historical review. BUMC Proc. 17, 400–406.

Rotz, L.D., Hughes, J.M., 2004. Advances in detecting and responding to threats from bioterrorism and emerging infectious disease. Nat. Med. 12, 130–136.

Selvolini, G., Giovanna, M., 2017. MIP based sensors: promising new tools for cancer biomarker determination. Sensors 4, 718–812.

Snowden, F.M., 2008. Emerging and re-emerging diseases: a historical perspective. Immunol. Rev. 225, 9–26.

Szinicz, L., 2005. History of chemical and biological warfare agents. Toxicology 214, 167–181.

Towner, J.S., Rollin, P.E., Bausch, D.G., Sanchez, A., Ksiazek, T.G., Lukwiya, M., Kaducu, R., Nichol, S., 2004. Rapid diagnosis of Ebola hemorrhagic fever by reverse transcription-PCR in an outbreak setting and assessment of patient viral load as a predictor of outcome. J. Virol. 78, 4330–4341.

Wang, D., Baudys, J., Krilich, J., Smith, T.J., Barr, J.R., Kalb, S.R., 2014. A two-stage multiplex method for quantitative analysis of *botulinum* neurotoxins type A, B, E, and F by MALDI-TOF mass spectrometry. Anal. Chem. (21), 10847–10854.

Wijesuriya, D.C., 1993. Biosensors based on plants and animal tissues. Biosens. Bioelectron. 8, 155–160.

Yan, Y., Luo, J.Y., Chen, Y., Wang, H.H., Zhu, G.Y., He, P.Y., Guo, J.L., Lei, Y.L., Chen, Z.W., 2017. A multiplex liquid-chip assay based on Luminex xMAP technology for simultaneous detection of six common respiratory viruses. Oncotarget 57, 96913–96923.

Microfluidics application for detection of biological warfare agents

6

Bhairab Mondal, N. Bhavanashri, S.P. Mounika, Deepika Tuteja, Kunti Tandi, H. Soniya

Shankaranarayana Life Sciences, Bengaluru, India

Introduction

A bioterrorism attack is the purposeful release of viruses, bacteria, toxins, or other harmful agents used to cause illness or death in people, animals, and/or plants. The benefits of this weapon are that these agents are inexpensive to obtain and easy to produce, can be easily disseminated, and can cause widespread fear and panic beyond the actual physical damage (Zhu et al., 2013). These agents are typically found in nature, but the pathogen could have been mutated or altered to increase its ability to cause disease, make it resistant to current medicines, or to increase the pathogen's ability to be spread around the environment. Biological agents can be spread by air, water, or food. Terrorists tend to use biological agents because they are extremely difficult to detect and do not cause illness for several hours to several days (Haes and Van Duyne, 2002).

However, military leaders have learnt that, as a military asset, bioterrorism has some important limitations—for example, it is difficult to employ a biological weapon in a way that only the enemy is affected and not friendly forces. A biological weapon is useful to terrorists mainly as a method of creating mass panic and disruption to a state or a country. However, technologists have warned of the potential power which genetic engineering might place in the hands of future bioterrorists. Bioterrorism is difficult to predict or prevent, and reliable platforms to rapidly detect and identify bioterrorism agents are therefore important in order to minimize the spread and wide use of these agents and to protect public health. To meet the challenges of bioterrorism, coordinated and concerted efforts of different agencies (the intelligence agency, the army, the BSF (Border Security Force), SSB (Sashastra Seema Bal), law enforcement machinery, health departments, and civil administration) are required. Therefore, the best approach for combating bioterrorism is to detect and treat affected individuals as quickly as possible post-release. Since pathogens employed as bio-weapons are typically highly toxic, detection methods should be rapid and unambiguous, and should be effective at very low analyte concentrations. These platforms must be sensitive, specific, and able to detect a variety

Handbook on Biological Warfare Preparedness. https://doi.org/10.1016/B978-0-12-812026-2.00006-2

of pathogens accurately, including modified or previously uncharacterized agents, directly from complex sample matrices. False-positives might trigger inappropriate and/or wasteful mobilization of resources. The current public discussion of the threat of bioterrorism is an opportunity to evaluate our collective capabilities and to assess weaknesses and vulnerabilities.

Various commercial tests that utilize biochemical, immunological, nucleic acid, and bioluminescence procedures are currently available to identify biological threat agents. Recently developed tests identify bioterrorism agents using DNA aptamers, biochips, evanescent wave biosensors, cantilevers, living cells, and other innovative technologies. This chapter describes current and developing technologies against bioterrorism and considers challenges to the rapid and accurate detection of biothreat agents. Although there is no ideal platform, many of these technologies have proved valuable for the detection and identification of bioterrorism agents.

This review summarizes microfluidics based current and developing technologies to monitor, detect, and characterize biothreat agents for both civilian and military biodefense. Sections also cover: micro-scale sample preparation methods; integrated lab-on-a-chip systems based on immunoassays; proteomics; polymerase chain reaction (PCR), quantitative PCR (qPCR), and other nucleic acid amplification methods; and DNA microarrays.

Biological weapons

Biological weapons (bio-weapons), pathogenic organisms and their toxic products constitute a particularly pernicious threat. They can be released into the air and into water systems or the spores can be disseminated (example: anthrax spore contamination in the U.S. postal system, 2001). The presence of such agents is not usually confirmed until they produce symptoms in compromised individuals. If the agent is infectious, the affected individuals may become new vectors for spreading the disease unless they are identified, isolated, and treated.

Terrorism with biological weapons is likely to remain rare. Bioterrorism differs from other types of terrorism (chemical, radiological, or nuclear). The area of miniaturized or microfluidic analysis systems, also called micro total analysis systems (μTAS) or lab-on-a-chip (LOC), has become increasingly popular. The ability of microfluidic systems to conduct measurements from small volumes of complex fluids with efficiency and speed, without the need for a skilled operator, has been regarded as the most powerful application of LOC technologies. Furthermore, portable LOC devices involve automated complex diagnostic procedures, normally performed in a centralized laboratory, and are able to provide healthcare workers and outpatients with important health-related information even in the most remote settings. Portable medical diagnostic tools to a patient-centric and home-testing approach and microfluidic LOC technology are of great importance in both developed and developing countries where more than half of deaths are attributed to infectious diseases.

Pathogens involved in bioterrorism

Bacterial, fungal, and viral pathogens have been identified as agents that have been, or could be, used as weapons of biological warfare and/or biological terrorism which have been recorded throughout history. These agents are relatively easily obtained, prepared, and dispersed either as weapons of mass destruction or for more limited terrorist attacks. As all these agents have the potential for aerosol dissemination, a large population are susceptible and few limited treatment and vaccination strategies exist. In 1954, *Brucella suis*, which is a category B agent, was first weaponized by the United States. Multiple viral agents have been classified by the Centers for Disease Control and Prevention (CDC) as potential weapons of mass destruction and biologic terrorism. Agents such as smallpox, viral hemorrhagic fever viruses, and agents of viral encephalitis are of concern because they are highly infectious and relatively easy to produce. European soldiers used smallpox as bio-weapons when variola-contaminated clothing and blankets were delivered to South American natives in the 15th century, and even during the French and Indian Wars (1754–1767), British soldiers used variola to initiate an outbreak of smallpox among American Indians sympathetic to Americans. With the cessation of routine smallpox vaccinations in the early 1980s, the deliberate release of smallpox would now be potentially catastrophic.

The CDC ranks the biological agents and diseases that have the potential to be used as weapons into three categories:

(a) *Category A* agents are characterized by ease of dissemination and transmission of disease with a high mortality rate, causing public panic and social disruption.
(b) *Category B* agents disseminate less easily, and have lower morbidity and mortality rates.
(c) *Category C* agents could be used for mass dissemination in the future because of their availability, ease of production, dissemination, and high morbidity and mortality rates (Tables 1 and 2).

Recent biological terrorism threats and outbreaks of microbial pathogens clearly emphasize the need for biosensors that can identify infectious agents quickly and accurately. The majority of rapid biosensors generate detectable signals when a molecular probe in the detector interacts with an analyte of interest. Analytes may be whole bacterial or fungal cells, virus particles, or specific molecules, such as chemicals or protein toxins, produced by the infectious agent. Peptides and nucleic acids are most commonly used as probes in biosensors because of their versatility informing various tertiary structures. The interaction between the probe and the analytes can be detected by various sensor platforms, including quartz crystal microbalances, surface acoustical waves, surface plasmon resonance, amperometrics, and magneto elastics. The field of biosensors is constantly evolving to develop devices that have higher sensitivity and specificity, and are smaller, portable, and cost-effective. The present article describes recent advances in biosensors for applications in the rapid

Table 1 Categories of biological agents according to the CDC

Category A agents	Category B agents	Category C agents
• *Bacillus anthracis* (anthrax) • *Clostridium botulinum* toxin (botulism) • *Francisella tularensis* (tularaemia) • Variola major (smallpox) • *Yesinia pestis* (plague) • Filo viruses • Ebola virus (Ebola hemorrhagic fever) • Marburg virus (Marburg hemorrhagic fever) • Arena viruses • Junin virus (Argentinian hemorrhagic fever) and related viruses • Lassa virus (Lassa fever)	• Alpha viruses • Eastern and western equine encephalomyelitis viruses (EEE, WEE) • Venezuelan equine encephalomyelitis virus (VEE) • *Brucella* species (brucellosis) • *Burkholderia mallei* (glanders) • *Coxiellaburnetii* (Q fever) • Epsilon toxin of *Clostridium perfringens* • Ricin toxin from *Ricinus communis* • Staphylococcal enterotoxin A subset of Category B agents includes pathogens that are food- or waterborne. These pathogens include but are not limited to: • *Cryptosporidium parvum* • *Escherichia coli* O157:H7 • *Salmonella* species • *Shigella dysenteriae* • *Vibrio cholerae*	• Hanta viruses • Multidrug-resistant tuberculosis • Nipah virus • Tick-borne encephalitis viruses • Tick-borne hamorrhagic fever viruses • Yellow fever

Table 2 Classification of agents of bioterrorism

Category A	Category B	Category C
The highest priority agents include organisms that pose a risk to national security because they: • are easily disseminated; • cause high mortality; • cause public panic and social disruption; and • require special action for public health preparedness	The second highest priority agents include those that: • are moderately easy to disseminate; • cause moderate morbidity; and • require enhanced disease surveillance and publichealth diagnostic capacity	The third highest priority agents include emerging pathogens that: • could be engineered for mass dissemination in the future; and • have the potential for high morbidity, mortality, and major health impact

detection of bioterrorism weapons. Although biosensors are exhibiting double-digit growth rates, they still have to overcome a number of challenges, including the following:

- Novel research focus to impending introduction of innovative applications and lesser towards fundamental research.
- Development of a single biosensor platform with multi-purpose diagnostics ability has restricted biosensor applications.
- Numerous problems encountered in successful commercialization of biosensors have encouraged conservative development strategies.
- Competition from non-biosensor technologies has hindered revenue progress.
- The low rate of technology transfer and a lower level of development have deterred the development of newer biosensors.

Planning and response

Planning may involve the development of biological identification systems. Until recently, most biological defense strategies have been geared to protecting soldiers on the battlefield rather than ordinary people in cities. Financial cutbacks have limited the tracking of disease outbreaks. Some outbreaks, such as food poisoning due to *E. coli* or *Salmonella*, could be of either natural or deliberate origin.

Preparedness

Biological agents are relatively easy to obtain by terrorists and are becoming more threatening. Laboratories are working on advanced detection systems to provide early warnings in contaminated areas and populations at risk, and to facilitate prompt treatment. Methods for predicting the use of biological agents in urban areas as well as assessing the area for the hazards associated with a biological attack are being established in major cities. In addition, forensic technologies are being developed to identify biological agents, their geographical origins, and/or their initial source. Efforts include decontamination technologies to restore facilities without causing additional environmental concerns. Early detection and rapid response to bioterrorism depend on close cooperation between public health authorities and law enforcement; however, such cooperation is currently lacking. National detection assets and vaccine stockpiles are not beneficial if local and state officials do not have access to them.

Challenges relating to biothreat detection are:

1. high sensitivity—detect very small amounts of pathogens, toxins, and chemical agents;
2. high selectivity—discriminate targets from other materials;
3. massively parallel to detect multiple pathogens, minimize false-positives, and have rapid response, without sample preparation;
4. transportable or handheld, robust, simple to operate;

5. inexpensive; and
6. adaptable to new biothreats, integrated chemical-biosensor, and allow the detection of single molecules.

Targets to achieve are:

1. single RNA molecule detection;
2. real-time monitoring of RNA hybridization at single molecule level;
3. real-time monitoring of protein binding to aptamers at single molecule level;
4. hybridization of synthetic target DNA with anti-anthrax; and
5. selectivity of protein detection, i.e., selectivity of human thrombin detection by anti-thrombin aptamers.

Available systems for bioterrorism

The rapid detection and identification of biothreat agents comprise a pressing issue in field of clinical and food-borne pathogen detection. It is well known that healthcare system can greatly benefit from faster, accurate diagnosis by significantly reducing healthcare costs. The developed world standards for target pathogen diagnosis, including culture-based, enzyme immunoassay, and polymerase chain reaction (PCR), often require 2–4 days. These platforms must not only be sensitive and specific, but must also be able to detect a variety of pathogens accurately, including modified agents directly from complex sample matrices. Various commercial tests utilizing biochemical, immunological, nucleic acid, bioluminescence aptamers, biochips, evanescent wave biosensors, cantilevers, living cells, and other innovative technologies are currently available to identify biological threat agents (Lim et al., 2005). This literature describes these current and developing technologies, and considers challenges to rapid, accurate detection of biothreat agents. Although there is no ideal platform, many of these technologies have proved invaluable for the detection and identification of biothreat agents. Therefore, robust and portable diagnostic devices capable of providing information on pathogens will also help in reduction of mortality rates, hospitalization, and timely isolation in the case of infectious pathogens. A variety of different biosensors have been developed, but there is still a need for miniaturized, low-cost, and disposable biosensors for rapid detection and accurate identification of a wide range of pathogens.

Immunoassays

Immunoassays include immuno-chromatographic lateral flow devices, enzyme-linked immunosorbent assay (ELISA), IMS-electrochemiluminescence, and magnetic force assays, which are routinely used to detect biosecurity threats.

1. Lateral flow assays are very simple and compatible with portable applications. A drop of test solution in buffer is added to a pad containing antibodies coupled to colloidal gold or other labels. The antibody and antigen, if present, bind and wick down the pad laterally and intercept detection lines that contain antibody-gold

complexes. The gold aggregates due to the bivalency of the antibodies and produces a line that is visible by eye (Koczula and Gallotta, 2016).

2. ELISA is the most commonly used form of immunoassay. This uses an antibody bound to a solid phase support, such as a microtiter plate, to capture analytes from liquids. After washing away unbound material, a secondary antibody is added which is labeled or coupled to an enzyme (e.g., horseradish peroxidase) for visual detection (Joos et al., 2002).

3. Electrochemiluminescence (ECL) assays contain ruthenium labels, which emit light when electrochemically reduced. In ECL detection, tripropylamine can be oxidized at the surface of an array of electrodes, and in turn it reduces the ruthenium, which then emits light. ECL has been further developed with microelectrodes manufactured using screen-printing of carbon inks onto microtiter plates (Liu et al., 2015).

4. Magnetic force assays: The strength of intermolecular interactions can be measured by the force required to disrupt a bond when the target is attached to a magnetic bead. This bead serves as the label which can be detected by microfabricated magneto-resistive transducers on microchips. Multiple analytes and multiple samples can be measured simultaneously, and magnetic force is tolerant of different types of analytes (Gijs et al., 2009).

Proteomic approaches

Proteomic approaches for biodefense rely on identification of proteins and peptides to evaluate and characterize potential biothreat agents. Biodefense includes separation of proteins and peptides by mass spectrometry platforms, two-dimensional gel analysis, and protein arrays (Pizarro et al., 2007). Proteomic approaches are being used to build a cyber-infrastructure of NIAID-funded centers that are applying these tools in biodefense to develop vaccines and proteomic targets. Protein arrays containing either antibodies to different epitopes or different proteins arrayed on a solid surface have been applied to characterize and type biothreats. A protein chip for the Array Tube platform was developed that uses a microtube-integrated protein chip to accomplish detection using the classical sandwich assay and horseradish peroxidase colorimetric substrate (Huelseweh et al., 2006).

Nucleic acid amplification and detection methods

The best-known and most used DNA amplification method is polymerase chain reaction (PCR). PCR uses thermal cycling to amplify DNA exponentially using a thermally stable DNA polymerase. Each cycle of amplification doubles the amount of template, thereby exponentially increasing the amount of target, and can amplify from as little as a single copy of DNA. The specificity of the amplification is determined by the pair of primers that initiate the amplification. PCR has become a standard clinical and research technique for nucleic acid testing and for biodefense. Many different variations of the basic PCR reaction have been developed, including qPCR, nested PCR, multiplexed PCR, and single nucleotide polymorphism PCR (Wiedbrauk, 2009).

Bio-warfare agents monitoring

There are two basic detection approaches for bio-warfare agents: detect-to-warn and detect-to-treat. Detect-to-warn systems aim to identify biothreats rapidly enough to provide sufficient warning to prevent exposure to the threat. Detect-to-treat systems aim to identify the causative agent for diagnostic purposes and thereby to direct healthcare workers to the most effective treatment as quickly as possible. For all systems, low false alarm rates (FAR) and affordable acquisition and operating costs are essential for widespread adoption (Hasan et al., 2005).

Civilian biodefense

Civilian biodefense is based upon surveillance to detect biothreat agents, response networks to warn and direct the treatment of the affected population, and the development of countermeasures. Bioterrorism incidents, involving releases of a bioagent in a form that can harm individuals or larger populations, include, e.g., mailing a toxin or bacteria, releasing aerosols in high-profile events, and attacks on the food supply. BioShield, a countermeasures program, is a 10-year, $5.6 billion U.S. program for the advanced development and purchase of medical countermeasures (Larsen and Disbrow, 2017). Acquisition programs have been announced to counter *Bacillus anthracis* (anthrax), Variola virus (smallpox), botulinum toxins, and radiological/nuclear agents.

Microfluidic applications in biodefense

The primary programs that have been implemented for civilian biodefense detection include screening of all postal mail at sorting centers, the Bio Watch program, selected localized screening in subways and other undisclosed locations, and the Laboratory Response Network (LRN). The currently deployed systems principally use full volume or meso-scale fluidics. Despite the need for more advanced detect-to-warn biodefense detection for the general public, these systems have remained largely undeveloped, due mainly to the complexity of integrating the complete process. The largest monitoring program is Bio Watch, a joint effort by the Department of Homeland Security (DHS), the Centers for Disease Control and Prevention (CDC), and the Environmental Protection Agency (EPA) (Shea et al., 2003).

Military biodefense

U.S. military biodefense programs aim to detect and identify biological warfare agents that an enemy might use to degrade forces, contaminate bases, and spread confusion throughout command and control systems. For example, the Portal Shield program is designed for facilities protection, the Joint Biological Agent Identification and Diagnostic System (JBAIDS) is designed for both detection and diagnostics of environmental and clinical samples, and the Joint Biological Point Detection System (JBPDS) is designed for detect-to-warn capabilities in the field (Carrico, 1998).

Portal Shield is an array-able sensor system developed to provide early warning of biological attacks for high-value, fixed-site assets, such as air bases and port facilities. Portal Shield is designed to detect and identify threats simultaneously within 25 min. It is programmable to survey continuously as well as performs random or directed sampling. It is fully modularized, self-contained, and can detect eight different agents. Using an array system, the false-positive rate diminishes towards zero.

The Joint Biological Agent Identification and Diagnostic System (JBAIDS) is the Department of Defense's first common platform for identification and diagnostic confirmation of biological agent exposure or infection. JBAIDS Block I is currently operational as a real-time PCR instrument with Food and Drug Administration (FDA)-approved assays. It utilizes manual sample preparation which results in complex operator procedures that may lead to human error, high operational costs, and support and logistical requirements that preclude remote operations. The Joint Biological Point Detection System (JBPDS) is planned to detect, identify, and warn against the presence of up to 24 bio-warfare agents at discrete points within a given field environment (Fitch et al., 2003). Future JBPDS improvements will reduce consumable use; weight and size are directly tied to the drastic reduction of reaction volume employed in microfluidic systems and the use of miniaturized microfluidic components. There are considerable issues in integrating all processing and analysis steps in a "hands-free" device. However, microfluidics requires a "hands-free" implementation once samples are loaded since typically there is no way that the user can intervene.

Microfluidics

Microfluidics is the science which studies the behavior of fluids through microchannels, and the technology of manufacturing microminiaturized devices containing chambers and tunnels through which fluids flow (Glückstad, 2004). It deals with very small volumes of fluids and the utmost importance of detecting low concentrations of small molecules and solutes in mixtures for toxicology, drug discovery, diagnostics, and anti-bioterrorism. Microfluidics systems work by using a pump and a chip. Inside the chip there are microchannels that allow the processing of the liquid such as mixing, chemical, or physical reactions. Microfluidics is often described as "lab on a chip" or "organ on a chip" technology (Fig. 1).

A biochip is a small device that directly connects an organism to an engineering system. It includes tiny built-in sensors which analyze biochemical targets in organisms such as cells, blood, or skin. Biochips utilize a wide array of technological advances such as microfluidics, microarrays, optics, or electronics. A microfluidic chip is a device that enables a tiny amount of liquid to be processed or visualized. The chip is usually transparent and its length or widths are from 1 cm (0.5″) to

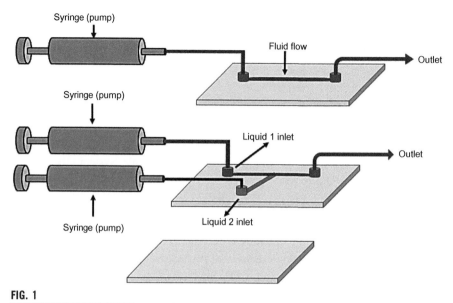

FIG. 1

Microfluidic working principle.

10 cm (4″). The chip thickness ranges from about 0.5 mm to 5 mm. Microfluidic chips have internal thin hair-like microchannels that are connected to outside by means of holes on the chip called inlet/outlet ports. Microfluidic chips are made of thermoplastics such as acrylic, glass, silicon, or a transparent silicone rubber called PDMS. Microfluidic chips are usually made by making thin grooves or small wells on surface of a layer, and then a second layer to form microchannels or chambers. Depending on material choice, the channels are made via soft lithography, hot embossing, injection molding, micro-machining, or etching. Three-dimensional printing may be used for producing microfluidic chips, although it has serious limitations in terms of minimum feature size, surface roughness, optical transparency, or choice of material. Microfluidics is also used to miniaturize or integrate conventional laboratory practices by making lab-on-a-chip device to save costs or reduce time (Castillo-León, 2015). It has applications in most experimental science and engineering, and can be used in a wide range of applications, such as: molecular and cell biology research, genetics, fluid dynamics, micro-mixing, point of care diagnostics, lab-on-a-chip, tissue engineering, organ-on-a-chip, drug delivery device, fertility testing and assistance, synthesis of chemicals or proteins, disease diagnostics, and gene sequencing, with many new applications emerging (Nguyen et al., 2019) (Fig. 2).

Microfluidic techniques are emerging as cost-efficient and disposable tools for rapid diagnosis of viral infection. Microfluidic devices are used for selective and sensitive detection based on immunoassays of related antigens or antibodies, nucleic acid amplification, and flow cytometry. At present, microfluidics provides efficient

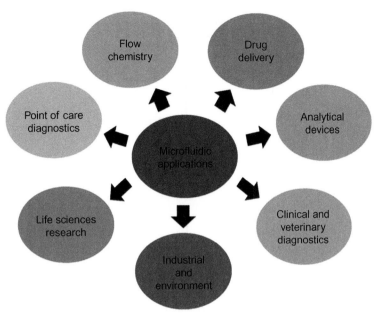

FIG. 2

Microfluidic applications.

tools for multiple research areas, and more specifically for biological analysis, with the following benefits:

- whole biological process integrated and simplified for the end-users;
- high-throughput, multiplexed and highly paralleled assays;
- faster analyses due to the shorter reactions and/or separation times;
- portable devices for point-of-care applications;
- low reagent consumptions;
- global cost reduction per analysis; and
- accurate measurement.

Types of microfluidics

Microfluidic platforms are of various types based on the liquid propulsion principle such as lateral flow tests, pressure-driven laminar, flow linear actuated devices, segmented flow microfluidics, microfluidic large-scale integration, centrifugal microfluidics, electrowetting, electrokinetics, surface acoustic waves, and dedicated systems for massively parallel analysis (Mark et al., 2010). The following table contains various types of microfluidic platforms along with their respective characterization (Fig. 3; Tables 3 and 4).

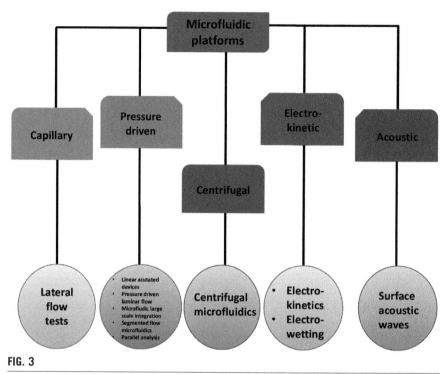

FIG. 3

Various microfluidic platforms.

Continuous flow microfluidics

The manipulation of liquid flow through fabricated microchannels without breaking continuity is defined as continuous flow microfluidics. Fluid flow is established by external sources such as micropumps or internal mechanisms such as electric, magnetic, or capillary forces. This assay has diverse applications including micro- and nanoparticles separators, particle focusing, and chemical separation as well as simple biochemical applications. The two major and opposing fluid handling tasks are mixing and separation (Chen et al., 2007a,b).

Mixing

The mixing of reactants is required to initiate the interactions such as protein folding and enzyme reactions. In most lab-on-a-chip platforms, sample preparation is required for a variety of biological and chemical assays (Mark et al., 2010). Diffusion-based mixing techniques fail to satisfy the recent demand for rapid and homogeneous mixing. One strategy for increasing mixing efficiency is employing external energy sources to create disturbances, such as acoustic, magnetic, or electrostatic (Nguyen et al., 2017). Using external actuations to increase the mixing efficiency could be expensive and challenging.

Table 3 The comparison between the three types of microfluidics

	Continuous-flow microfluidics	Droplet-based microfluidics	Digital microfluidics
Operating method	Motion of continuous fluid in micro-channels	Motion of droplets in micro-channels using streams of immiscible fluids	Motion of discrete droplets on an array of planar electrodes
Flow actuation	Mechanical (syringe) pumps, pneumatic pressure, electrokinetic	Mechanical (syringe) pumps, pneumatic pressure	Electrowetting, on dielectric, dielectrophoresis
Advantages	Ease of fabrication and operation, suitable for applications that require a continuous flow with relatively high sampling volume, and being compatible with most of current screening and sensing mechanisms	Ease of fabrication and operation, suitable for a applications that require isolated reaction sites to avoid cross-contamination	Lower sample consumption, scalability, better localization, reconfigurability, and portability
Disadvantages	High sample volume consumption compared to other microfluidic systems, possible contamination, and not being scalable due to fabrication and physical limitations	No control over individual droplets, challenging to create droplets of different sizes using the same setup, and challenging to implement stable gas–liquid systems	Complicated fabrication procedure, and bio-adsorption and evaporation

Separation

Separation also plays an important role in sample preparation for both analytical chemistry and biological applications. In the last two decades, significant advances have been made in the development of continuous-flow microfluidic separation (Kang et al., 2008). With continuous injection and collection of samples, a high separation throughput can be achieved. The separation of particles and cells can employ a variety of external forces such as hydrodynamic, electrophoretic, dielectrophoretic, magnetophoretic, acoustic, and inertial force (Lenshof et al., 2017).

Digital microfluidics

The combination of microfluidics with the science of emulsion and with the manipulation of individual droplets digital microfluidics (DMF) has been developed. It involves the manipulation of small, discrete droplets, usually in the microliter scale or smaller (Gokmen et al., 2009). The main tasks of DMF involve dispensing droplets, moving droplets, merging droplets, or mixing contents within a droplet

Table 4 Microfluidic platforms classified according to the main liquid propulsion principle (Mark et al., 2010)

Microfluidic platform	Characterization
Lateral flow tests	A lateral flow test is also known as test strips. Liquid movement is controlled by the wettability and feature size of the porous or microstructured substrate, where liquids are driven by capillary forces. The readout of a test is typically done optically and is quite often implemented as colorimetric detection
Linear actuated devices	By the method of mechanical displacement of a liquid, linear actuated devices control liquid movement, e.g., by a plunger. Liquid control is mostly limited to a one-dimensional liquid flow in a linear manner.
Pressure-driven laminar flow	Based on pressure gradients, pressure-driven laminar flow platform is characterized by liquid transport mechanisms. Typically this leads to hydrodynamically stable laminar flow profiles in microchannels. There is a broad range of different implementations in pressure sources such as using gas expansion principles, syringes, pumps or micropumps, pneumatic displacement of membranes, etc.
Microfluidic large-scale integration	A microfluidic channel circuitry with chip-integrated microvalves based on flexible membranes between a liquid-guiding layer and a pneumatic control-channel layer. The microvalves are closed or open corresponding to the pneumatic pressure applied to the control-channels like micropumps, mixers, multiplexers, etc. can be built up with hundreds of units on one single chip
Segmented flow microfluidics	This assay is carried out by the use of small liquid plugs and/or droplets immersed in a second immiscible continuous phase (gas or liquid) as stable micro-confinements within closed microfluidic channels. They can be transported by pressure gradients and can be merged, split, sorted, and processed without any dispersion in microfluidic channels
Centrifugal microfluidics	In centrifugal microfluidics, all processes are controlled by the frequency protocol of a rotating microstructured substrate. The relevant forces for liquid transport are centrifugal force, Euler force, Coriolis force, and capillary force. Assays are implemented as a sequence of liquid operations arranged from radially inward positions to radially outward positions
Electrokinetics	The microfluidic unit operations are controlled by electric fields acting on electric charges, or electric field gradients acting on electric dipoles. There are several electrokinetic effects depending on buffers and/or sample such as electroosmosis, electrophoresis, dielectrophoresis, and polarization
Electrowetting	Electrowetting platforms use droplets immersed in a second immiscible continuous phase (gas or liquid) as stable micro-confinements. The voltage between a droplet and the electrode underneath the droplet defines its wetting behavior. By changing voltages between neighboring electrodes, droplets can be generated, transported, split, merged, and processed

Table 4 Microfluidic platforms classified according to the main liquid propulsion principle (Mark et al., 2010)—cont'd

Microfluidic platform	Characterization
Surface acoustic waves	The surface acoustic wave's platform using droplets residing on a hydrophobic surface in a gaseous environment (air). The microfluidic unit operations are mainly controlled by acoustic shock waves traveling on the surface of the solid support. Most of the unit operations such as droplet generation, transport, mixing, etc. are freely programmable
Dedicated systems for massively parallel analysis	Within the category of dedicated systems for massively parallel analysis, we discuss specific platforms that do not comply with our definition of a generic microfluidic platform. The characteristics of those platforms are not given by the implementation of the fluidic functions but by the specific way to process up to millions of assays in parallel

(Choi et al., 2012). Numerous techniques have been developed to perform these tasks, as follows.

Droplet-based DMF

DMF devices can have a basic open planar form, where the plate is usually engineered to provide an energy gradient to drive the droplet as the droplet is placed on a solid planar surface (Li et al., 2016). To facilitate control of the sandwiched droplet, in some cases a top plate is added. With proper design, droplets can be moved across the plate in two dimensions and can also be further controlled by constructing channels between the plates, thus restricting the droplet to a one-dimensional movement (Zhu et al., 2013).

Electrowetting-on-dielectric (EWOD) technique

Electrowetting-on-dielectric (EWOD) is one of the most popular techniques in DMF, where a droplet is placed between two plates, one of which contains a dielectric layer and voltage difference across the droplet generates asymmetric droplet contact angles, thus creating a driving force (Shen et al., 2014). Switching the voltage difference in a timely manner moves the droplet. Optoelectrowetting is a modified version of the EWOD technique, where the voltage switching is accomplished optically (Pamula et al., 2009).

Dielectrophoretic technique

The dielectrophoresis technique similarly uses electrostatic force, but the droplet itself acts as a dielectric. The device harnesses the user's mechanical input and converts it into electrostatic energy, which is then used to move the droplets suspended in fluids (Nguyen et al., 2017).

Magnetic-based techniques

A magnetic field can be applied to move droplets containing magnetite via magnetowetting instead of an electric field. The magnetic field generates a body force throughout the entire droplet. Displacing a permanent magnet under a ferrofluid droplet creates asymmetric contact angles and moves the droplet (Biswas et al., 2016).

Other techniques

Droplets can be manipulated using other means such as surface acoustic waves (SAW) or thermocapillary forces. Acoustic energy is generated using a piezoelectric element and transferred to the droplet (Bordbar et al., 2018). As the SAW hit the droplet, energy is transferred onto the droplet, which causes it to de-pin from the surface and move. More energetic SAW can even cause droplets to nebulize. Unlike EWOD, most SAW devices need only one plate (Nguyen et al., 2017).

Liquid-marble-based DMF

Another emerging field in DMF is the use of liquid marble (LM) as the discrete platform. The LM is a small droplet encapsulated by a hydrophobic coating, which consists of a porous particle layer (Jin et al., 2017). The hydrophobic and porous shell removes the need for surface treatment, as the droplet is physically isolated from its surroundings. An added benefit is that an LM is able to float on a liquid surface and seemingly skid around with low friction (Ooi et al., 2018). Among the most popular techniques is manipulating a LM containing magnetite using a permanent magnet. Furthermore, a LM can be driven by thermo- or soluto-capillary forces.

Advanced digital microfluidic platforms

Recent advances in manipulating micro droplets predominantly involve EWOD-based devices. Researchers have pioneered the use of DMF in the immune precipitation process. This concept was accomplished using an existing DMF device which combines both EWOD and magnetic manipulation of the droplet. DMF has also been used for the first time in solid-phase micro extraction, as well as in high field nuclear magnetic resonance spectroscopy (Nguyen et al., 2017). Nanostructure initiator mass spectrometry (NIMS) arrays can be integrated into a EWOD device to conduct enzyme screening, which increases the throughput of the process (Heinemann et al., 2017). Liquid marble can also be used as a microbioreactor, as it can accommodate liquid volumes across several orders of magnitude and can still be easily handled. Recently, a spinning liquid marble has been used to improve mixing (Nguyen et al., 2017).

Various available platforms for bioterrorism

For treatment of infectious disease, it is important to detect and identity pathogens in the early stages. Infectious disease accounts for the major causes of illness and mortality worldwide. Several detection platforms have been developed which provides a fundamental tool in different fields including clinical diagnostics, pathology, drug discovery, clinical research, disease outbreaks, and food safety (Foudeh et al., 2012). For decades, several conventional methods have been used based on cell culture, nucleic acid amplification, and ELISA; these are often laborious and take from several hours to days to perform. For point of care (POC) applications, the detection platform should also be simple to use and interpret, affordable, robust, stable under a wide range of operating conditions (such as temperature, humidity), and preferably portable and disposable (Wang et al., 2013). They should provide sensitivity and specificity. The ability to perform multiplex tests is one of the major important prerequisite for pathogen detection devices, such as in case of lower respiratory infections. To overcome the existing limitations of the conventional methods, lab-on-a-chip (LOC) and microfluidic systems are gaining attention (Poritz et al., 2011). Microfludics offers more rapid and sensitive detection even at lower concentrations of sample input and gives accurate results. These new generation devices combine sample processing with detection of pathogens on a single microchip. The geometries and dimensions of channels help in the isolation and detection of pathogens (Baker et al., 2009). Advantages like high surface to volume ratio, ability to handle small volumes of reagents from nano to picoliters, and faster rate of mass and heat transfer make them a suitable detection system. Table 5 sets out the pathogen detection methods based on microfluidics (Fig. 4) (Foudeh et al., 2012).

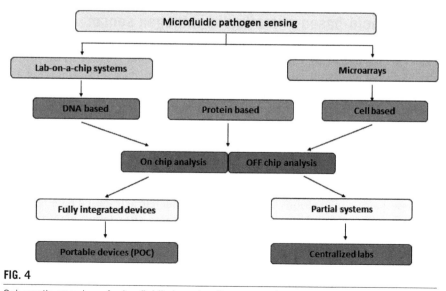

FIG. 4

Schematic overview of microfluidic based pathogen sensing systems.

Figure adopted and modified from Mairhofer et al., 2009

Table 5 Common biological recognition elements used in microfluidic devices

Biological recognition element	Advantages	Disadvantages
Enzymes	High sensitivity towards their targets Suitable for oxidation reduction reactions	Possibility of losing their activity upon immobilization Most suitable for small analytes, e.g., glucose, urea, and lactate
Antibodies	Rapid analysis for direct immunoassays Suitable for bio-affinity interaction, e.g., antibody–antigen interaction Suitable for the detection of large targets, e.g., bacteria and pathogens	Requires labeling for indirect immune assays which can result in the increase of cost and time required for analysis Not suitable for detection of small targets using direct and sandwich immunoassays Not suitable for oxidation reduction reactions
Aptamers	Highly sensitive and selective Suitable for the detection of a wide range of analytes Long-term stability, inexpensive and rapid synthesis Flexibility to be modified with labels without losing their performance or binding properties	Higher toxicity than antibodies Faster excretion due to their small size Weaker binding to analytes

Nucleic acid-based microfluidic pathogen sensor

A microarray consists of an arrayed series of thousands of microscopic spots of DNA oligonucleotides covalently attached on a solid support to determine the relative abundance of nucleic acid sequences in the target (Mairhofer et al., 2009). Sequencing and genotyping of nucleic acid involves advancement in microchannel and capillary electrophoreses (Medintz et al., 2001). Complete analysis of nucleic acid involves steps such as cell concentrating, capture, cell lysis, nucleic acid purification, amplification, and final detection. Microfluidic integrated microarrays have been used in the identification of *Bacillus* species, *Yesinia enterocolitica*, influenza, and fungal pathogens (Dutse and Yusof, 2011). Nucleic acid detection is highly sensitive and specific due to target amplification and base-pairing interactions. High-throughput nucleic acid-based pathogen detection can be achieved by direct target probing or after target amplification (Iqbal et al., 2000). The limitation associated with the direct target probing is the variation in detection limit from 10^5 to 10^6 target molecules, which results in low sensitivity and requires sophisticated signal enhancement techniques. Bead-based methods have shown to reduce the diffusion time and increase biorecognition events (Bielecki et al., 2012). The sensitivity and specificity can be increased further using magnetic forces which have the ability to differentiate between specific and nonspecific (Foudeh et al., 2012). Target amplification techniques are highly specific and can be achieved through polymerase chain reaction (PCR), ligase chain reaction (LCR), or nucleic acid sequence-based amplification (NASBA). A commercial microfluidic-based PCR device has been developed, called BioMark 48.48 Dynamic Array developed by the Fluidigm Corporation (San Francisco, CA), which is capable of running up to 2304 reactions per chip while only 96 total liquid loading steps are required. Another example of a microfluidic-based PCR device is the GeneXpert® system, produced by Cepheid (Sunnyvale, CA), which detects *Mycobacterium bovis* (Chang et al., 2013). The system consists of a filter to capture organisms from a clinical sample, which is then followed by cell lysis. Real-time PCR is carried out of released nucleic acids to identify pathogens (Chen et al., 2007a,b). Micro-PCR chips can be classified into three categories: (i) stationary-chamber micro PCR-chips as nano/picoliter reservoir for conventional thermocycling; (ii) continuous-flow micro-PCR chips where different temperature zones are established at different locations and the sample is moved between individual temperature zones for cycling; and (iii) droplet-based PCR systems where amplification reactions are conducted in water-in-oil droplets for each amplicon (Mairhofer et al., 2009). Non-specific adsorption due to the large surface-to-volume ratios present in microchannels is one of the limitations known to inhibit PCR reactions. A variety of specific surface modification procedures or bulk modification methods for polymers have recently been implemented to overcome this limitation (Auroux et al., 2004).

In many biosensing applications, the starting samples are usually tissue, blood, environmental, or food samples, and these need to undergo careful sample preparation for sensitive detection due to trace or low-abundance species. A sample (pre)-preparation step is often required when using environmental or otherwise complex samples in order to identify pathogens in food samples, whole blood, urine, wastewater, etc. (Haes and Van Duyne, 2002). In the case of LOC devices, sample pre-treatment is used routinely combined with DNA/RNA isolation procedures. Methods used for isolation include pathogen capture using antibody labeled magnetic beads or electro-kinetic capture of bacterial cells such as the di-electrophoretic capture of malarial-parasitized cells (Rombach, 2016). Cell lyses and PCR analysis can be accomplished chemically or optically. Examples of optical approaches include the Laser-Irradiated Magnetic Bead system (LIMBS), which combines optical forces with magnetic beads for direct cell lyses and DNA capture. Optical methods utilize opto-thermal properties of nanoparticles to transform near infrared light energy into thermal energy for pathogen lyses (Gorkin III, 2010).

After nucleic acid isolation, direct target detection or micro-sized PCR, also called a PCR microfluidic chip, is seen as the next step in the development of integrated micro-total analysis system (Dutse and Yusof, 2011). Fluid control and flow stability in a microfluidic-based system are critical for successful nucleic acid detection. The introduction and maneuvering of any fluids must be done with extreme care to prevent bubble formation within the channels or chambers. Reverse-transcriptase PCR, real-time reverse transcription PCR, limited dilution PCR, and real-time PCR have been successfully used for nucleic acid-based microfluidic pathogen sensing (Lui, 2010). Other amplification-based methods include the application of immobilized primers for bacterial DNA detection, combination of on-chip PCR followed by microarray-based fluorescence detection, and the application of field-effect transistors for label-free detection of bacterial DNA. Various isolation and amplification protocols have been combined in LOC devices. Although fluorescence detection is mostly used in the field of DNA detection, a variety of electrochemical and magneto resistive sensors have also been successfully integrated in microfluidic based nucleic acid detection devices (Lisowski and Zarzycki, 2013).

Microfluidic cell-based pathogen sensing

Identification, differentiation, and quantization of clinical relevant cellular systems can be done by cell-based systems (Perez and Nolan, 2006). The progression of HIV infection can be checked by using cytometers to count erythrocytes. A variety of pathogens can be detected based on chemiluminescence immunoassay (EIA) using 10-channel capillary chip-based microfluidic coated with capture antibodies (Mairhofer et al., 2009). In this system, controlled fluid flow through capillaries and microchannels can be achieved through hydrodynamic (e.g., pressure driven) or electrokinetic flow switching and dielectrophoresis. Electric fields can also be used in microfluidics which works as cell separation system capable of trapping bacteria or discriminating between dead and live yeast. Despite of all these advances, there is an urgent need for miniature, low-cost, and portable sensors which can detect and accurately identify bacteria in complex matrices (Fang et al., 2018). Immunoassay-based microfluidics are also used for cell-based pathogen detection. The presence of a high surface to volume ratio in microchannels helps to functionalize surfaces selectively with capture agents (Zhou et al., 2010). Here microfluidic devices take advantage of the significantly increased probability of pathogen interaction and cell capture at modified/activated sensor surfaces along the flow pathway that allows for the identification of small amounts of pathogens in a short period of time (Mairhofer et al., 2009). Cell-based microfluidics based on antigen–antibody interaction for pathogen detection has been demonstrated using fluorescence, chemiluminescence, optical leaky waveguides, SPR, impedance, love acoustic waves, and conducting polymers. Immunoassays offer a high degree of selectivity in many instances; additional signal amplification is required to detect small amounts of pathogens (Fig. 5). This can be achieved either by enzymatic signal amplification or through pre-concentrations steps including dielectrophoresisultrasonic deposition of cells, magnetic beads, and membrane filters (Nasseri et al., 2018).

Bacteria lysis and immunoreaction

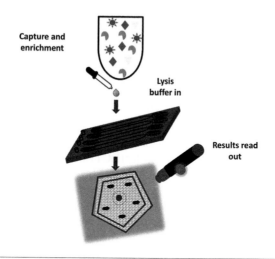

FIG. 5

Schematic illustration of the microfluidic system for on-chip bacterial enrichment and detection.

Figure adapted and modified from Zhang et al., 2018

Microfluidics in combination with mass spectrometry

Mass spectrometry (MS) can be employed in bacterial pathogen detection based on spectra pattern matching against a library of matrix-assisted laser desorption/ionization-time of flight (MALDI-TOF) mass spectra from standard strains or relies on proteomic strategies (Demirev and Fenselau, 2008). The advantage of incorporating MALDI-TOF MS in bacterial identification excludes the use of expensive reagents for gene amplification and biochemical experiments can be avoided (Geoghegan and Kelly, 2005). MS can identify much wider types of bacteria in comparison to PCR and LAMP. The advantage of MALDI-TOF MS based bacterial identification is that it is cost-effective. Microfluidic chips have been coupled with mass spectrometers via various interfaces, e.g., microchip electrospray ionization and droplets collection from microchips on MALDI target plates. A microfluidic chip was developed by Bian et al., which could catch and enrich bacteria in air with high efficiency. The chip was based on liquid chromatography-mass spectrometry (LC-MS) to identify multiple bacterial species, i.e., *Listeria monocytogenes, Vibrio parahemolyticus,* and *Escherichia coli.* Cho et al. utilized MALDI-TOF MS in combination with a microchip integrated with magnetic beads-based affinity chromatography to detect RNA polymerase of hepatitis C virus from patient serum. The ligand which was immobilized on the magnetic beads to capture the proteins was aptamer. The captured protein was eluted by UV irradiation. The eluted target proteins were then digested and analyzed by MALDI-TOF MS (Cho et al., 2004).

Microfluidic pathogen detection systems-based antibody and aptamer sensor

Microfluidics based on antigen-antibody interaction is most commonly used in pathogen detection. Most of the microfluidics employs combination of electrochemical and optical or label-free detection techniques, nanotechnology advanced detection systems and antibody microarray systems (Myers and Lee, 2008). Electrochemicals coupled with fluorescence-based microfluidic biosensors have been used in the detection of Cholera toxin subunit B (CTB) (Bunyakul et al., 2009). In this system, a combination of CTB-antibodies and Ganglioside GM1, which is the natural target of the CTB, were used as a specific recognition system. Another microfluidic for detecting different *Bacillus* species was developed based on the direct-charge transfer (DCT) immunosensor which relies on antibody recognition, in combination with conducting polymers (e.g., polyaniline) as transducers. This microfluidic is based upon the principle of antibody-antigen-antibody sandwich interaction and DCT to generate a resistance signal which can detect concentrations as low as 100 CFU/mL. (Mairhofer et al., 2009). Multiple pathogens can also be detected by using quantum dot barcodes conjugated to targeting antibodies within an electrokinetically driven microfluidics and photon counting detection system (Rivet et al., 2011). Alternatively, the application of magnetic beads and fluidic force discrimination (FFD) for antibody-based pathogen detection has shown multiplexed detection capability using two proteins: ricin A chain (RCA) and staphylococcal enterotoxin B (SEB) (Fig. 6). Microfluidic devices are

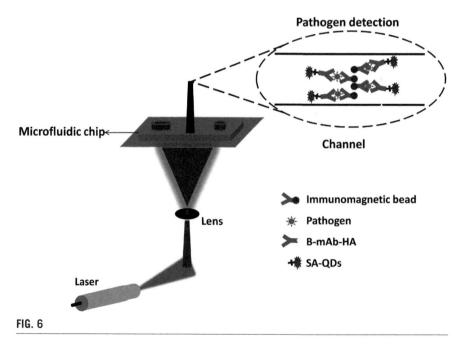

FIG. 6

Working principle of the on-chip magnetic immunofluorescence assay for the detection.

Figure adopted and modified from Zhang et al., 2018

attractive platforms for immunoassays as they reduce the time and labor difficulties of conventional ELISA. The Triage system developed by Biosite Inc. is an ELISA-based microfluidic device which detects *Clostridium difficile* (Mairhofer et al., 2009).

When compared to antibodies, aptamers can be generated by an *in-vitro* selection process referred to as systematic evolution of ligands by exponential enrichment (SELEX) (Ye et al., 2012). They represent as a promising alternative to antibodies as recognition agents since the generation of poly- or monoclonal antibodies is often time-consuming and challenging. Aptamers can be used as a biorecognition element, which involves the use of single-walled carbon nanotubes (SWNT) as transducers. They have remarkable advantages including feasibility of commercial scale-up and storage stability; affinity retention and the ability to differentiate between structurally similar analytes are also reported as advantages that offer great potential for pathogen and bimolecular screening (Ye et al., 2012). Aptamers can easily be modified at their 5′ or 3′ terminus with thiols, amines, or epoxy groups, which helps in immobilization onto a microfluidic chamber.

Microfluidic protein/enzyme based pathogen sensing

Other methods used for pathogen detection rely on immunological methods which employ specific affinities of protein-carbohydrate, protein–protein, or protein-DNA interactions (Templin et al., 2003). Detection systems based on antigen (Ag)-antibody (Ab) interaction are most widely used. A portable detection chip integrated with antibodies was developed to detect enterohemorrhagic *E. coli* (EHEC); this involved a magneto-resistive immunosensor in a four-channel configuration (Mairhofer et al., 2009). The problem associated with the use of antibodies is the difficulty in production of quality-assured antibodies. However, recombinant antibody-fragments (rAbs) have comparable specificity and can be derived through low-cost production. The polymeric materials most commonly used in microfluidic pathogen sensing systems are poly-(methyl methacrylate) (PMMA), polycarbonate (PC), and poly(dimethylsiloxane) (PDMS). Sensor surfaces can be made functional by attaching affinity tags covalently like poly-amino acids, protein G/A, biotin and recombinant fusion proteins, or simply by physisorption (Gervais et al., 2011). The non-specificity can be removed by using the bilayer membranes (SBMs). The self-assembled protein can be arrayed through contact-printing of complementary DNA onto glass slides followed by translating the target proteins with mammalian reticulocyte. The limitation associated with this detection system is that it requires protein to be in native conformation, which allows optimal protein-target interaction. If the binding is incorrectly done, it reduces the reproducibility and sensitivity. Use of additional enzymes or liposomes encapsulating fluorescence dyes or electroactive compounds has been demonstrated to increase sensor sensitivity. This system also requires that the antigen-antibody system should be recycled. Several polymers have shown themselves to be useful alternatives to antibodies due to their inherent robustness and reproducibility. Alternatively, enzyme-substrate reactions have the advantage of auto-regeneration of the binding site without any specificity loss over a large number of cycles.

Microfluidics in combination with fluorescence spectrometry

Optical detection strategies are most widely employed platforms for microfluidic pathogen detection. Fluorescence spectroscopy has several advantages like high sensitivity, allowing for single molecule detection high selectivity and easy integration of fluorescent tags within the microfluidic system. In this system, bacteria and fluidic flow can be visualized in microfluidic devices. A micro device integrated with immunofluorescence has been developed that detects AIV infection. This system contains AIV H9N2 captured specifically by antibodies coupled with magnetic beads. The antigen-antibody interaction can then be detected by visualizing the beads by fluorescence spectroscopy. The LOD for AIV H9N2 was down to 3.7×104 copies/μL with only $2\,\mu$L of sample consumption. Another example of a fluorescence-based microfluidic detection system was developed by Xiang et al. (2006) that is capable of detecting *E. coli* O157:H7 at a limit of detection of $0.3\,$ng/μl. A membrane-based immunoassay system was developed by Floriano et al. (2005) that incorporated fluorescent detection of *Bacillus globigii* at a detection limit of as little as 500 spores.

Microfluidics in combination with electrochemistry

Electrochemistry in combination with microfluidic devices are one of the best biosensing platforms for the detection of pathogenic bacteria. By integrating electrodes in microfluidic chips, electrochemical measurements can be performed on-chip, which helps in the sensitive detection of microbes (Fig. 7). A fully automated microfluidic-based electrochemical biosensor was developed for *E. coli* detection from a water sample. The biosensor contained 8 Au electrodes. Bacterial cells could be concentrated on the surface of the Au electrodes by immunoaffinity. Labeled antibodies like horseradish peroxidase (HRP) were then introduced to form a sandwich structure with the captured bacterial cells for electrochemical measurement of HRP-TMB (3,3′,5,5′-tetramethylbenzidine) interaction. Cross-reactivity was checked with *Shigella*, *Salmonella* spp., *Salmonella typhimurium*, and *Staphylococcus aureus* on an *E. coli* specific antibody surface that confirmed the high specificity of the developed immunoassays. The system achieved waterborne pathogenic *E. coli* detection, with a LOD as low as $50\,$CFU/mL. Ölcer et al. (2015) developed a sensing microsystem that consisted of Au electrode arrays and Au nanoparticles for real-time amperometric measurements. The platform was applied to detect cyanobacterial gene fragments.

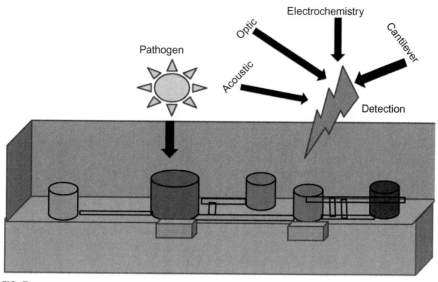

FIG. 7

Microfluidics in combination with electrochemistry.

Figure adopted and modified from Lin et al., 2014

Conclusion

Bioweapons are extremely damaging and efficient. The progress of biotechnology also opens new routes for weaponizing pathogens. Bacterial pathogens have been expected to be the most likely agents of a modern era biological attack, and this prediction has recently proven to be correct. Clinicians should be familiar with the features of the diverse diseases due to these pathogens, to enable early recognition and reporting and appropriate management. Because, as the recent US epidemic has proven, the manifestations of intentional release are not fully predictable, this review has focused on all clinical manifestation of infection with these critical pathogens. Unfortunately, potentially useful pathogens may extend beyond this list of critical pathogens, but it is imperative to review disease manifestations caused by those pathogens that have been identified.

This review has attempted to provide a survey of available microfluidics devices and developing technologies for biothreat agent detection. Many other technologies have not been included because insufficient published data were available to ascertain their accuracy and reliability. While an ideal platform has yet to be developed, many systems described in this review have proved invaluable in identifying biothreat agents rapidly and accurately. Although the risk of bioterrorism remains, detection technologies will continue to be improved to meet the challenges of this threat. Security and biodefense are emerging as a strong market for new applications.

References

Auroux, P.A., Koc, Y., Manz, A., Day, P.J.R., 2004. Miniaturised nucleic acid analysis. Lab Chip 4 (6), 534–546.

Baker, C.A., Duong, C.T., Grimley, A., Roper, M.G., 2009. Recent advances in microfluidic detection systems. Bioanalysis 1 (5), 967–975.

Bielecki, Z., Janucki, J., Kawalec, A., Mikołajczyk, J., Pałka, N., Pasternak, M., Pustelny, T., Stacewicz, T., Wojtas, J., 2012. Sensors and systems for the detection of explosive devices-an overview. Metrol. Measur. Syst. 19 (1), 3–28.

Biswas, S., Pomeau, Y., Chaudhury, M.K., 2016. New drop fluidics enabled by magnetic-field-mediated elastocapillary transduction. Langmuir 32 (27), 6860–6870.

Bordbar, A., Taassob, A., Zarnaghsh, A., Kamali, R., 2018. Slug flow in microchannels: numerical simulation and applications. J. Ind. Eng. Chem. 62, 26–39.

Bunyakul, N., Edwards, K.A., Promptmas, C., Baeumner, A.J., 2009. Cholera toxin subunit B detection in microfluidic devices. Anal. Bioanal. Chem. 393 (1), 177–186.

Carrico, J.P., 1998, August. Chemical-biological defense remote sensing: what's happening. In: Electro-Optical Technology for Remote Chemical Detection and Identification III. Vol. 3383. International Society for Optics and Photonics, pp. 45–57.

Castillo-León, J., 2015. Microfluidics and lab-on-a-chip devices: history and challenges. In: Lab-on-a-Chip Devices and Micro-Total Analysis Systems. Springer, Cham, pp. 1–15.

Chang, C.M., Chang, W.H., Wang, C.H., Wang, J.H., Mai, J.D., Lee, G.B., 2013. Nucleic acid amplification using microfluidic systems. Lab Chip 13 (7), 122–1242.

Chen, L., Lee, S., Choo, J., Lee, E.K., 2007a. Continuous dynamic flow micropumps for microfluid manipulation. J. Micromech. Microeng. 18 (1), 013001.

Chen, L., Manz, A., Day, P.J., 2007b. Total nucleic acid analysis integrated on microfluidic devices. Lab Chip 7 (11), 1413–1423.

Cho, S., Lee, S.H., Chung, W.J., Kim, Y.K., Lee, Y.S., Kim, B.G., 2004. Microbead-based affinity chromatography chip using RNA aptamer modified with photocleavable linker. Electrophoresis 25 (21–22), 3730–3739.

Choi, K., Ng, A.H., Fobel, R., Wheeler, A.R., 2012. Digital microfluidics. Annu. Rev. Anal. Chem. 5, 413–440.

Demirev, P.A., Fenselau, C., 2008. Mass spectrometry for rapid characterization of microorganisms. Annu. Rev. Anal. Chem. 1, 71–93.

Dutse, S.W., Yusof, N.A., 2011. Microfluidics-based lab-on-chip systems in DNA-based biosensing: An overview. Sensors 11 (6), 5754–5768.

Fang, X., Zheng, Y., Duan, Y., Liu, Y., Zhong, W., 2018. Recent advances in design of fluorescence-based assays for high-throughput screening. Anal. Chem. 91 (1), 482–504.

Fitch, J.P., Raber, E., Imbro, D.R., 2003. Technology challenges in responding to biological or chemical attacks in the civilian sector. Science 302 (5649), 1350–1354.

Floriano, P.N., Christodoulides, N., Romanovicz, D., Bernard, B., Simmons, G.W., Cavell, M., McDevitt, J.T., 2005. Membrane-based on-line optical analysis system for rapid detection of bacteria and spores. Biosens. Bioelectron. 20 (10), 2079–2088.

Foudeh, A.M., Didar, T.F., Veres, T., Tabrizian, M., 2012. Microfluidic designs and techniques using lab-on-a-chip devices for pathogen detection for point-of-care diagnostics. Lab Chip 12 (18), 3249–3266.

Geoghegan, K.F., Kelly, M.A., 2005. Biochemical applications of mass spectrometry in pharmaceutical drug discovery. Mass Spectrom. Rev. 24 (3), 347–366.

Gervais, L., De Rooij, N., Delamarche, E., 2011. Microfluidic chips for point-of-care immunodiagnostics. Adv. Mater. 23 (24), H151–H176.

Gijs, M.A., Lacharme, F., Lehmann, U., 2009. Microfluidic applications of magnetic particles for biological analysis and catalysis. Chem. Rev. 110 (3), 1518–1563.

Glückstad, J., 2004. Microfluidics: sorting particles with light. Nat. Mater. 3 (1), 9.

Gokmen, M.T., Van Camp, W., Colver, P.J., Bon, S.A., Du Prez, F.E., 2009. Fabrication of porous "clickable" polymer beads and rods through generation of high internal phase emulsion (HIPE) droplets in a simple microfluidic device. Macromolecules 42 (23), 9289–9294.

Gorkin III, R.A., 2010. Enabling Technologies for Nucleic Acid Sample-to-Answer Centrifugal Microfluidics. University of California, Irvine, CA.

Haes, A.J., Van Duyne, R.P., 2002. A nanoscale optical biosensor: sensitivity and selectivity of an approach based on the localized surface plasmon resonance spectroscopy of triangular silver nanoparticles. J. Am. Chem. Soc. 124 (35), 10596–10604.

Hasan, J., Goldbloom-Helzner, D., Ichida, A., Rouse, T., Gibson, M., 2005. Technologies and techniques for early warning systems to monitor and evaluate drinking water quality: A state-of-the-art review. (No. EPA/600/R-05/156). Environmental Protection Agency Washington DC Office of Water.

Heinemann, J., Deng, K., Shih, S.C., Gao, J., Adams, P.D., Singh, A.K., Northen, T.R., 2017. On-chip integration of droplet microfluidics and nanostructure-initiator mass spectrometry for enzyme screening. Lab Chip 17 (2), 323–331.

Huelseweh, B., Ehricht, R., Marschall, H.J., 2006. A simple and rapid protein array based method for the simultaneous detection of biowarfare agents. Proteomics 6 (10), 2972–2981.

Iqbal, S.S., Mayo, M.W., Bruno, J.G., Bronk, B.V., Batt, C.A., Chambers, J.P., 2000. A review of molecular recognition technologies for detection of biological threat agents. Biosens. Bioelectron. 15 (11–12), 549–578.

Jin, J., Ooi, C., Dao, D., Nguyen, N.T., 2017. Coalescence processes of droplets and liquid marbles. Micromachines 8 (11), 336.

Joos, T.O., Stoll, D., Templin, M.F., 2002. Miniaturised multiplexed immunoassays. Curr. Opin. Chem. Biol. 6 (1), 76–80.

Kang, L., Chung, B.G., Langer, R., Khademhosseini, A., 2008. Microfluidics for drug discovery and development: from target selection to product lifecycle management. Drug Discov. Today 13 (1–2), 1–13.

Koczula, K.M., Gallotta, A., 2016. Lateral flow assays. Essays Biochem. 60 (1), 111–120.

Larsen, J.C., Disbrow, G.L., 2017. Project BioShield and the biomedical advanced research development authority: a 10-year progress report on meeting US preparedness objectives for threat agents. Clin. Infect. Dis. 64 (10), 1430–1434.

Lenshof, A., Johannesson, C., Evander, M., Nilsson, J., Laurell, T., 2017. Acoustic cell manipulation. In: Microtechnology for Cell Manipulation and Sorting. Springer, Cham, pp. 129–173.

Li, Y., Baker, R.J., Raad, D., 2016. Improving the performance of electrowetting on dielectric microfluidics using piezoelectric top plate control. Sens. Actuators B 229, 63–74.

Lim, D.V., Simpson, J.M., Kearns, E.A., Kramer, M.F., 2005. Current and developing technologies for monitoring agents of bioterrorism and biowarfare. Clin. Microbiol. Rev. 18 (4), 583–607.

Lin, Y.S., Yang, C.H. and Huang, K.S., 2014. Biomedical devices for pathogen detection using microfluidic chips. Current Proteomics. 11 (2), 116–120.

Lisowski, P., Zarzycki, P.K., 2013. Microfluidic paper-based analytical devices (µPADs) and micro total analysis systems (µTAS): development, applications and future trends. Chromatographia 76 (19–20), 1201–1214.

Liu, Z., Qi, W., Xu, G., 2015. Recent advances in electrochemiluminescence. Chem. Soc. Rev. 44 (10), 3117–3142.

Lui, C., 2010. Integrated Biosensor Systems: Automated Microfluidic Pathogen Detection Platforms and Microcantilever-Based Monitoring of Biological Activity.

Mairhofer, J., Roppert, K., Ertl, P., 2009. Microfluidic systems for pathogen sensing: a review. Sensors 9 (6), 4804–4823.

Mark, D., Haeberle, S., Roth, G., Von Stetten, F., Zengerle, R., 2010. Microfluidic lab-on-a-chip platforms: requirements, characteristics and applications. In: Microfluidics Based Microsystems. Springer, Dordrecht, pp. 305–376.

Medintz, I.L., Paegel, B.M., Blazej, R.G., Emrich, C.A., Berti, L., Scherer, J.R., Mathies, R.A., 2001. High-performance genetic analysis using microfabricated capillary array electrophoresis microplates. Electrophoresis 22 (18), 3845–3856.

Myers, F.B., Lee, L.P., 2008. Innovations in optical microfluidic technologies for point-of-care diagnostics. Lab Chip 8 (12), 2015–2031.

Nasseri, B., Soleimani, N., Rabiee, N., Kalbasi, A., Karimi, M., Hamblin, M.R., 2018. Point-of-care microfluidic devices for pathogen detection. Biosens. Bioelectron. 117, 112–128.

Nguyen, N.T., Hejazian, M., Ooi, C., Kashaninejad, N., 2017. Recent advances and future perspectives on microfluidic liquid handling. Micromachines 8 (6), 186.

Nguyen, N.T., Wereley, S.T., Shaegh, S.A.M., 2019. Fundamentals and Applications of Microfluidics. Artech House.

Ölcer, Z., Esen, E., Ersoy, A., Budak, S., Kaya, D.S., Gök, M.Y., Barut, S., Üstek, D., Uludag, Y., 2015. Microfluidics and nanoparticles based amperometric biosensor for the detection of cyanobacteria (*Planktothrix agardhii* NIVA-CYA 116) DNA. Biosens. Bioelectron. 70, 426–432.

Ooi, C.H., Jin, J., Sreejith, K.R., Nguyen, A.V., Evans, G.M., Nguyen, N.T., 2018. Manipulation of a floating liquid marble using dielectrophoresis. Lab Chip 18 (24), 3770–3779.

Pamula, V.K., Pollack, M.G., Paik, P.Y., Ren, H., Fair, R.B., Advanced Liquid Logic Inc, 2009. Methods for manipulating droplets by electrowetting-based techniques. U.S. Patent 7,569,129.

Perez, O.D., Nolan, G.P., 2006. Phospho-proteomic immune analysis by flow cytometry: from mechanism to translational medicine at the single-cell level. Immunol. Rev. 210 (1), 208–228.

Pizarro, S.A., Lane, P., Lane, T.W., Cruz, E., Haroldsen, B., VanderNoot, V.A., 2007. Bacterial characterization using protein profiling in a microchip separations platform. Electrophoresis 28 (24), 4697–4704.

Poritz, M.A., Blaschke, A.J., Byington, C.L., Meyers, L., Nilsson, K., Jones, D.E., Thatcher, S.A., Robbins, T., Lingenfelter, B., Amiott, E., Herbener, A., 2011. FilmArray, an automated nested multiplex PCR system for multi-pathogen detection: development and application to respiratory tract infection. PLoS One 6 (10), e26047.

Rivet, C., Lee, H., Hirsch, A., Hamilton, S., Lu, H., 2011. Microfluidics for medical diagnostics and biosensors. Chem. Eng. Sci. 66 (7), 1490–1507.

Rombach, M., 2016. Pre-Storage of Reagents for Nucleic Acid Analysis in Unit-Use Quantities for Integration in Lab-on-a-Chip Test Carriers (Doctoral dissertation). Albert-Ludwigs-Universität Freiburg imBreisgau.

Shea, D.A., Lister, S.A., Foreign Affairs, Defense, and Trade Division and Domestic Social Policy Division, 2003, November. The BioWatch Program: Detection of Bioterrorism. Congressional Research Service [Library of Congress].

Shen, H.H., Fan, S.K., Kim, C.J., Yao, D.J., 2014. EWOD microfluidic systems for biomedical applications. Microfluid. Nanofluid. 16 (5), 965–987.

Templin, M.F., Stoll, D., Schwenk, J.M., Pötz, O., Kramer, S., Joos, T.O., 2003. Protein microarrays: promising tools for proteomic research. Proteomics 3 (11), 2155–2166.

Wang, S., Inci, F., De Libero, G., Singhal, A., Demirci, U., 2013. Point-of-care assays for tuberculosis: role of nanotechnology/microfluidics. Biotechnol. Adv. 31 (4), 438–449.

Wiedbrauk, D.L., 2009. Nucleic acid amplification and detection methods. In: Clinical Virology Manual. fourth ed. American Society of Microbiology, pp. 156–168.

Xiang, Q., Hu, G., Gao, Y., Li, D., 2006. Miniaturized immunoassay microfluidic system with electrokinetic control. Biosens. Bioelectron. 21 (10), 2006–2009.

Ye, M., Hu, J., Peng, M., Liu, J., Liu, J., Liu, H., Zhao, X., Tan, W., 2012. Generating aptamers by cell-SELEX for applications in molecular medicine. Int. J. Mol. Sci. 13 (3), 3341–3353.

Zhang, D., Bi, H., Liu, B. and Qiao, L., 2018. Detection of pathogenic microorganisms by microfluidics based analytical methods, pp 5512–5520.

Zhou, J., Ellis, A.V., Voelcker, N.H., 2010. Recent developments in PDMS surface modification for microfluidic devices. Electrophoresis 31 (1), 2–16.

Zhu, Y., Zhang, Y.X., Cai, L.F., Fang, Q., 2013. Sequential operation droplet array: an automated microfluidic platform for picoliter-scale liquid handling, analysis, and screening. Anal. Chem. 85 (14), 6723–6731.

Further reading

Das, S., Kataria, V.K., 2010. Bioterrorism: a public health perspective. Med. J. Armed Forces India 66 (3), 255–260.

Lenshof, A., Laurell, T., 2010. Continuous separation of cells and particles in microfluidic systems. Chem. Soc. Rev. 39 (3), 1203–1217.

Luka, G., Ahmadi, A., Najjaran, H., Alocilja, E., DeRosa, M., Wolthers, K., Malki, A., Aziz, H., Althani, A., Hoorfar, M., 2015. Microfluidics integrated biosensors: A leading technology towards lab-on-a-chip and sensing applications. Sensors 15 (12), 30011–30031.

Collection, storage, and transportation of samples for offsite analysis

7

Anju Tripathi, Kshirod Sathua, Vidhu Pachauri, S.J.S. Flora

National Institute of Pharmaceutical Education and Research-Raebareli, Lucknow, India

Introduction

Nowadays, devastating infectious agents have become the prime choice for bioterrorists for incapacitating human beings. To identify the origin and causes of these diseases and outbreaks in a community is a challenging task for epidemiologists, laboratory scientists, physicians, and other health care professionals. Moreover, for sorting out the root of these problems, the use of advanced biochemical and molecular techniques is very important (Perera and Weinstein, 2000; Rothman et al., 2001) and for getting error-free, repeatable biochemical or molecular results, proper collection, storage, and transportation of samples plays a vital role. If there is unavailability of resources/equipment, lack of competent technical/health care staff in the place of attack or epidemic area, then during that time, offsite analysis of samples is necessary (Bonassi and Au, 2002; Vaught and Henderson, 2011).

Knowledge of continually developing biochemical/molecular diagnostic tools must be implemented into study designs and procedures. Hence, there must be certain provision to be made for the collection, storage, and transportation of specimens. Moreover, selection and validation of these procedures can also affect the outcome of molecular epidemiology studies (Landi and Caporaso, 1997).

Collection of a wide variety of specimens is needed during sample collection. Hence, collection is a critical step for any research/technical personnel. Ideally based on International Society for Biological and Environmental Repositories (ISBER) and International Agency for Research on Cancer (IARC) guidelines, different specimens like blood and blood fractions, tissue, urine, saliva/buccal cells, bronchoalveolar lavage, exhaled air, hair, nail clippings, semen, feces, and cytological fluids like pleural fluid and synovial fluid need to be collected for identification, detection, and diagnosis of any devastating outbreak (Campbell et al., 2012). Sample collection also depends on intended use and study goal. Involvement of advanced technology and well skilled technical personnel is highly necessary for obtaining the best quality or quantity sample; for example, trained phlebotomists should take blood samples. Moreover sterile, evacuated glass or plastic ware should be used for blood sample collection. After successful collection, whether fractionation of blood is to be done

Handbook on Biological Warfare Preparedness. https://doi.org/10.1016/B978-0-12-812026-2.00007-4

depends on further use or any special purpose. For investigation of modern analytical techniques like array of proteins and peptides, separation of plasma or serum is considered the prime choice. Hence, selection of anticoagulant is very important for sorting out such problems. Nowadays, special protease inhibitors containing readymade blood collection tubes are also available for proteomic analysis (Nishad and Ghosh, 2016). Various blood collection procedures, anticoagulant, additives, and their use have been given in Table 1 and Fig. 1.

When any toxicant, either of biological origin or of chemical origin, enters into the body, it will undergo various biotransformation reactions and finally produce toxic metabolites as products or by products. These toxic metabolites are mostly excreted in urine and feces. So urine and feces samples are considered to be convenient specimens for analyzing in a variety of studies (Begou et al., 2019). For successful collection, choosing an appropriate preservative like EDTA and sodium metabisulfite is an important criterion. Besides these, for collection of tissue specimens, different methods like biopsy, surgery, and autopsy (Mager et al., 2007; van der Linden et al., 2014) should be performed under the strict ethical and legal guidance of ISBER and IARC. As an excellent source of DNA, collection of saliva/buccal cells is often preferred over blood collection (Theda et al., 2018). Moreover, for trace metal analysis, collection of nail and hair samples is considered as a prime source (Koseoglu et al., 2017).

After the completion of successful collection, depending on the intended use, sample can be analyzed immediately or it can be stored for subsequent use. For specimens like plasma, serum, and other cellular specimens, maintenance of storage temperature is very important. Various storage temperatures at which samples can be securely stored for long periods are -20, -80, and $-196°C$ (Lloyd et al., 2016; Pegg, 2009). Standard conditions for different sample analyses are described in Table 2.

If there are no onsite facilities available, the sample is needed for future use, or any newer biological agents have been used by bioterrorists, recognition of which is not possible at the attack location, immediate transportation of specimens to a designated laboratory is necessary (World Health Organization, 1997). In other words, for offsite analysis, transportation of specimen is of the utmost importance after collection. A safe mode of transportation and the best possible leak-proof packaging with proper labeling as per transport guidelines is required for successfully arrival of samples at the designated laboratory (Gekas, 2017).

Recently advances in technology like gene sequencing can be beneficially used for identifying antigenic/pathogenic gene, virulent genes, or genes responsible for drug resistance (Klemm and Dougan, 2016). This can help mankind to design new diagnoses and drugs, which can be helpful during a biological warfare (BW) attack. Hence, there must be a proper database of gene sequences and repositories of microbial and environmental samples to keep a record and track of previous outbreaks. For these, a proper management cascade from sample collection, storage, and transportation, to data analysis and data base preparation is highly necessary. On the other hand, panic caused by BW may disturb the whole setup of medical facilities available in that particular area. Hence, there is a shortage of clinical facilities, pathological staff, medicines, safety consumables, etc. To overcome such situations, offsite analysis of samples plays a vital role.

Table 1 Collection of some common specimen types for various identification, detection, and molecular studies

Sample collection	Fractions of samples	Additives/ preservative used	Applications	Limitations
Blood	Plasma	Anticoagulant (heparin, EDTA), protease inhibitor	Biochemical estimation, proteomics	Low DNA yield
Blood	Serum	No anticoagulant	Biochemical estimation, sources of DNA, proteomics, multiple analysis	Low DNA yield
Blood	Whole blood	Anticoagulant (heparin, EDTA), protease inhibitor	Genomic studies, sources of DNA, RNA	Wounded skin while handling infected animal tissue
Blood	Buffy coat	Anticoagulant (heparin, EDTA)	Sources of DNA, lymphocytes, cell lines	Limited yield if not properly collected
Blood	Blood clot	None	Sources of DNA	Cost effectiveness, extraction difficulties
Urine	Whole urine	EDTA and sodium metabisulfite as preservatives	Estimation of toxic metabolites	
Tissue	Biopsy, surgery, autopsy	Protease inhibitor	For forensic identification	Need for highly skilled personnel, temperature maintenance
Saliva/ buccal sample	Whole saliva/ buccal specimen		Excellent source of DNA	
Nail and hair	Nail and hair clippings	None	Trace metal analysis	For measurement of long-term exposure

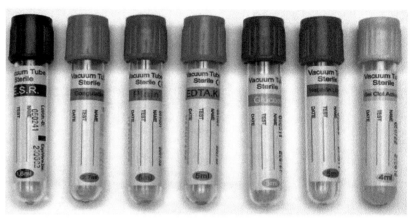

FIG. 1

Different color blood sample collection tubes indicating different volumes with proper labeling. The color of collection tube is based upon the different anticoagulants and their intended use.

Table 2 Standard conditions for sample analysis during storage/transport

Sample	Condition
Urine	1 h at room temp
Stool	1 h for hard stool
Fluid stool	½ h since collection
CSF	1–2 h room temp
Blood	1–2 days (with particular type of colored vial)

Offsite analysis

Sending samples to well-established labs when it is not possible to analyze at a time or place due to any reason, for example, unavailability of resources/equipment, staff, or contamination spread, is called offsite analysis. Biological samples are sometimes sent for offsite analysis for research purposes and to form repository stocks (Vaught and Henderson, 2011). This is helpful for preparing databases related to a particular disease or pathogen, and its vigor, symptoms, mortality and survival rates, and acute/chronic outbreaks. The biological samples taken for offsite analysis may be sent to any referral labs/research centers that could be helpful for referral, diagnosis, or preparing a database. In developing countries like India, there are about 26 permanent institutes/centers, 57 field stations, and 31 population-based cancer registries working under the MCI (Medical Council of India) and ICMR (Indian Council of Medical Research). These standardize treatment protocols across various medical institutes and make medical facilities available, affordable, and accessible to the every person in an area, commonly known as "Last mile delivery of services amongst tribal populations" (Sharma et al., 2013; Sudan et al., 2018). Moreover, offsite analysis strengthens the capacity and scope to carry out multidisciplinary research on a single sample; e.g., a single blood sample can be used for culture, sensitivity, biochemical, stain/morphology, and genomics. Offsite analysis generalizes huge volumes of data through intramural and extramural research programs.

Sample collection

Collection of biological samples for offsite analysis is a critical step for any research/technical personnel, as collection of a wide variety of specimen types, depending on the study goals, may be necessary for identification, detection, and diagnosis of any devastating outbreaks (Perera, 2000). For ideal collection of biological samples, one should follow various guidelines provided by different governing agency like International Society for Biological and Environmental Repositories (ISBER), International Agency for Research on Cancer (IARC), etc. (Perera, 2000). The wide varieties of specimen types include:

(1) blood and blood fractions;
(2) tissue;
(3) urine;
(4) saliva/buccal cells;
(5) placental tissue;
(6) cord blood;
(7) bone marrow;
(8) bronchoalveolar lavage;
(9) exhaled air;
(10) hair;
(11) nail clippings;
(12) semen;
(13) feces; and
(14) cytological fluids, like pleural fluid and synovial fluids.

These specimen types must be collected and stored as per the guidelines, and preserved in such a manner that their stability is maintained for future analyses.

Specimen collection procedures

Although collection of specimens will vary according to the specimen types and the intended analysis, all protocols and procedure should be ideally designed and well documented. For validation of new collection protocols and methods, performing pilot studies is essential (Holland et al., 2003; Perera, 2000). The following section focuses on collection of some common specimen types that have been used for various identification, detection, and molecular epidemiological studies.

Collection of blood samples

In order to obtain the best-quality or best-quantity sample, the collection of blood should be done by well-trained technical persons like phlebotomists, and always following standard recommended protocols by national or international organizations (Vaught, 2006). Generally, interchangeable evacuated glass or plastic tubes are used for blood sample collection. For avoiding any cross contamination and/or to differentiate various additives, the stopper of the collection tubes are coded with different colors (Landi and Caporaso, 1997).

Prior to storage or analysis, fractionation of blood may need to be done, resulting in separation of various components like:

(1) mononuclear leukocytes;
(2) neutrophils;
(3) erythrocytes;
(4) plasma; and
(5) serum.

When collecting fractionized portions, one needs to maintain the utmost care and conditions as each single part has its own importance and is separated for a special purpose. For example, peripheral blood mononuclear cells like mononuclear leukocytes need to be maintained in a viable state, erythrocytes are used as adducts of hemoglobin, and plasma is collected from an anticoagulant blood sample, whereas serum samples are isolated without addition of anticoagulant by maintaining different conditions. Serum allows the analyzing of antibodies, nutrients, and lipoproteins. Despite many advantages and disadvantages, based on the guidelines of the Human Proteomic Organization (HUPO) both serum as well as plasma may be useful for proteomic analysis (Vaught and Henderson, 2011). However, for investigation of an array of proteins and peptides, plasma is considered the best choice, as there is the possibility of loss of many proteins and peptides during blood coagulation processes. Hence, based on the intended form of analysis, blood samples should be collected in tubes either anticoagulated or coagulated (Vaught, 2006).

The choice of anticoagulant is very important for sorting out problems of certain laboratory applications (Landi and Caporaso, 1997) as shown in Fig. 1. Nowadays, special protease inhibitors containing readymade collection tubes are available for proteomic analysis. Sometimes, for preservation of analytes, stabilizing agents are necessary; hence, special caution should be taken during collection of blood samples and the stabilizing agents must be added as soon as possible after collection. The time lag between the blood collection to storage and removal from storage unit to subsequent processing is very important, hence, special precaution should be taken in accordance with guidelines, depending on the intended analyses. Some other factors like temperatures of blood specimen processing and freeze–thaw cycle play crucial roles for getting effective results. In addition, enzymatic degradation affects many biomarkers. Particularly RNA and proteins are very sensitive to this, hence, maintaining their integrity is of the utmost importance during specimen processing and collection. Currently, RNase inhibitors are commercially available for preserving RNA integrity. Hence collection of blood samples with special collection systems and in accordance with guidelines ideally allows for the most efficient, effective results (Vaught and Henderson, 2011).

Urine sample collection

Many varieties of metabolites, of either biological or chemical origin, are excreted in urine. Hence, it becomes a convenient specimen for analyzing a variety of studies (Vaught and Henderson, 2011). Depending on the analytical goals and study design, collection of urine samples can be performed under several conditions, described as follows (International Society for Biological and Environmental Repositories (ISBER), 2008; Hallmans and Vaught, 2011):

- First morning sample collection for performing various laboratory assays.
- Collection of random urine specimens for cytology studies.
- Fractional urine specimen collection for studying of urine analyte levels with analyte concentration in blood.
- Timed urine sample collections for studying excretion patterns of various analytes.
- Maintenance of collected urine specimen on ice or in a refrigerator.

Collection of urine samples is done, preferably, in collection vessels as shown in Fig. 2. The size of vessel is generally larger than other liquid collection specimens, having a 50–3000 mL capacity. The amount of sample to be collected depends on measurement of analyte. Different types of preservative may be used for maintenance of sample; choice of appropriate preservative should be in accordance with

FIG. 2

Urine specimen collection vessels with appropriate labeling.

the test methodologies and transport conditions. EDTA and sodium metabisulfite are widely used preservatives during urine collection.

Tissue sample collection

Tissue samples are mainly collected in three different ways: biopsy, surgery, and autopsy. As per ISBER and IARC guidelines, "tissue sample must be collected under strict ethical and legal guidelines and the collection of samples for research must never compromise the diagnostic integrity of a specimen." The ideal collection of tissue specimen, based upon ISBER and IARC guidelines, is as follows (International Society for Biological and Environmental Repositories (ISBER), 2008; Campbell et al., 2012).

- Tissue specimen should be collected by a trained pathologist during a surgical or autopsy procedure.
- Collected tissue specimen should be properly labeled with tissue type, organ type, and time of collection.
- After successful collection, the specimen should be immediately placed in ice-cool saline and then transported to a tissue repository for further processing.
- There should be minimization of time lag between collection, stabilization (freezing and fixing), and processing of tissue specimens. The best approach for preservation of a tissue sample is preservation within 1 h or less (Eiseman et al., 2003).
- The tissue specimen should be frozen either directly or immersed in the freezing medium.
- A surgical tissue sample may be collected as per IRB approval and, depending upon the intended use, either transported or frozen immediately by methods such as snap freezing specimen in Dewar flask of liquid nitrogen or dry ice immediately at the time of collection (Leonel et al., 2019).
- During the collection of autopsy samples, it is very important to know about the time interval between death, collection, and processing of specimen.
- It should be possible to fix a tissue sample in formalin or alcohol and embed it in paraffin, in case adequate freezing or storage facilities are not available.

Saliva/buccal specimen collection

Collection of saliva/buccal cells is often preferred over blood collection. Buccal cell collection is a safe and convenient method of collection and is an excellent source of DNA (García-Closas et al., 2001). Various protocols exist for collecting of specimens, including, swabbing, cyto-brushing, and mouth washing. Among these, the mouthwash protocol is popularly used for saliva/buccal specimen collection and produces a good quality and quantity of DNA for genetic analysis. Recently, a new saliva specimen collection method has been developed by DNAGenotek that contains Oragene, which can preserve saliva at room temperature. It is preferably used for epidemiological studies. Even nowadays, a treated filter paper cards method is used to collect DNA from blood and buccal swabs, which is helpful for genetic studies as well as other clinical applications. The US Centers for Disease Control and Prevention (CDC) use this blood spot card for neonatal screening (Mei et al., 2001, 2010)

Nail and hair sample collection

Nail and hair clippings are used for trace metal analysis, and will provide a clue for longer-term measure of exposure. Collection, storage, and transportation of these samples are very simple. These samples can also be used as a source of DNA (International Society for Biological and Environmental Repositories (ISBER), 2008; Daniel III et al., 2004).

Precautions to take during sample collection (International Society for Biological and Environmental Repositories (ISBER), 2008)

During sample collection, the following precautions should be taken by technical personnel:

- Biological samples must be collected under the guidelines of Good Clinical Laboratory Practice (GCLP) guidelines by the technician or individual who is collecting the sample.
- Medical staff should be protected and there should be clear follow-up, such as types of sample to be investigated, quantity of sample to be collected, and after collection, types of storage conditions to be maintained.
- Samples should be properly labeled using a form or slip with important data:
  - name;
  - no OPD/patient/numerical identities;
  - type of specimen; and
  - date of collection.
- There must be a form along with the collected sample that should contain the following information for offsite analysis:
  - name;
  - date;
  - place;
  - test to be performed; and
  - reporting/finding.
- If the samples are taken from casualties, the sample label, as well as the form, should include:
  - brief description of illness/autopsy/biopsy finding;
  - place, date, and time of collection;
  - pathologist/trained medical/forensic staff; and
  - individual's address.
- Routine laboratory check steps should be added with each sample of either blood, urine, or sputum.

General guidelines for clinical specimen collection (Campbell et al., 2012; De Souza and Greenspan, 2013)

- All biological samples should be handled with adequate precautions as they are capable of transmitting infections.
- At the time of collection, the sample amount must be adequate, because an inadequate amount of sample may yield a false negative result.
- All the samples must be collected in a sterile environment and the container should also be free of aerosols.
- Any bodily fluid, such as pus, urine, sputum, or CSF, should be taken from the syringe directly and must be kept in a specific vial. It must not be carried out in the form of a swab from one place to another.
- If swab collection is recommended by a medical practitioner, the swab should be kept in a suitable liquid media, otherwise the culture may dry out.
- A sample received a long time after collection is of no use, and so prolonged delay should be avoided for, e.g., urine at room temperature.
- A stool sample in which trophozoites are to be detected must be processed within 1 h if the stool texture is soft; if it is fluid only 30 min is the time the trophozoites could be seen.
- For gonorrhea samples, an anal swab can be transported without media for half an hour; if the sample has to be processed after a delay or offsite, we must use a transport medium.
- If any sample container is leaking, collection should be avoided, and if received in leaking condition offsite, the staff should ask that the collection be repeated.
- The offsite laboratory verifies the collection/verification sheet for further processing. The collected sample will be processed only when the requisition form is completely filled.
- Collection of samples in which anaerobic microbes are expected and need to be cultured should be done by the aspiration method with a sterile syringe and needle. The syringe can be capped and submitted, or it can be transferred into a vacontainer anaerobic specimen collector.
- For anaerobic samples, the technician must specify the source of the sample, i.e., whether it is from a wound, boil, normal tissue/vein, etc.
- Where there is a request for an anaerobic culture there must be an aerobic culture also, as if we have to culture the sputum for aerobic microbes, the sample collection method will be different. For example, for aerobic culture one has to take percutaneous transtracheal aspirate, but for anaerobic one needs expectorated sputum tracheal tube suctioning bronchoscopic aspirate or wash, throat, and mouth swab.
- In order to check UTI anaerobically, percutaneous suprapubic aspirate of urine is required. In order to take a sample for anaerobic culture, the technician must perform decontamination of the vagina and collect the sample in a syringe IV catheter passing through the cervical opening.

- Taking samples for anaerobic culture from the vagina/cervix is wrong without decontamination of vagina, cervix, feces, rectum, or superficial wound.
- We must use vacutainer for the anaerobic specimen collection.
- For any liquid sample collection, the air trapped in the syringe should be expelled by holding the syringe and needle upright.
- Nasopharyngeal sample collection should be done very carefully by well-trained staff and if there is suspicion of any viral infection, the swab collection should be done through both the nostrils for offsite analysis, and the specimen to be transported immediately on ice, with the sample to be carried either in normal saline or in M4 viral transport media.

Packing of clinical samples (Campbell et al., 2012)

During collection, storage, and transportation of specimens, one must follow the general packaging guidelines given by the WHO. As per guidelines, the samples should be collected aseptically and wrapped in four layers in the following manner:

1. Watertight/airtight container/vials. If multiple samples are being handled together, then each vial/container should be individually wrapped and separated to prevent contact between them.
2. Absorbent material/covering material. Cellulose wadding, cotton balls, paper towels, or super-absorbent packets must be used to wrap the individual vial/container of sample.
3. Secondary airtight and watertight container. This may be any bigger container/plastic zip lock bag, plastic canister, or screw capped can.
4. Outer packing. This must be of rigid and hard construction and may be made of fiberboard, wood, metal, or plastic.

Storage of samples

After successful collection, samples and their aliquots may be stored under different conditions, depending on the intended laboratory analyses. Storage of a specimen is a critical step as incorrect storage may directly affect the result. Maintenance of storage temperature is a vital factor for commonly used specimens like plasma, serum, and other cellular specimens. Plasma and serum may be securely stored at $-80°C$, whereas other cellular specimens like lymphocytes should be stored in liquid nitrogen at $-196°C$ for a longer period (Jang et al., 2017). Different storage temperature refrigerator systems for successful storage of specimens are shown in Fig. 3. For the preservation of samples, liquid nitrogen is the best option as it is less expensive and without any temperature deviation. It also allows the sample to be stored for a longer time (Fig. 4). However, if freezer systems are not available, then an alternative approach is collection of blood spots on filter cards and storage at room temperature for subsequent analysis (Elsener and Reisch, 2016; Whitney et al., 2018). The following factors need to be taken into consideration before storage of samples:

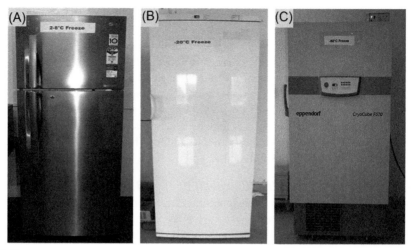

FIG. 3

Different temperature refrigerator systems for storage of specimens after collection, depending upon intended use. (A) 2–8°C refrigerator; (B) −20°C refrigerator; and (C) −80°C refrigerator.

FIG. 4

(A) Cryocan containing liquid nitrogen (−196°C) for long-term storage of specimens. (B) Precautions to be taken during storage of specimen in cryocan containing liquid nitrogen.

- Standard operating procedures should be followed with proper documentation during rapid transfer of material, preferably for specimens that needs low temperature storage.
- There should be sufficient back-up storage facilities; these must be connected to a generator system that can provide an immediate power supply during an electric outage.
- Automated freezer systems are preferable for convenient storage.
- There must be an adequate supply of liquid nitrogen for the maintenance of cryo storage specimens. Cryo vials must be able to withstand liquid nitrogen

temperatures and screw caps of cryo vials must be leak-proof to avoid cross contamination (Shu et al., 2015), as shown in Fig. 4.

- A liquid nitrogen tank should contain an alarm system that will activate if the level goes down.
- There should be sufficient dry ice for emergency backup.
- A special unit for the maintenance and repair of storage equipment is required.
- Before use, all equipment should be properly validated.
- The storage vessels must be properly labeled and should not deteriorate even after long term storage.

Maintenance of storage systems

Maintenance of storage systems is vital as insufficient maintenance may increase the chances of alteration of results. Hence, to prevent the variability of results and for repeatability of data, validation of freezers and other storage equipment is necessary (Pegg, 2009). In addition, the biorepository should develop a protocol to assure the functional status of equipment (De Souza and Greenspan, 2013). Furthermore, there should be special procedures to assure that the storage equipment is properly validated. As per ISBER regarding best maintenance practice, "any device that provides a readout, data, or has a meter movement, is considered an instrument, and requires calibration" (Vaught et al., 2010; Vaught and Henderson, 2011). For awareness, there must be a preventive maintenance program at regular intervals. Maintenance of temperature for any storage systems is a vital criterion for safe storage of specimens. Hence, monitoring of freezer temperature must be continuous. For a small biorepository, manual logging of temperature twice daily is adequate; however, for a larger biorepository, there should be automated monitoring temperature systems for avoiding any malfunctions. Monitoring of both levels and temperature is necessary for liquid nitrogen freezers. Hence, monitoring on a regular basis is very important for avoiding any detriment to samples.

Precautions to be taken before storage of specimens

The following precautions should be taken before storage of specimens:

- Containers used for blood, urine, and sputum collection should be sterile and leak-proof.
- The specimen should not be accepted by the offsite analysis staff if the sample container is not well labeled, not tightly packed, or without a verification sheet.
- A specimen collected through intact skin must be collected and stored in such a way that the risk factors could be maximally minimized.
- Collected samples must be kept in sterile, screw cap, and leak proof containers. If the containers are with lid only, it must be capable of stopping aerosols.
- Specimens/samples like fresh pus, fluid, or tissue must be aspirated aseptically and stored in sterile containers.

- If a swab is to be taken, there must be a transport media with it for prevention of dryness and losing its viability. Hence it must be transported within 2–8 h only.
- All samples must be stored in a sterile container and packed in zip lock bags for storage and transporting to offsite analysis.
- There are a few samples, like stool/feces, which are to be transported for analysis within a certain time and in certain amounts, i.e., a fecal specimen should be at least 50 g if solid and 20 mL if fluid for onsite analysis. If these samples are to be transported for offsite analysis, they should be transferred immediately to a Cairy Blair vial.

Transportation of collected samples

Detection of BW agent is necessary to provide effective treatment to victims and decontamination of attacked location (Kamboj et al., 2006). A number of instruments and kits are available for identification of biological warfare agents, but they are not effective if engineered strains or new biological agents are used in bioterrorist events, and therefore, recognition is not possible at the attacked location. In that case, samples must be collected and transported to a designated laboratory for identification (Peruski et al., 2002; Thavaselvam and Vijayaraghavan, 2010). The carrying boxes must be safe in the transport vehicle and travel should be directly from the location of origin to the drop-off location (Holland et al., 2003). The principle of safe transport by this means is the same as for air or international transport: the material should not have any possibility of escaping from the package under normal conditions of transport. The following practices should be observed:

- Biological samples must be enclosed in containers that are watertight and leak-proof.
- If the sample container is a tube, it must be tightly capped and placed in a rack to maintain it in an upright position.
- Specimen containers and racks should be placed in robust, leak-proof plastic or metal transport boxes with secure, tight-fitting covers.
- The transport box should be secured in the transport vehicle.
- Each transport box should be labeled appropriately, consistent with its contents.
- Specimen data forms and identification data should accompany each transport box.
- A spill kit containing absorbent material, a chlorine disinfectant, a leak-proof waste disposal container, and heavy-duty reusable gloves should be kept in the transport vehicle (World Health Organization, 1997).

Before the shipment of biological and environmental samples in the vehicle, ensure the safe packaging of the samples (González-Gross et al., 2008). The vehicle and persons that are involved in the transportation purpose must meet the following criteria:

- The vehicle must be in good condition and fully air-conditioned.
- The driver should be a responsible person and have a valid driver's license.
- The person must be clear about category A and category B agents.

- In case of problems with the vehicle, a backup vehicle should be available quickly.
- All travel routes should be planned in advance.
- The designated person must be aware and follow the guidelines of motor vehicles laws of that particular city or region.
- Notify the Office of Environmental Health and Safety before transporting material.

During transportation of biological samples, the motor vehicle should be used for that purpose only, and there is to be no passenger or food transport. In the event of a motor vehicle accident, the transporter should follow the steps below:

- Call for emergency support, if required.
- Let all emergency responders know that you are transporting potential biohazards.
- Inform recipient of sample status.
- Make an arrangement for alternate transport vehicle if you are not able to arrive at destination (Lippi and Mattiuzzi, 2016).

Conclusion

Today, it is possible to analyze large numbers of samples in a time efficient manner by using advanced tools and high-throughput technology. To obtain accuracy and reproducible results, proper handling of biological samples is of the utmost importance. Hence, collection, storage, and transportation of biological samples are challenging tasks for epidemiologists, laboratory scientists, physicians, and other healthcare professionals. The method of collection by competent persons, maintenance of conditions during storage, and proper transportation of specimens from an affected area to a designated laboratory with proper guidelines helps error-free diagnosis of any disease/outbreak.

Acknowledgments

I would like to express special thanks and gratitude to my respected guide, Dr. S.J.S. Flora, who gave me the golden opportunity to write this chapter, which also helped me acquire knowledge about many things. I also wish to acknowledge the Department of Pharmacology and Toxicology, National Institute of Pharmaceutical Education and Research (NIPER-Raebareli).

References

Begou, O., Deda, O., Agapiou, A., Taitzoglou, I., Gika, H., Theodoridis, G., 2019. Urine and faecal samples targeted metabolomics of carobs treated rats. J. Chromatogr. B 1 (1114–1115), 76–85.
Bonassi, S., Au, W.W., 2002. Biomarkers in molecular epidemiology studies for health risk prediction. Mutat. Res. Rev. Mutat. Res. 511 (1), 73–86.

Campbell, L.D., Betsou, F., Garcia, D.L., Giri, J.G., Pitt, K.E., Pugh, R.S., et al., 2012. Development of the ISBER best practices for repositories: collection, storage, retrieval and distribution of biological materials for research. Biopreserv. Biobanking 10 (2), 232–233.

Daniel III, C.R., Piraccini, B.M., Tosti, A., 2004. The nail and hair in forensic science. J. Am. Acad. Dermatol. 50 (2), 258–261.

De Souza, Y.G., Greenspan, J.S., 2013. Biobanking past, present and future: responsibilities and benefits. AIDS (London, England) 27 (3), 303.

Eiseman, E., Bloom, G., Brower, J., Clancy, N., Olmsted, S.S., 2003. Case Studies of Existing Human Tissue Repositories: "Best Practices" for a Biospecimen Resource for the Genomic and Proteomic Era. Rand Corporation.

Elsener, D., Reisch, D., 2016. Storage Unit and Transfer System for Biological Samples. Google Patents.

García-Closas, M., Egan, K.M., Abruzzo, J., Newcomb, P.A., Titus-Ernstoff, L., Franklin, T., et al., 2001. Collection of genomic DNA from adults in epidemiological studies by buccal cytobrush and mouthwash. Cancer Epidemiol. Biomarkers Prev. 10 (6), 687–696.

Gekas, V., 2017. Transport Phenomena of Foods and Biological Materials. Routledge.

González-Gross, M., Breidenassel, C., Gómez-Martínez, S., Ferrari, M., Beghin, L., Spinneker, A., et al., 2008. Sampling and processing of fresh blood samples within a European multi-center nutritional study: evaluation of biomarker stability during transport and storage. Int. J. Obes. 32 (S5), S66.

Hallmans, G., Vaught, J.B., 2011. Best practices for establishing a biobank. Methods Mol. Biol. 675, 241–260.

Holland, N.T., Smith, M.T., Eskenazi, B., Bastaki, M., 2003. Biological sample collection and processing for molecular epidemiological studies. Mutat. Res., Rev. Mutat. Res. 543 (3), 217–234.

International Society for Biological and Environmental Repositories (ISBER), 2008. Collection, storage, retrieval and distribution of biological materials for research. Cell Preserv. Technol. 6 (1), 3–58.

Jang, T.H., Park, S.C., Yang, J.H., Kim, J.Y., Seok, J.H., Park, U.S., et al., 2017. Cryopreservation and its clinical applications. Integr. Med. Res. 6 (1), 12–18.

Kamboj, D.V., Goel, A.K., Singh, L., 2006. Biological warfare agents. Def. Sci. J. 56 (4), 495–506.

Klemm, E., Dougan, G., 2016. Advances in understanding bacterial pathogenesis gained from whole-genome sequencing and phylogenetics. Cell Host Microbe 19 (5), 599–610.

Koseoglu, E., Koseoglu, R., Kendirci, M., Saraymen, R., Saraymen, B., 2017. Trace metal concentrations in hair and nails from Alzheimer's disease patients: relations with clinical severity. J. Trace Elem. Med. Biol. 39, 124–128.

Landi, M., Caporaso, N., 1997. Sample collection, processing and storage. IARC Sci. Publ. (142)223–236.

Leonel, E.C.R., Lucci, C.M., Amorim, C.A., 2019. Cryopreservation of human ovarian tissue: a review. Transfus. Med. Hemother. 1–9.

Lippi, G., Mattiuzzi, C., 2016. Biological samples transportation by drones: ready for prime time? Ann. Transl. Med. 4 (5).

Lloyd, R.M., Burns, D.A., Huong, J.T., 2016. Methods for Collection, Storage and Transportation of Biological Specimens. Google Patents.

Mager, S., Oomen, M.H., Morente, M.M., Ratcliffe, C., Knox, K., Kerr, D.J., et al., 2007. Standard operating procedure for the collection of fresh frozen tissue samples. Eur. J. Cancer 43 (5), 828–834.

Mei, J.V., Alexander, J.R., Adam, B.W., Hannon, W.H., 2001. Use of filter paper for the collection and analysis of human whole blood specimens. J. Nutr. 131 (5), 1631S–1636S.

Mei, J.V., Zobel, S.D., Hall, E.M., De Jesús, V.R., Adam, B.W., Hannon, W.H., 2010. Performance properties of filter paper devices for whole blood collection. Bioanalysis 2 (8), 1397–1403.

Nishad, S., Ghosh, A., 2016. Dynamic changes in the proteome of human peripheral blood mononuclear cells with low dose ionizing radiation. Mutat. Res., Genet. Toxicol. Environ. Mutagen. 797, 9–20.

Pegg, D.E., 2009. Principles of cryopreservation. In: Preservation of Human Oocytes. CRC Press, pp. 33–45.

Perera, F.P., 2000. Molecular epidemiology: on the path to prevention? J. Natl. Cancer Inst. 92 (8), 602–612.

Perera, F.P., Weinstein, I.B., 2000. Molecular epidemiology: recent advances and future directions. Carcinogenesis 21 (3), 517–524.

Peruski, A.H., Johnson, L.H., Peruski, L.F., 2002. Rapid and sensitive detection of biological warfare agents using time-resolved fluorescence assays. J. Immunol. Methods 263 (1), 35–41.

Rothman, N., Wacholder, S., Caporaso, N., Garcia-Closas, M., Buetow, K., Fraumeni Jr., J., 2001. The use of common genetic polymorphisms to enhance the epidemiologic study of environmental carcinogens. Biochim. Biophys. Acta Rev. Cancer 1471 (2), C1–C10.

Sharma, J.D., Kataki, A.C., Vijay, C., 2013. Population-based incidence and patterns of cancer in Kamrup Urban Cancer Registry, India. Natl. Med. J. India 26 (3), 133–141.

Shu, Z., Gao, D., Pu, L.L., 2015. Update on cryopreservation of adipose tissue and adipose-derived stem cells. Clin. Plast. Surg. 42 (2), 209–218.

Sudan, P., Mehendale, S., Jain, N., Kant, R., 2018. Indian Council of Medical Research: A Beacon of Medical Research in India.

Thavaselvam, D., Vijayaraghavan, R., 2010. Biological warfare agents. J. Pharm. Bioallied Sci. 2 (3), 179–188.

Theda, C., Hwang, S.H., Czajko, A., Loke, Y.J., Leong, P., Craig, J.M., 2018. Quantitation of the cellular content of saliva and buccal swab samples. Sci. Rep. 8 (1), 6944.

van der Linden, A., Blokker, B.M., Kap, M., Weustink, A.C., Riegman, P.H., Oosterhuis, J.W., 2014. Post-mortem tissue biopsies obtained at minimally invasive autopsy: an RNA-quality analysis. PLoS One 9 (12), e115675.

Vaught, J.B., 2006. Blood collection, shipment, processing, and storage. Cancer Epidemiol. Biomarkers Prev. 15 (9), 1582–1584.

Vaught, J.B., Henderson, M.K., 2011. Biological sample collection, processing, storage and information management. IARC Sci. Publ. 163, 23–42.

Vaught, J.B., Caboux, E., Hainaut, P., 2010. International efforts to develop biospecimen best practices. Cancer Epidemiol. Biomarkers Prev. 19 (4), 912–915.

Whitney, S.E., Wilkinson, S., Muller, R., 2018. Compositions for Stabilizing DNA, RNA, and Proteins in Blood and Other Biological Samples During Shipping and Storage at Ambient Temperatures. Google Patents.

World Health Organization, 1997. Guidelines for the Safe Transport of Infectious Substances and Diagnostic Specimens. World Health Organization, Geneva.

Medical management of diseases associated with biological warfare

8

Jayant Patwa, S.J.S. Flora

National Institute of Pharmaceutical Education and Research-Raebareli, Lucknow, India

Introduction

Biological weapons pose a potential threat to military and civilian populations. Biological warfare (BW) agents include bacteria, fungi, viruses, and toxins that affect human beings, animals, plants, and can be used in bioterrorist activity and biological warfare attacks (Agarwal et al., 2004). BW agents are a serious concern nowadays, due to emerging bioterrorist incidents all over the world. Although various treatment strategies have been developed in recent years to treat or neutralize such kinds of biological warfare agents, these are not enough for a complete cure (Woods, 2005). This chapter provides a broad overview of antibacterial, antifungal, antiviral agents, vaccines, and monoclonal antibodies that can be used to treat infections caused by biological warfare agents. Specific antimicrobial, antifungal, and antiviral agents are being used to treat particular BW agents. Antimicrobial drug resistance is a major hurdle in the treatment of bacterial biological warfare agents. Drug resistance in the pathogens may be either intrinsic or acquired type. In the case of intrinsic drug resistance, the microorganism may not have of an appropriate target, or possess natural barriers that prevent the agent from reaching its target. For example, it has been observed that Staphylococcus aureus has intrinsic resistance with the second and third-generation cephalosporins. Acquired resistance generally occurs due to frequent/high doses of antibiotics exposure, or may be due to genetic changes occurring in an organism (Munita and Arias, 2016).

Genetic modification in the pathogens may be the prime reason for drug resistance, for example, a number of pathogenic microbes have shown resistance to quinolone antibiotics because genetic modifications take place in the following genes: gyrA, parC, and gyrB of Neisseria gonorrhoeae (Lindback et al., 2002). As BW agents are generally genetically modified strains, they have inheritance resistance to a number of antibiotics. Genetically modified strains can be prepared either by genetic editing or selected in vitro by serial passages, for example, fluoroquinolone resistance can be developed in the laboratory by in vitro continuous

Handbook on Biological Warfare Preparedness. https://doi.org/10.1016/B978-0-12-812026-2.00008-6

passages on culture media along with regularly increasing the concentration of fluoroquinolones in the media (Price et al., 2003). Mono-drug resistance is quite common in BW agents, but multidrug resistance is a major concern in the effective treatment of BW disease. Stepanov and his co-worker prepared a multidrug-resistant strain of B. anthracis (anthrax). In his experiment, he used Russian vaccine strain STI-1, genetically modified by the introduction of a pTEC plasmid. The newly prepared STI-AR strain was resistant to the following antibiotics: rifampin, chloramphenicol, penicillin, lincosamides, tetracycline, and macrolides (Stepanov et al., 1996).

Passive immunization may be a promising approach to provide immunity against biological agents. Active immunization involves administering an immunogen into the body to produce an adaptive immune response; the response takes days/weeks to develop immunity, but may be long-lasting, even lifelong (Casadevall and Relman, 2010). At the time of biological warfare or bioterrorist activity, active immunization may not be useful, or may be limited because effective vaccination often takes more time and multiple doses to boost the immunogenic response. Prophylactic immunization may perhaps be a helpful approach to provide protection against biological warfare agents, but it has the drawback that, during an attack, vaccination of a large population might not be possible (Saleh et al., 2017). Vaccine administration also has some adverse effects that would not be acceptable during a bioterrorist or biological warfare attack situation. Furthermore, vaccines fail to induce protective immunity in all recipients, particularly in immune-compromised persons. In recent years, certain vaccines and antibodies have been developed against BW agents. Apart from these therapies, monoclonal antibodies and antitoxins have been developed to provide passive immunity against the immunogenic reaction induced by biological agents (Casadevall, 2002).

Treatment protocols include several principles

1. Identification of biological warfare causality
2. Triage
3. EMT (emergency medical treatment)
4. Decontamination
5. Medical evacuation
6. Specific therapy
7. Prevention

Treatment for bacteria BW agents
Anthrax

Anthrax is caused by the Bacillus anthracis organism, most frequently found in wild and domestic mammals (Pile et al., 1998). It occurs mainly in three forms—cutaneous, gastrointestinal, and inhalation—which depend on exposure to causative organisms; while skin exposed to contaminated product leads to cutaneous infection, consumption of contaminated food causes gastrointestinal disease, and breathing in of aerosolized spores during the handling of contaminated materials leads to the inhalation form (Bell et al., 2002). Warning signs start between 1 day and 2 months following the infection being contracted. Characteristic symptoms including fever/chills, fatigue/malaise, cough, nausea/vomiting, diaphoresis, chest pain, myalgia, confusion, and headache (Inglesby, 1999).

Mode of transmission
Anthrax spores can easily be delivered through the aerosol form or by contaminated food and water (Beeching et al., 2002). Normally Bacillus anthracis spores have the capacity to enter the human body by means of the GI tract (ingestion), skin (damaged skin), or lungs (inhalation), and cause particular clinical indications, dependent on the site of entry. In general, a contaminated human will be isolated so that other humans will not be infected, although anthrax does not commonly spread from a contaminated individual to an uninfected individual (Beyer and Turnbull, 2009).

Treatment
Three types of treatment strategies are used for treatment of anthrax infection: antibiotics, monoclonal antibodies, and for prophylactic purposes, vaccines are also available against the anthrax infection (Hull et al., 2005). Antibiotics generally suitable to manage anthrax infections include large doses of oral and intravenous antibiotics, such as fluoroquinolones (ciprofloxacin), erythromycin, doxycycline, vancomycin, or penicillin. Generally, ciprofloxacin 400 mg and doxycyline 100 mg are preferred for the treatment of anthrax in adults; clindamycin and rifampin are good choices for additional antibiotics (Brook, 2002).

Vaccine
A vaccine is a suspension form of biological products including whole organism, DNA, toxoid, and subunit conjugates, which provides acquired immunity against a specific disease. Generally, two forms of vaccine are available: inactivated vaccine and attenuated vaccine. The first effective vaccine was developed by Louis Pasteur in 1881. In the late 1930s, the Soviet Union developed the first vaccine against human anthrax disease, followed later by the UK and US in the 1950s. The current FDA-approved US vaccine was formulated in the 1960s. The two varieties of anthrax vaccine now available include acellular, which is a US form, and live vaccine, which is Russian (Scorpio et al., 2006; Cybulski et al., 2009). BioThrax, is currently approved by the US FDA, and was previously given in a six dose primary sequence at 0, 2, and 4 weeks, and 6, 12, and 18 months, with annual booster doses to prolong immunity. However, in 2008, omission of the week 2 dose was approved by the FDA, resulting in the five dose series currently recommended. The currently available anthrax vaccines elicit considerable local and systemic reactogenicity (local and systemic adverse reaction includes erythema, induration, soreness, fever) and serious adverse responses take place in about 1% of recipients (Malkevich et al., 2013).

Monoclonal antibody

A monoclonal antibody (mAb or moAb) is a protein molecule that is produced by a specific immune cell. Functions include recognition of organisms and neutralization of foreign organisms like bacteria, viruses, and their toxins. Raxibacumab (brand name ABthrax) and Obiltoxaximab are approved to neutralize the toxins produced by B. *anthracis* (Yamamoto et al., 2016). Raxibacumab is a human monoclonal antibody anticipated for prophylactic purposes and treatment of anthrax. Its efficacy has been shown in monkeys and rabbits. Adverse effects associated with the use of raxibacumab may include, but are not limited to, rash extremity pain, and itching (Tsai and Morris, 2015).

Case study of anthrax treatment

At the time of the anthrax attack in 2001 in the USA, 65 species of anthrax were isolated and each species was related to the 2001 outbreak. All isolated species were sensitive to the quinolones, tetracycline, rifampin, vancomycin, monobactom antibiotics imipenem, meropenem, clindamycin chloramphenicol, and the aminoglycosides. The isolates have moderate range vulnerability to the macrolides antibiotics but they were resistant to extended-spectrum cephalosporins, including ceftriaxone, which is a third-generation cephalosporin agent, as well as to co-trimoxazole (trimethoprim- sulfamethoxazole) (Bell et al., 2002).

Plague

Plague is caused by the bacterium *Yersinia pestis*, which is a Gram-negative bacterium, and more than 200 species have been identified (Whitby et al., 2002). Plague usually occurs in three different forms: bubonic, septicemic, and pneumonic. The consequences of the bubonic form include swelling in the lymph nodes, whereas in the septicemic form, black pigmentation occurs in tissue, and in the pneumonic form, symptoms included difficulties in breathing, and cough and chest pain may occur. In the midst of World War II, weaponized plague was made by the Japanese army. During the Japanese control of Manchuria, Unit 731 intentionally infected Manchurian civilians, as well as Korean and Chinese prisoners of war with the plague bacterium (Inglesby et al., 2000).

Mode of transmission

In biological warfare and bioterrorist events, plague can be spread through airborne transmission (aerosolization) along with contaminated food and water (Beeching et al., 2002). Naturally, plague is most commonly transmitted from rodent to human by the bite of an infected flea, especially the oriental rat flea, Xenopsylla cheopis. Possible modes of transmission include:

- droplet contact—sneezing or coughing on someone else;
- direct physical contact—contact with a contaminated individual, including sexual contact;

- indirect contact—usually by touching contaminated soil or a contaminated surface;
- airborne transmission—if the bacteria can persist in the air for long periods;
- fecal-oral transmission—usually from contaminated food or water sources; and
- vctor borne transmission—carried by insects or other animals.

Treatment

Protein synthesis inhibitor antibiotics commonly preferred against plague infection include streptomycin, tetracycline, and chloramphenicol (Anisimov and Amoako, 2006). Among the newer generation of antimicrobial agents, doxycycline and gentamicin have shown to be effective against plague in monotherapeutics regimes (Mwengee et al., 2006). Drug-resistance is a major issue with antibiotics in the case of plague infections. In 1995, one case of drug-resistance was reported in Madagascar; further outbreaks were reported in November 2014 and October 2017 in Madagascar (Drancourt and Raoult, 2017).

Vaccine

Plague vaccine is a vaccine that is effective against *Yersinia pestis* infection. Since 1980, the killed bacteria were used as vaccines which were less effective against the pneumonic plague. To overcome this problem, a live attenuated vaccine has been developed to reduce the threat of plague infection (Sun et al., 2011). In recent years, a number of potential vaccine candidates have been identified including live, attenuated, and subunit vaccines.

Monoclonal antibodies

LcrV or F1 have been identified for the development of monoclonal antibodies to provide passive immunity in plague victims. Earlier preclinical studies demonstrated that anti-LcrV or F1 monoclonal antibodies (mAbs) administration is able to passively protect mice against plague challenge. LcrV and F1 monoclonal antibody (MAbs 7.3 and F1-04-A-G1) administered by the intratracheal route in mice have shown protection from pneumonic plague (Sun and Singh, 2019).

Brucellosis

Brucellosis is generally recognized by other names, such as Mediterranean fever, Malta fever, and undulant fever. Brucellosis is caused by a *brucella* species, which is a gram-negative rod shaped (*Cocco bacilli*) bacteria (Aparicio, 2013). They work as facultative intracellular parasites causing illness, which is generally permanent. Several species have been identified, but humans are sensitive to only four species: B. abortus, B. canis, *B. melitensis*, *B. suis*, and B. abortus; among these, *B. melitensis* is a particularly dangerous and invasive species. In the middle of the 20th century, brucella strains were weaponized by many countries. According to the literature,

B. suis was the first species to be weaponized, by the USA in 1954. Brucellosis is very infectious. It is proposed that 10–100 microorganisms would be adequate to deliver contaminating spray for individuals (Pappas et al., 2006).

Mode of transmission

Transmission of Brucellosis is possible through a spray in biological warfare and bioterrorist activity. Brucellosis is a zoonotic disease that is spread by consumption of unpasteurized milk or uncooked meat from contaminated animals, or close contact with their secretions. The incubation period of brucellosis is typically 1–2 months, but it may occur in around 5–60 days (Vigeant et al., 1995).

Treatment

Broad ranges of antibiotics are effective against the *brucella* bacteria, for example, tetracyclines, rifampin, and the aminoglycosides streptomycin and gentamicin. However, it has been recommended that combination of more than one antibiotic is required to achieve the desired effects, because the bacteria are nurtured inside cells. The gold standard regimen for adults includes 1 g streptomycin for 14 days by intramuscular route along with doxycycline 100 mg twice a day consecutively for 45 days though oral route. If streptomycin is not available or the patient has susceptibility with streptomycin, it can be replaced with gentamicin 5 mg/kg once a day by intramuscular route for 7 days (Roushan et al., 2006). Apart for this, an alternate regimen is doxycycline and rifampin two times daily for at least 6 weeks. This combination has benefits of oral administration. In case of neuro-brucellosis treatment, combination of doxycycline and rifampin along with co-trimoxazole has been used. In this triple drug combination, doxycycline is capable of crossing the blood-brain barrier (BBB), but in addition rifampin and co-trimoxazole are necessary to avoid relapse condition. Ciprofloxacin and co-trimoxazole regimen is related with the high risk of relapsing condition. Indeed, even with ideal antibrucellic treatment, relapses still happen in 5% to 10% of patients having Malta fever (Solera, 2000).

Vaccine

Vaccination could be a promising approach for prophylactic treatment of Brucellosis, but there is no licensed vaccine available for humans, although animal vaccine is available. Still, researchers are vigorously trying to develop a brucellosis vaccine. RB51 vaccine is an attenuated live bacterial vaccine that was licensed conditionally by the Centers for Veterinary Biologics, Veterinary Services, Animal and Plant Health Inspection Service, USDA, on February 23, 1996, for immunization of cattle in the United States (Tittarelli et al., 2008).

Cholera

Cholera is a devastating disease caused by the bacterium *Vibrio cholera*, which is commonly used in bio-terrorist and biological warfare activities for mass destruction. Symptoms include watery diarrhea, vomiting, and in addition, muscle cramps may occur; consequently, dehydration and electrolyte imbalance occur due to severe diarrhea. It has been estimated that 3–5 million people are affected by a form of cholera, globally, and approximately 28–130 thousand people die every year from the disease. In 2010, it was classified in the pandemic category in undeveloped countries (Harris et al., 2012). Children are more susceptible to cholera. It was used during both world wars, and many people died because of dysentery, which is a common symptom of cholera (Riedel, 2004).

Mode of transmission

Only humans are affected through cholera infection and it does not require any insect vector or animal. Transmission of cholera infection generally happens through ingestion of food or water contaminated directly or indirectly by stool or vomitus of infected individuals (Beeching et al., 2002). The incubation time is on or after a couple of hours of infection to 5 days; it is generally 2–3 days. Mechanism of cholera induced pathological changes has been well elucidated. It has been suggested that cholera bacteria secrete a toxin that binds with a specific receptor present in the intestine, which is responsible for the alteration of pathology changes (Campbell Mcintyre et al., 1979).

Treatment

Treatment regimens include replacement therapy, electrolytes, along with antibiotics treatment. Replacement therapy and electrolytes are needed due to extensive fluid evacuation, leading to electrolytes imbalance. It has been reported that oral rehydration therapy (ORT) successfully recovers the cholera pathological changes in most cases. ORT is very effective, safe, and easy to administer. The rehydration solution is commercially available; if it is not available or hard to get, it can be prepared by using I L of boiled water, 1/2 tea spoon of salt, 6 tea spoons of sugar and includes squashed banana as potassium supplement and taste improvement. Pathological condition can be restored without the use of antibiotics if proper hydration is maintained (Guerrant et al., 2003). The World Health Organization (WHO) only recommends antibiotics for those who have severe dehydration condition. Doxycycline is considered to be a first-line treatment for cholera infection; however, resistance occurs in a few strains of *V. cholera*. Several other antibiotics are also effective against *V. cholera*, such as erythromycin, cotrimoxazole, chloramphenicol, tetracycline, and furazolidone. Ciprofloxacin is a fluoroquinolone; it may be effective against cholera but resistance has been reported in lature. It has been reported that azithromycin and tetracycline may work better than doxycycline or ciprofloxacin (Woods, 2005).

Vaccine

The first vaccines were made in the late 1800s and oral vaccines were first presented during the 1990s. The available cholera oral vaccines are effective and safe, although sometimes mild adverse effects may occur, including abdominal pain and diarrhea in some individuals. It is considered that cholera vaccines are safe for use by pregnant women. The period of each vaccine's effectiveness is around 2 years in adults, but in children, nearly 6 months in the 2–5 years age group (Martin et al., 2014). The Dukoral monovalent vaccine was developed by Sweden, and, used along with formalin, heat-killed whole cell of V. Cholera O1 with genetically modified cholera toxin B subunits; this regimen was licensed in 1991 specifically for travelers. The Shanchol and mORCVAX bivalent vaccines, which consist of the O1 and O139 serogroups, primarily, was authorized in Vietnam in 1997. At present, both oral and injectable vaccines are available.

Two forms of oral vaccine exist: inactivated and attenuated. Two varieties of the inactivated oral inoculation are currently being used: WC-rBS and BivWC (Masuet Aumatel et al., 2011). WC-rBS (trade name "Dukoral") is a kind of monovalent inactivated vaccine that encompasses killed entire cells of V. cholerae O1 and additionally contains recombinant cholera toxin B subunit. BivWC (marketed under the names "Shanchol" and "mORCVAX") is a bivalent inactivated vaccine containing killed whole cells of V. cholerae O1 and V. cholerae O139 (Desai et al., 2016). In 2016, the US FDA approved an oral vaccine (CVD 103-HgR or Vaxchora), derived from a serogroup O1 classical Inaba strain, which is a live, attenuated oral vaccine. Injectable cholera vaccine is uncommon in use, but is generally effective for people living where cholera is more frequent. It has been reported that they show protection up to 2 years after a single vaccination, and for 3–4 years with annual booster immunization they reduce the risk of death from cholera by 50% in the first year after vaccination (Sinclair et al., 2011).

Melioidosis

Melioidosis bacterium is an important biological warfare agent, classified as a category B biological warfare agent by CDC. Melioidosis, also known as Whitmore's disease, is caused by a gram-negative bacterium, *Burkholderia pseudomallei*. It mainly occurs in endemic areas including Australia, Thailand, and Vietnam. It can be recognized by symptoms including chest pain, bone or joint pain, cough, skin infections, lung nodules, and pneumonia.

Mode of transmission

The aerosolization properties of melioidosis bacteria make its utilization attractive for bioterrorists and biological warfare events. Transmission of melioidosis can take place through inhalation, wounds, and consumption of contaminated food or water. The incubation period is 1–21 days.

Treatment

Treatment of melioidosis includes two phases: initial acute-phase and eradication phase therapies. The initial acute-phase therapy includes Ceftazidime 50 mg/kg (upto 2g) intravenous every 8h, or 6g/day by continuous infusion. The second choice is Meropenem 25 mg/kg (up to 1g) also active through intravenous route in every 8h. Amoxicillin-Clavulanate (co-amoxiclav) may be effective if the abovementioned drugs are not available. Eradication phase therapy includes treatment with co-trimoxazole, and doxycycline is prescribed to be utilized for 12–20 weeks to diminish the rate of repetition. Chloramphenicol is never again routinely suggested for this reason. Co-amoxiclav is an option for those patients who can't take co-trimoxazole and doxycycline (e.g., pregnant women and children under the age of 12), yet is not as useful (Dance, 2014).

Vaccine

Live attenuated vaccines (LAV) are perceived for their ability to evoke intense humoral and cell-mediated immune responses that can provide long lasting protection against the antigen. A number of vaccine targets like DNA and antigenic protein have been identified and investigated; but as of 2018, no vaccine has been approved for melioidosis. Several targets selected virulence factors of *B. pseudomallei* are listed in Table 1 (Warawa and Woods, 2002).

Tularemia

According to the CDC (Centers for Disease Control and Prevention) *Fransisella tularensis* is a potential biological warfare agent and it was a part of various biological warfare programs of Japan, the Soviet Union, and the United States. The bacterium *Fransisella tuarensis* is basically responsible for tularemia infection; it is also known as rabbit fever. It has been reported in literature that approximately 200 cases were reported every year between 1970 and 2015. Symptoms generally seen in the tularemia-infected individual include fever (this is quite common), skin ulcers, and enlarged lymph nodes; sometimes pneumonia and throat infections are also observed in patients (Dennis et al., 2001; Katz et al., 2002).

Mode of transmission

Tularemia is an attractive biological warfare agent due to some specific reasons, i.e., aerosolization of the bacteria is easy. It is a potent biological warfare agent. Ten to fifty bacteria are enough to infect victims. In comparison with other biological warfare agents, it is non-persistent and decontamination is very easy. Normally it is spread through arthropod vectors including haemaphysalis, amblyomma, ixodes, and dermacentor, and hares, rodents, and rabbits frequently serve as reservoir hosts (Beeching et al., 2002).

Treatment

Aminoglycoside antibiotics are generally preferred to treat tularemia infection. Aminoglycosides include streptomycin and gentamicin. Earlier, doxycyline was used. Even quinolone antibiotics may be useful (Hepburn and Simpson, 2008).

Table 1 Vaccine candidates for *B. pseudomallei*.

Adherence	Invasion	Endocytic escape	Intracellular survival	Others
pilA	irlR	bopA	purM	tssM
boaA	bipD	bsaQ	sodC	wcb
boaB	bopE	bsaZ	katG	operon
bpaC	bipB	bsaU	ahpC	BLF1
fliC	bipC	CHBP	rpoE	
			virAG	
			locus	

Newer antibiotics like ketolides and tigecycline, which is a protein synthesis inhibitor, have shown effect in in vitro experiments against *F. tularensis* Subsp. *holarctica*. Recent findings suggest that LL-37 is also effective against tularemia infection. LL-37, which is kind of antimicrobial peptide, has been investigated in the marine model (Boisset et al., 2014).

Vaccine and monoclonal antibodies

As yet, a vaccine has only been available to protect laboratorians routinely working with *Francisella tularensis*. This immunization is at present under re-evaluation by the US Food and Drug Administration (FDA) and is not commonly accessible in the United States. An attempt was made by researchers in recent years to prepare monoclonal antibodies against the tularemia infection such as:

1. monoclonal antibodies against the LPS of *F. tularensis* LVS;
2. immune sera; and
3. anti-MPF IgM and IgG antibodies.

However, such monoclonal antibodies are ineffective in in vivo models (Table 2) (Savitt et al., 2009; Boisset et al., 2014).

Table 2 Summary of treatment for bacterial biological warfare agents.

S. no.	Bacterial warfare agent	Antibiotics	Vaccine	Monoclonal antibodies
1	Anthrax	Ciprofloxacin, Erythromycin, Doxycycline, Vancomycin, Penicillin, Clindamycin, and Rifampin	BioThrax	Raxibacumab, Obiltoxaximab
2	Plague	Streptomycin, Tetracycline Chloramphenicol, and Gentamicin	Plague vaccine (no trade name)	No
3	Brucellosis	Rifampin, Tetracyclines, and Aminoglycosides (Streptomycin and Gentamicin)	No	No
4	Cholera	Erythromycin, Cotrimoxazole, Chloramphenicol, Tetracycline, and Furazolidone	Dukoral and ShanChol	No
5	Meliodosis	Ceftazidime, Meropenem, Amoxicillin-Clavulanate (co-amoxiclav), and Doxycycline	No	No
6	Tularemia	Streptomycin, Gentamicin, Doxycyline, Ketolides, and Tigecycline	No	No

Treatment for virus BW agents
Smallpox

Variola minor and Variola major, which belong to the genus Orthopoxvirus, the family Poxviridae, and subfamily Chordopoxvirinae, are the two virus variants that are responsible for smallpox infection in humans; although smallpox has been eradicated from the world (Mlinaric-Galinovic et al., 2003). The emerging symptoms in infected individuals include fever and vomiting, and subsequently, formation of sores in the mouth and skin rash. The smallpox virus is considered a high risk for the following reasons:

- A large population of the world is unimmunized as the process of vaccination against smallpox was stopped in 1980.
- High infectivity through aerosol.
- The relative ease of culturing the virus.

In 1980, the World Health Organization announced that smallpox had been eliminated from the Earth, and the immunization process was stopped for the general population. For that reason, the entire unimmunized human population has become prone to the infection of Variola virus, if they encounter it through bioterrorism or by any other means. There is no powerful treatment for smallpox infection and the vaccine is not available, except in the USA (Whitley, 2003).

Mode of transmission

Transmission of smallpox generally occurs through aerosol in bioterrorist and biological warfare events. It has been reported that during the French and Indian Wars (1754–1767) by British armed forces in North America, smallpox virus contaminated blankets (that had been acquired from infected patients) were circulated to American Indians. Consequently, 50% mortality was observed in the effected tribe due to the epidemics (Henderson et al., 1999). Naturally, Variola virus is most frequently transmitted through the airborne droplets of saliva that is aroused form the infected person during the close contact or face-to-face contact with susceptible individuals (Beeching et al., 2002; Milton, 2012).

Treatment

Some antiviral drugs are available that may be helpful to treat or prevent smallpox from getting worse. Tecovirimat (TPOXX) is the only drug to treat smallpox infection; it was approved by the FDA in July 2018. Preclinical studies results revealed that tecovirimat is effective against the virus, which is similar to smallpox infection. The efficacy of tecovirimat has not been evaluated in smallpox-infected subjects, but it has been tested in healthy individuals. Investigation outcomes disclose that it is safe and causes only minor side effects. Cidofovir showed effectiveness during the last large smallpox epidemics; the drug must be administered through the IV (intravenous) route, and nephrotoxicity may be a possible adverse drug effect (Berhanu et al., 2015; Stower, 2018).

Vaccine

ACAM2000, which was developed by Acambis, is the only US FDA approved (on 31 August 2007) vaccine that is effective against smallpox. It encompasses the live attenuated Vaccinia virus, cloned from the calf lymph (Kennedy et al., 2009). It must be administered by the percutaneous route. Earlier, Dryvax vaccine was licensed for smallpox disease in the United States, which is a freeze-dried calf lymph smallpox vaccine. It is the world's oldest vaccine for smallpox; it has since been replaced by the ACAM2000.

Crimean-Congo hemorrhagic fever

Crimean-Congo hemorrhagic fever (CCHF) is a deadly fever that is caused by the CCHF virus. The virus belongs to the genus Nairovirus in the Bunyaviridae family, and is responsible for the disease in humans; it has a lethality rate of 15%–70% (Ergonul et al., 2006). CCHFV belongs to single-stranded RNA virus, and contains 3-segmented genomic designs: small (S), medium (M), and large (L), according to size. The CCHF virions are spherical—approximately 100 nm in diameter and contain a lipid covering that is host derived. The CCHF virus has been widely spread in areas like Africa, Eastern Europe, Asia, and the Middle East (Hawman and Feldmann, 2018).

Mode of transmission

In biological warfare and bioterrorist activities it can be transmitted in aerosolized form. In nature, the vectors are the *Hyalomma* species of ticks, distribution of which is widespread throughout the world. CCHF is transferred through tick bites or direct contact with the blood. Hospital-acquired infections are also seen in different countries, such as Iran, UAE, Sudan, Afghanistan, and Pakistan. CCHF is spread via thick fluids, as was studied by different scientists; further studies are required to investigate CCHF virus are present in intracellular and extracellular fluids, of subsisted, seen in semen also. Likewise, studies are required into the clinical course of CCHF in women who are pregnant, as well as study of the outcomes for pregnancy and the infant, fertility in women who suffer from CCHF, and the virus present in body fluids and breast milk in pregnant women (Gordon et al., 1993).

Treatment

In the acute infection of CCHF, ribavirin is preferred. Ribavirin (1-b-d-ribofuranosyl-1,2,4-triazole-3-carboxamide) belongs to the nucleoside analogue, which has been shown to be effective only in in vitro experiments against viral activity on both DNA as well as RNA viruses. Ribavirin is effective with INF-α used to treat hepatitis-c virus. Ribavirin is also used in syncytial virus (Fisgin et al., 2009). Japan's regulatory agency has approved Favipiravir for treatment of Ebola and pathogenic RNA virus with different possible treatments like immunotherapy with antibodies for CCHF. The antibodies were derived from CCHF patients who had recovered. The MxA protein used to treat some Bunyaviridae family is also CCHF virus (Hawman and Feldmann, 2018).

Vaccine

Since the 1970s, several CCHF vaccination programs around the world have been terminated due to it high toxic effects. The main accessible, and most likely to some degree useful, CCHF vaccine has been an inactivated antigen preparation then used in Bulgaria; but the reality is, there is no FDA approved vaccine available to date (Dowall et al., 2017).

Current research

There are no specific therapies approved for CCHF. The only available treatment of CCHF is Ribavirin, which is a broad spectrum therapy. It inhibits RNA synthesis of virus and have shown antiviral activity in vivo using mouse model. The clinical studies conducted between 2000 and 2015 in Iran showed 80% efficacy and has now been recommended by WHO against CCHF infection. The targets for vaccine in CCHF virus is divided into three segments consisting of large (L), medium (M), and small (S) of RNA which is coded by nucleoprotein virus. The structure of the RNA virus is the nucleoprotein and glycoproteins which are identified as targets for antigenic vaccine for CCHF virus. The monoclonal antibodies have targeted Gc, but not Gn (Al-Abri et al., 2017).

Ebola hemorrhagic fever

The Centers for Disease Control has kept Ebola as a Category A biological warfare agent. Ebola (EBOV) is also known as Ebola hemorrhagic fever (EHF). It is an RNA virus belonging to the *Filoviridae* family (Bray, 2003). This viral disease is widely recognized in the world, despite rare occurrence. However, EBOV is a life threatening pathogen, the rate of fatalities being around 25%–90%. Signs and symptoms typically begin between 2 days and 3 weeks after the exposure to the virus. Common symptoms include headaches, fever, muscular pain, and sore throat. Rashes, diarrhea, and vomiting are commonly observed in infected persons, along with improper functioning of liver and kidney. The disease has a huge risk of casualty, killing between 25% and 90% of those infected, with an average of about 50% (Moole et al., 2015).

Mode of transmission

Aerosol transmission of Ebola virus is possible, and this unique property encourages its use in bioterrorist and biological warfare activities. Incubation periods of EHF virus are usually 3–10 days, but initial symptoms generally occur within 2 days following exposure to the virus. The Ebola viruses is spread by direct contact with infected body fluid and different organ fluid secretions that may be fatal; the disease is transmissible by infected blood. The transmission of Ebola virus thus mainly occurs through person-to-person contacts, initially showing symptoms which may lead to life threatening EVD case. In Ebola virus does not spread through airborne because of mortal in nature, this virus comes under class-4 as a biologic pathogen (Jaax et al., 1995).

Treatment

There is no FDA approved drug against EHF to date; however, patients can symptomatically improve by different antiviral drugs, and there may chances of survival of patients. Probability of improvement and recovery are in the early stages of infection. The symptoms of EHF are treated by:

- providing electrolytes by IV infusion (body salts);
- maintaining oxygen and blood pressure through the infection; or
- treatment of different infection caused by the Ebola virus.

Vaccine

Zaire Ebola virus glycoprotein (ChAd3-EBO-Z) and recombinant vesicular stomatitis virus (rVSV-ZEBOV) vaccine are two potential vaccine candidates against the Ebola virus. Both vaccines have so far exhibited adequate safety and tolerability profiles (Moekotte et al., 2016; Regules et al., 2017).

Monoclonal antibodies

As systematically reviewed recently by Gonzalez-Gonzalez and colleagues, to date, around 20 monoclonal antibodies have been identified and characterized, of which some 10 were found promising to progress to testing in non-human primate models. These included mAb114, MB-003, ZMAb, ZMapp, and MIL77E. Currently, researchers are trying to develop a DNA vaccine that gives genetic immunity against the Ebola virus, in which they introduce some nucleoprotein and envelop protein in the host cell, which provides an immune boost for the host (Moekotte et al., 2016).

Venezuelan equine encephalitis

During the Cold War, both the United States biological weapons program and the Soviet biological weapons program researched and weaponized VEE (Croddy and Perez-Armendariz, 2002). Venezuelan equine encephalitis virus belongs to Alphavirus in the Togaviridae family. They come under positive RNA genomic and the size of genome is 11.5 kb. Infected individuals may experience flu like symptoms, like high fevers and headaches (Weaver et al., 2004).

Mode of transmission

The transmission of VEE virus is by mosquito bite, although it is also spared through parenteral injection, nasal instillation, direct contact with broken skin, and sometimes by contaminated animals, particularly in laboratory animals. The incubation period of VEE virus is about 1 week after a bite by an infected mosquito, and symptoms may appear within 24 h. The VEE virus cannot be transmitted person to person, as far as is known at the time of writing, but it can be transmitted by the bite of an infected individual. Mainly, some species of mosquito transmit this infection in humans, like the *Photophore* and *Ochlerotatus* genuses.

Treatment

Venezuelan equine encephalitis can be treated by melatonin (MLT) to improve patient immunity against the virus. In a preclinical study, mice were immunized by vaccine with TC-83 VEE virus and giving daily melatonin treatment (2–5 mg/kg) for three days before immunization; after 10 days, the amount of IgM antibody was determined at different days (5, 10, 15, 20) after that immunization, and also determined the IL-10 levels after 15 days of vaccination. Treatment may increase or decrease according to IgM and IL-10 level (Kinney et al., 1989).

Vaccine

The first vaccine for VEE was developed in 1938. 1AB strains were isolated from mouse brain and different organs to make Formalin-inactivated preparations, at that time. This vaccine was effective in different animal, but with a high risk of residual live virus. The disappearance of subtype IAB VEEV since 1973, due to the replacement of live-attenuated TC-83 strain for inactivated vaccines, supports this conclusion. The commercial equine vaccines currently marketed in the US consist of inactivated TC-83, which are combined with the other principal alpha viral encephalitis, eastern and western equine encephalitis viruses (Paessler and Weaver, 2009).

Current research

Currently, attempts are being made to develop a live attenuated vaccine for VEE virus. There is an ongoing preclinical study for development of a new treatment that is much safer, with minimum risk of vaccine-born outbreaks, such as the case in late 2017. However, there is a requirement for a preclinical study to insure the safety of the vaccine in humans. The preclinical data suggest that the immune response with VEE virus include whole body infection except brain which needs to be controlled. Recently work has been going on for a development against encephalitis virus that helps to remove infection from the brain (Table 3) (Paessler and Weaver, 2009).

Table 3 Summary of treatment for viral biological warfare agents.

S. No.	Viral warfare agent	Antiviral	Vaccine
1	Smallpox	Tecovirimat	Dryvax, ACAM2000
2	CCHF	Ribavirin	No
3	EHS	No	No
4	VEE	No	TC-83, C-84 (candidate only)

Treatment for toxins BW agents

Saxitoxin

Saxitoxin is a marine alkaloid that is synthesized in marine dinoflagellates (*Alexandrium* sp., *Gymnodinium* sp., *Pyrodinium* sp.) and freshwater cyanobacteria (*Anabaena* sp., some *Aphanizomenon* spp., *Cylindrospermopsis* sp., *Lyngbya* sp., *Planktothrix* sp.). It can also be synthesized in the laboratory. Saxitoxins have been investigated by the United States natural and compound weapons programs. It has been reported that the CIA used saxitoxins for suicide capsules and other covert uses in the 1950s. It is best known as paralytic shellfish toxin (PST) and acts as a potent neurotoxin in humans: it is 1000 times more toxic than sarin. The LD 50 of saxitoxin is 5.7 µg per kg body weight. It is potent neurotoxin that exerts toxic response by selectively blocking the voltage gated sodium channel in the brain that is responsible for the proper neuronal function; this leads to paralysis (Cusick and Sayler, 2013).

Mode of transmission

Saxitoxin has an attractive ability to aerosolize that encourages its use in bioterrorist activity but it can also be administered in food and water. After absorption, it reaches the brain where it selectively blocks sodium channels in the neurons (Anderson, 2012).

Treatments

There is no specific treatment for saxitoxin poisoning; symptomatic treatment may be helpful. Neostigmine and edrophonium have been used to improve muscle weakness following tetrodotoxin intoxication, which is similar to saxitoxin intoxication. Nonetheless, no clinical trials have evaluated the use of these drugs upon saxitoxin exposure (Bane et al., 2014).

Botulinum toxin

The botulinum toxin (BTX) is considered to be a potent neurotoxic protein that is secreted by the bacterium *Clostridium botulinum* and related species. It is a potent neurotoxin, with an estimated human median lethal dose (LD50) of 1.3–2.1 ng/kg via intravenous or intramuscular route and 10–13 ng/kg when inhaled. The botulinum toxin has been recognized as a potential agent for use in bioterrorist and biological warfare activity (Nigam and Nigam, 2010). It can be absorbed through the respiratory tract, mucous membranes, eyes, or non-intact skin. The botulinum toxin exerts its action at the neuromuscular junction, where it blocks the release of the neurotransmitters like acetylcholine from axon endings, consequently causing flaccid paralysis, characterized by double or blurred vision, drooping eyelids, difficulty swallowing, slurred speech, and shortness of breath (Arnon et al., 2001).

Mode of transmission

The botulinum toxin can be spread though aerosol form. It can also be used in covert operations by injection into the victim. Naturally, it can be transmitted through contaminated food and water consumption.

Treatment for botulinum toxin

Heptavalent BAT (botulism antitoxin) is available to neutralize the toxic effect of botulinum toxin. The heptavalent BAT (A, B, C, D, E, F, and G) was prepared by Emergent Bio Solutions Canada Inc. It is the only approved antitoxin by CDS against all seven types of botulinum nerve toxins (types A, B, C, D, E, F, and G) (Arnon et al., 2001). In the early hours, administration of HBAT is considered critical as the antitoxin can able to neutralize only circulating toxin, not toxin that has become bound to nerve terminals. It can be effective in both life-threatening botulism conditions as well as botulinum nerve toxins being used in a bioterrorist attack. In 2010, the CDC approved BAT for Botulinum toxin and in 2013, the FDA licensed it for commercial purpose (Hill et al., 2013).

Ricin

Basically, ricin is a lectin which is a carbohydrate binding protein obtained from the seed of the castor oil plant; the biological source is Ricinus communis. It is a highly potent toxin with an LD 50 around 22 micrograms/kg of body weight via injection or inhalation route; however, it gets inactivated in the oral route due to the acidic environment of the stomach. During World War I, ricin was first investigated by the US for military purposes. At that time it was used either as a toxic dust or as coating for bullets (Moshiri et al., 2016).

Mode of transmission

Ricin transmission generally occurs by inhalation, ingestion, and parenteral exposure. Inhalation exposure may take place through an aerosol, powder or dust Ingestion of contaminated food, water, and consumer products. Parenteral exposure can occur by direct injection.

Treatment for ricin poisoning

There is no antidote available for ricin poisoning; only symptomatic treatment that can be helpful by giving supportive medical care so that poisoning effects may be minimized in the victims. Several factors have to consider before providing supportive care. Basically, it depends upon the route of exposure by which victims were poisoned (for example poisoning can take place through ingestion, inhalation, eye, or skin exposure). Care may include helping victims breathe, giving them intravenous fluids and medications to treat adverse conditions like low blood pressure and seizure, flushing with activated charcoal may be helpful if the ricin has been very recently ingested. It can be possible to remove ricin from the stomach and prevent further absorption, and if eye irritation occurs in patients, washing out their eyes with clean water (Table 4) (Zhang et al., 2015).

Table 4 Summary of treatment for toxin biological warfare agents.

S. no.	Toxins warfare agent	Drugs used for treatment	Antitoxin
1	Saxitoxin	Neostigmine and edrophonium (muscle weakness)	No
2	Botulinum toxin	No	Heptavalent BAT (botulism antitoxin)
3	Ricin	Supportive medical care	No

Conclusion

Despite the availability of various antibiotics and antiviral drugs, there is still a need for a promising treatment therapy against biological warfare agents. Many challenges exist with antimicrobial therapy, including drug resistance, are a major issue with the antimicrobial agents due to development of engineered stains. Although in recent years, numbers of vaccine and monoclonal antibodies has been discovered that have shown effectiveness toward BW agents, but they are not enough. Hence, in future for effective treatment, the approach will likely have to include not only novel antimicrobials agents, but also include vaccines and monoclonal antibodies.

References

Agarwal, R., Shukla, S.K., Dharmani, S., Gandhi, A., 2004. Biological warfare—an emerging threat. J. Assoc. Physicians India 52, 733–738.

Al-Abri, S.S., Al Abaidani, I., Fazlalipour, M., Mostafavi, E., Leblebicioglu, H., Pshenichnaya, N., et al., 2017. Current status of Crimean-Congo haemorrhagic fever in the World Health Organization eastern Mediterranean region: issues, challenges, and future directions. Int. J. Infect. Dis. 58, 82–89.

Anderson, P.D., 2012. Bioterrorism: toxins as weapons. J. Pharm. Pract. 25 (2), 121–129.

Anisimov, A.P., Amoako, K.K., 2006. Treatment of plague: promising alternatives to antibiotics. J. Med. Microbiol. 55, 1461–1475. Pt 11.

Aparicio, E.D., 2013. Epidemiology of brucellosis in domestic animals caused by Brucella melitensis, Brucella suis and Brucella abortus. Rev. Sci. Tech. 32 (1), 53–60.

Arnon, S.S., Schechter, R., Inglesby, T.V., Henderson, D.A., Bartlett, J.G., Ascher, M.S., et al., 2001. Botulinum toxin as a biological weapon: medical and public health management. JAMA 285 (8), 1059–1070.

Bane, V., Lehane, M., Dikshit, M., O'Riordan, A., Furey, A., 2014. Tetrodotoxin: chemistry, toxicity, source, distribution and detection. Toxins 6 (2), 693–755.

Beeching, N.J., Dance, D.A.B., Miller, A.R.O., Spencer, R.C., 2002. Biological warfare and bioterrorism. BMJ 324 (7333), 336–339. (Clinical research ed.).

Bell, D.M., Kozarsky, P.E., Stephens, D.S., 2002. Clinical issues in the prophylaxis, diagnosis, and treatment of anthrax. Emerg. Infect. Dis. 8 (2), 222–225.

Berhanu, A., Prigge, J.T., Silvera, P.M., Honeychurch, K.M., Hruby, D.E., Grosenbach, D.W., 2015. Treatment with the smallpox antiviral tecovirimat (ST-246) alone or in combination with ACAM2000 vaccination is effective as a postsymptomatic therapy for monkeypox virus infection. Antimicrob. Agents Chemother. 59 (7), 4296–4300.

Beyer, W., Turnbull, P., 2009. Anthrax in animals. Mol. Asp. Med. 30 (6), 481–489.

Boisset, S., Caspar, Y., Sutera, V., Maurin, M., 2014. New therapeutic approaches for treatment of tularaemia: a review. Front. Cell. Infect. Microbiol. 4, 40.

Bray, M., 2003. Defense against filoviruses used as biological weapons. Antivir. Res. 57 (1–2), 53–60.

Brook, I., 2002. The prophylaxis and treatment of anthrax. Int. J. Antimicrob. Agents 20 (5), 320–325.

Campbell Mcintyre, R., Tira, T., Flood, T., Blake, P., 1979. Modes of transmission of cholera in a newly infected population on an atoll: implications for control measures. Lancet 313 (8111), 311–314.

Casadevall, A., 2002. Passive antibody administration (immediate immunity) as a specific defense against biological weapons. Emerg. Infect. Dis. 8 (8), 833–841.

Casadevall, A., Relman, D.A., 2010. Microbial threat lists: obstacles in the quest for biosecurity? Nat. Rev. Microbiol. 8 (2), 149–154.

Croddy, E., Perez-Armendariz, C., 2002. Chemical and Biological Warfare: A Comprehensive Survey for the Concerned Citizen. Springer Science & Business Media.

Cusick, K.D., Sayler, G.S., 2013. An overview on the marine neurotoxin, saxitoxin: genetics, molecular targets, methods of detection and ecological functions. Marine Drugs 11 (4), 991–1018.

Cybulski Jr., R.J., Sanz, P., O'Brien, A.D., 2009. Anthrax vaccination strategies. Mol. Asp. Med. 30 (6), 490–502.

Dance, D., 2014. Treatment and prophylaxis of melioidosis. Int. J. Antimicrob. Agents 43 (4), 310–318.

Dennis, D.T., Inglesby, T.V., Henderson, D.A., Bartlett, J.G., Ascher, M.S., Eitzen, E., et al., 2001. Tularemia as a biological weapon: medical and public health management. JAMA 285 (21), 2763–2773.

Desai, S.N., Pezzoli, L., Alberti, K.P., Martin, S., Costa, A., Perea, W., et al., 2016. Achievements and challenges for the use of killed oral cholera vaccines in the global stockpile era. Hum. Vaccin. Immunother. 13 (3), 579–587.

Dowall, S.D., Carroll, M.W., Hewson, R., 2017. Development of vaccines against Crimean-Congo haemorrhagic fever virus. Vaccine 35 (44), 6015–6023.

Drancourt, M., Raoult, D., 2017. Investigation of pneumonic plague, Madagascar. Emerg. Infect. Dis. 24 (1), 183.

Ergonul, O., Tuncbilek, S., Baykam, N., Celikbas, A., Dokuzoguz, B., 2006. Evaluation of serum levels of interleukin (IL)–6, IL-10, and tumor necrosis factor–α in patients with Crimean-Congo hemorrhagic fever. J. Infect. Dis. 193 (7), 941–944.

Fisgin, N.T., Ergonul, O., Doganci, L., Tulek, N., 2009. The role of ribavirin in the therapy of Crimean-Congo hemorrhagic fever: early use is promising. Eur. J. Clin. Microbiol. Infect. Dis. 28 (8), 929–933.

Gordon, S.W., Linthicum, K.J., Moulton, J., 1993. Transmission of Crimean-Congo hemorrhagic fever virus in two species of Hyalomma ticks from infected adults to cofeeding immature forms. Am. J. Trop. Med. Hyg. 48 (4), 576–580.

Guerrant, R.L., Carneiro-Filho, B.A., Dillingham, R.A., 2003. Cholera, diarrhea, and oral rehydration therapy: triumph and indictment. Clin. Infect. Dis. 37 (3), 398–405.

Harris, J.B., LaRocque, R.C., Qadri, F., Ryan, E.T., Calderwood, S.B., 2012. Cholera. Lancet (Lond., Engl.) 379 (9835), 2466–2476.

Hawman, D.W., Feldmann, H., 2018. Recent advances in understanding Crimean-Congo hemorrhagic fever virus. F1000Res. 7, . (F1000 Faculty Rev-1715).

Henderson, D.A., Inglesby, T.V., Bartlett, J.G., Ascher, M.S., Eitzen, E., Jahrling, P.B., et al., 1999. Smallpox as a biological weapon: medical and public health management. Working Group on Civilian Biodefense. JAMA 281 (22), 2127–2137.

Hepburn, M.J., Simpson, A.J., 2008. Tularemia: current diagnosis and treatment options. Expert Rev. Anti-Infect. Ther. 6 (2), 231–240.

Hill, S.E., Iqbal, R., Cadiz, C.L., Le, J., 2013. Foodborne botulism treated with heptavalent botulism antitoxin. Ann. Pharmacother. 47 (2), e12.

Hull, A.K., Criscuolo, C.J., Mett, V., Groen, H., Steeman, W., Westra, H., et al., 2005. Human-derived, plant-produced monoclonal antibody for the treatment of anthrax. Vaccine 23 (17–18), 2082–2086.

Inglesby, T.V., 1999. Anthrax: a possible case history. Emerg. Infect. Dis. 5 (4), 556–560.

Inglesby, T.V., Dennis, D.T., Henderson, D.A., Bartlett, J.G., Ascher, M.S., Eitzen, E., et al., 2000. Plague as a biological weapon: medical and public health management. JAMA 283 (17), 2281–2290.

Jaax, N., Jahrling, P., Geisbert, T., Geisbert, J., Steele, K., McKee, K., et al., 1995. Transmission of Ebola virus (Zaire strain) to uninfected control monkeys in a biocontainment laboratory. Lancet 346 (8991–8992), 1669–1671.

Katz, L., Orr-Urteger, A., Brenner, B., Hourvitz, A., 2002. Tularemia as a biological weapon. Harefuah 141, 120. Spec No: 78-83.

Kennedy, R.B., Ovsyannikova, I., Poland, G.A., 2009. Smallpox vaccines for biodefense. Vaccine 27 (Suppl 4), D73–D79.

Kinney, R.M., Johnson, B.J., Welch, J.B., Tsuchiya, K.R., Trent, D.W., 1989. The full-length nucleotide sequences of the virulent Trinidad donkey strain of Venezuelan equine encephalitis virus and its attenuated vaccine derivative, strain TC-83. Virology 170 (1), 19–30.

Lindback, E., Rahman, M., Jalal, S., Wretlind, B., 2002. Mutations in gyrA, gyrB, parC, and parE in quinolone-resistant strains of Neisseria gonorrhoeae. APMIS 110 (9), 651–657.

Malkevich, N.V., Basu, S., Rudge Jr., T.L., Clement, K.H., Chakrabarti, A.C., Aimes, R.T., et al., 2013. Effect of anthrax immune globulin on response to BioThrax (anthrax vaccine adsorbed) in New Zealand white rabbits. Antimicrob. Agents Chemother. 57 (11), 5693–5696.

Martin, S., Lopez, A.L., Bellos, A., Deen, J., Ali, M., Alberti, K., et al., 2014. Post-licensure deployment of oral cholera vaccines: A systematic review. Bull. World Health Organ. 92 (12), 881–893.

Masuet Aumatell, C., Ramon Torrell, J.M., Zuckerman, J.N., 2011. Review of oral cholera vaccines: efficacy in young children. Infect. Drug Resist. 4, 155–160.

Milton, D.K., 2012. What was the primary mode of smallpox transmission? Implications for biodefense. Front. Cell. Infect. Microbiol. 2, 150.

Mlinaric-Galinovic, G., Turkovic, B., Brudnjak, Z., Gjenero-Margan, I., 2003. The variola virus as a biological weapon. Lijec. Vjesn. 125 (1–2), 16–23.

Moekotte, A.L., Huson, M.A., van der Ende, A.J., Agnandji, S.T., Huizenga, E., Goorhuis, A., et al., 2016. Monoclonal antibodies for the treatment of Ebola virus disease. Expert Opin. Investig. Drugs 25 (11), 1325–1335.

Moole, H., Chitta, S., Victor, D., Kandula, M., Moole, V., Ghadiam, H., et al., 2015. Association of clinical signs and symptoms of Ebola viral disease with case fatality: a systematic review and meta-analysis. J. Comm. Hosp. Int. Med. Perspect. 5 (4), 28406.

Moshiri, M., Hamid, F., Etemad, L., 2016. Ricin toxicity: clinical and molecular aspects. Rep. Biochem. Mol. Biol. 4 (2), 60–65.

Munita, J.M., Arias, C.A., 2016. Mechanisms of antibiotic resistance. Microbiol. Spectr. 4 (2), https://doi.org/10.1128/microbiolspec.VMBF-0016-2015.

Mwengee, W., Butler, T., Mgema, S., Mhina, G., Almasi, Y., Bradley, C., et al., 2006. Treatment of plague with gentamicin or doxycycline in a randomized clinical trial in Tanzania. Clin. Infect. Dis. 42 (5), 614–621.

Nigam, P.K., Nigam, A., 2010. Botulinum toxin. Indian J. Dermatol. 55 (1), 8–14.

Paessler, S., Weaver, S.C., 2009. Vaccines for Venezuelan equine encephalitis. Vaccine 27, D80–D85.

Pappas, G., Panagopoulou, P., Christou, L., Akritidis, N., 2006. Brucella as a biological weapon. Cell. Mol. Life Sci. 63 (19–20), 2229–2236.

Pile, J.C., Malone, J.D., Eitzen, E.M., Friedlander, A.M., 1998. Anthrax as a potential biological warfare agent. Arch. Intern. Med. 158 (5), 429–434.

Price, L.B., Vogler, A., Pearson, T., Busch, J.D., Schupp, J.M., Keim, P., 2003. In vitro selection and characterization of Bacillus anthracis mutants with high-level resistance to ciprofloxacin. Antimicrob. Agents Chemother. 47 (7), 2362–2365.

Regules, J.A., Beigel, J.H., Paolino, K.M., Voell, J., Castellano, A.R., Hu, Z., et al., 2017. A recombinant vesicular stomatitis virus Ebola vaccine. N. Engl. J. Med. 376 (4), 330–341.

Riedel, S., 2004. Biological warfare and bioterrorism: a historical review. Proc. (Baylor Univ. Med. Cent.) 17 (4), 400–406.

Roushan, M.R.H., Mohraz, M., Hajiahmadi, M., Ramzani, A., Valayati, A.A., 2006. Efficacy of gentamicin plus doxycycline versus streptomycin plus doxycycline in the treatment of brucellosis in humans. Clin. Infect. Dis. 42 (8), 1075–1080.

Saleh, E., Swamy, G.K., Moody, M.A., Walter, E.B., 2017. Parental approach to the prevention and management of fever and pain following childhood immunizations: a survey study. Clin. Pediatr. 56 (5), 435–442.

Savitt, A.G., Mena-Taboada, P., Monsalve, G., Benach, J.L., 2009. Francisella tularensis infection-derived monoclonal antibodies provide detection, protection, and therapy. Clin. Vaccine Immunol. 16 (3), 414–422.

Scorpio, A., Blank, T.E., Day, W.A., Chabot, D.J., 2006. Anthrax vaccines: Pasteur to the present. Cell. Mol. Life Sci. 63 (19–20), 2237–2248.

Sinclair, D., Abba, K., Zaman, K., Qadri, F., Graves, P.M., 2011. Oral vaccines for preventing cholera. Cochrane Database Syst. Rev. (3)Cd008603.

Solera, J., 2000. Treatment of human brucellosis. J. Med. Liban. 48 (4), 255–263.

Stepanov, A.V., Marinin, L.I., Pomerantsev, A.P., Staritsin, N.A., 1996. Development of novel vaccines against anthrax in man. J. Biotechnol. 44 (1–3), 155–160.

Stower, H., 2018. An antiviral for smallpox. Nat. Med. 24 (8), 1088.

Sun, W., Singh, A.K., 2019. Plague vaccine: recent progress and prospects. npj Vaccines 4 (1), 11.

Sun, W., Roland, K.L., Curtiss 3rd, R., 2011. Developing live vaccines against plague. J. Infect. Dev. Ctries. 5 (9), 614–627.

Tittarelli, M., Bonfini, B., De Massis, F., Giovannini, A., Di Ventura, M., Nannini, D., et al., 2008. Brucella abortus strain RB51 vaccine: immune response after calfhood vaccination and field investigation in Italian cattle population. Clin. Dev. Immunol. 2008, 584624.

Tsai, C.-W., Morris, S., 2015. Approval of Raxibacumab for the treatment of inhalation anthrax under the US Food and Drug Administration "animal rule". Front. Microbiol. 6, 1320.

Vigeant, P., Mendelson, J., Miller, M.A., 1995. Human to human transmission of Brucella melitensis. Can. J. Infect. Dis. 6 (3), 153–155.

Warawa, J., Woods, D.E., 2002. Melioidosis vaccines. Expert Rev. Vaccines 1 (4), 477–482.

Weaver, S.C., Ferro, C., Barrera, R., Boshell, J., Navarro, J.-C., 2004. Venezuelan equine encephalitis. Annu. Rev. Entomol. 49 (1), 141–174.

Whitby, M., Ruff, T.A., Street, A.C., Fenner, F., 2002. Biological agents as weapons 2: anthrax and plague. Med. J. Aust. 176 (12), 605–608.

Whitley, R.J., 2003. Smallpox: a potential agent of bioterrorism. Antivir. Res. 57 (1–2), 7–12.

Woods, J.B., 2005. Antimicrobials for biological warfare agents. In: Biological Weapons Defense. Springer, pp. 285–315.

Yamamoto, B.J., Shadiack, A.M., Carpenter, S., Sanford, D., Henning, L.N., O'Connor, E., et al., 2016. Efficacy projection of Obiltoxaximab for treatment of inhalational anthrax across a range of disease severity. Antimicrob. Agents Chemother. 60 (10), 5787–5795.

Zhang, T., Yang, H., Kang, L., Gao, S., Xin, W., Yao, W., et al., 2015. Strong protection against ricin challenge induced by a novel modified ricin A-chain protein in mouse model. Hum. Vaccin. Immunother. 11 (7), 1779–1787.

Protective equipment for protection against biological warfare agents

Virendra V. Singh, Mannan Boopathi, Vikas B. Thakare,
Duraipandian Thavaselvam, Beer Singh
Defence Research and Development Establishment, DRDO, Gwalior, India

Introduction

With increasing science and technological innovation, sophistication of equipment, and the proliferation of knowledge through the Internet across the world, the production, storage, and dissemination of biological warfare agents (BWAs) has become cost effective, easier to operate, and methods have become easier to execute its attack (Lederberg, 1999; Eitzen and Takafuji, 1997; Stockholm International Peace Research Institute (SPIRI), 1971; Kaufmann et al., 1997; Christopher et al., 1997; Poupard and Miller, 1992; Atlas, 2002). BWAs can be potentially misused to kill, incapacitate, or seriously impede an individual as well as entire cities or places (Lederberg, 1999; Eitzen and Takafuji, 1997; Stockholm International Peace Research Institute (SPIRI), 1971; Kaufmann et al., 1997; Christopher et al., 1997; Poupard and Miller, 1992; Atlas, 2002). BWAs are considered weapons of mass destruction, and a small amount of infectious material is sufficient to cause epidemics of the disease (Lederberg, 1999; Eitzen and Takafuji, 1997; Stockholm International Peace Research Institute (SPIRI), 1971; Kaufmann et al., 1997; Christopher et al., 1997; Poupard and Miller, 1992; Atlas, 2002). The effects of BWAs are not instantaneous, requiring a few hours to a week before the symptoms appear in the affected population, which leads to delay in reacting to a BWAs attack (Stockholm International Peace Research Institute (SPIRI), 1971; Kaufmann et al., 1997; Christopher et al., 1997).

The history of BWAs began during the fourteenth century, where Tartars Forces used plague infected dead bodies. Instances are recorded in the American Revolutionary War (1776–1781) and American Civil War during 1863 (Christopher et al., 1997, 1999; Poupard and Miller, 1992; Atlas, 2002; Wheelis, 1991). During World War I (1914–1918), BWAs was more advanced and the Germans developed some potential BWA (Christopher et al., 1997, 1999; Poupard and Miller, 1992; Atlas, 2002; Wheelis, 1991). Despite the signing of the 1925 Geneva Protocol, which banned the use of bacteriological weapons, BWAs development and use continued following the First World War. The Geneva Protocol (1925) was the first multilateral agreement that extended prohibition of BWAs (Christopher et al., 1997, 1999;

Handbook on Biological Warfare Preparedness. https://doi.org/10.1016/B978-0-12-812026-2.00009-8

Poupard and Miller, 1992; Atlas, 2002; Wheelis, 1991). During World War II, the Japanese used BWAs against Chinese prisoners (Christopher et al., 1997, 1999; Poupard and Miller, 1992; Atlas, 2002; Wheelis, 1991). In 1994, a Japanese cult, Aum Shinrikyo, attempted an aerosolized release of anthrax in Tokyo, followed by some minor incidents with ricin and anthrax (Christopher et al., 1997, 1999; Poupard and Miller, 1992; Atlas, 2002; Wheelis, 1991). The Biological weapon convention (BWC) was opened to signature in 1972 with the agreement coming into force in 1975 (Christopher et al., 1997, 1999; Poupard and Miller, 1992; Atlas, 2002; Wheelis, 1991). More than 175 state parties have ratified the BWC and agreed to prohibit the development, production, and stockpiling of BWAs. However, this convention is still not foolproof (having no provisions for monitoring and inspection of BWAs, and enforcement of the treaty), which resulted in the incident of 2001 (Kaufmann et al., 1997; Christopher et al., 1997, 1999; Poupard and Miller, 1992; Atlas, 2002; Wheelis, 1991; Klietmann and Ruoff, 2001; Davis, 1999; Meselson et al., 1994). BWAs emerged as a twenty-first century threat and it gained increasing attention of entire world community after the terrorist attack of September 11, 2001 on the World Trade Centre followed by letters harboring anthrax spores (Shah and Wilkins, 2003).

Appropriate protective measures are required to overcome the toxic and virulent effects of BWAs. Protection is an in-depth field and it is the first and foremost important field which prevents the exposure of BWAs and allows the combatant to be effective with a minimal degradation in their operational performance. The mechanism of exposure and the route of entry of the BWAs will govern the level of appropriate protection sought for first responders or any individual. With the help of early warning detection systems, one can enable the personnel to take appropriate protective gear to prevent the exposure and mortality.

This chapter describes documented literature on currently available and also ongoing protection technologies, their types, and new trends, and also highlights the opportunities and challenges of this important field. The scenarios where bio protection is required are vast, starting from scenarios such as battlefield, epidemic, confined health working environments, and also bio-laboratories. Because of differences in threat perception and intensity, the BWAs individual protective equipment (IPE) required also varies as per the requirement. Hence, this chapter mainly deals with battlefield scenarios for military application. However, cross-references of other applications are also discussed in brief. Particular attention has been given to individual and collective protection for the combatant. The recent trends in protection have shifted to the development of multifunctional, lightweight, hybrid materials, and nano-coated, in situ self-decontaminating, active material based protective gear, which could help to defend soldiers and the public at large from BWAs scenarios. Efforts have been devoted by researchers around the globe to make protective equipment intelligent and smarter; this is achieved by imparting functionality to the textile materials. Newer technological innovation towards material development and integration of material science and textile science will lead to rapid translation of the research activity into real practical defense applications, addressing the escalating threat of BWAs.

Classification of biological warfare agents and their symptoms

In nature, a large number of microorganisms exist; however, some of them cause the disease to humans and they cannot all be classified as BWAs. BWAs have been defined as living organisms, including viruses or infectious material derived from them, those cause disease or death in humans, animal or plants, when used for hostile purposes (Lederberg, 1999; Eitzen and Takafuji, 1997; Stockholm International Peace Research Institute (SPIRI), 1971; Kaufmann et al., 1997; Christopher et al., 1997, 1999). Table 1 shows the list of some potential BWAs and their symptoms (Shah and Wilkins, 2003; Crudy, 2001; Norris and Fowler, 1997; Linsel, 2001).

Routes of exposure and modes of delivery

BWAs are unconventional weapons and can be delivered by a number of means based on the requirement. Regardless of mode of delivery, there are three major potential routes for BWAs exposure: skin (dermal), gastrointestinal, and pulmonary. Pulmonary exposure is the major route of entry when compared to other routes (Wheelis, 1991; Klietmann and Ruoff, 2001; Davis, 1999; Meselson et al., 1994; Shah and Wilkins, 2003; Crudy, 2001; Norris and Fowler, 1997; Linsel, 2001; Prasad et al., 2008; Sidell et al., 1997a; Noll, 2005). The dissemination of BWAs can be covert or overt with the aim to disrupt society or to gain international supremacy and they differ from other conventional weapons. A small amount of BWAs has the ability to induce disease; however, the mode of delivery significantly affects the mass destructive potential of BWAs (Wheelis, 1991; Klietmann and Ruoff, 2001; Davis, 1999; Meselson et al., 1994; Shah and Wilkins, 2003; Crudy, 2001; Norris and Fowler, 1997; Linsel, 2001; Prasad et al., 2008; Sidell et al., 1997a; Noll, 2005). Dissemination of BWAs have special challenges to ensure a balance between the virulence and the stability of the agent as it is strongly affected by atmospheric conditions such as sunlight, rain, altitude, wind speed, etc., which may alter the BWAs dissemination in target areas. A potential BWA may be dispersed in the following forms:

(i) *Aerosol*: Aerosolized particles of diameter 1–5μm are the most effective form of BWAs as per aerodynamics, causing maximum destruction to large numbers of population. The largest aerosol particles are not preferred for warfare scenario due to their heaviness, as they either settle onto environmental surfaces or are deposited in the upper respiratory tract and eliminated by normal mucociliary activity. This aerosol form is more commonly used within bioterrorism and also by military groups (Klietmann and Ruoff, 2001; Crudy, 2001; Norris and Fowler, 1997; Linsel, 2001).

(ii) *Food and waterborne*: This causes infection through ingestion. BWA may be used for sabotage, which involves contaminating food, water, or medical

Table 1 List of potential BWAs and their symptoms

BWA	Aerodynamic diameter (µm)	Disease	Agent type	Possible release	Symptoms
Bacillus anthracis	0.2–10 (Bacteria) 0.5–1.5 (Bacterial spore)	Anthrax	Bacteria	Aerosol	Aerosol exposure: chills, high fever, vomiting, painful lymph nodes, headache, joint ache, internal and external bleeding lesions Cutaneous: swelling, bleeding, and tissue death
Yersinia pestis	0.2–10 (Bacteria) 0.5–1.5 (Bacterial spore)	Plague	Bacteria	Aerosol or infected fleas	Enlarged lymph nodes in groin, fever, affects spleen and lungs
Francisella tularensis	0.2–10 (Bacteria) 0.5–1.5 (Bacterial spore)	Tularemia	Bacteria	Aerosol	Fever, Chills, headache, enlarged lymph nodes
Burkholderia mallei	0.2–10 (Bacteria) 0.5–1.5 (Bacterial spore)	Glanders	Bacteria	Contamination of food and water	Skin lesions, ulcers in skin, mucous membranes, Pneumonia, pulmonary abscess
Vibrio cholerae	0.2–10 (Bacteria) 0.5–1.5 (Bacterial spore)	Cholera	Bacteria	Aerosol	Nausea, vomiting, diarrhea, dehydration, toxemia, and collapse
Variola major	0.02–0.3	Smallpox	Virus	Aerosol	High fever, headache, rashes which lead to blisters
Ebola	0.02–0.3	Hemorrhagic fever	Virus	Aerosol	High fever, delirium, severe joint pain, bleeding, and convulsions followed by death
Staphylococcal Enterotoxin B	–	Food poisoning	Toxin	Contamination of food and water	Sudden chills, fever, headache, cough, nausea, vomiting, and diarrhea
Ricin communis	–	Ricin toxin poisoning	Toxin	Aerosol	Weakness, fever, cough, pulmonary edema, severe respiratory distress, enlarged lymph nodes with organ involvement and death
Clostridium botulinum	–	Botulism	Toxin	Aerosol	Nausea, diarrhea, weakness, dizziness, dry mouth and throat, blurred vision, and respiratory paralysis
Mycotoxin	2–8 (Fungal spores)	Mycotoxicoses	Fungi	Aerosol	Pain, redness and vesicles, sloughing of epidermis, throat pain, sneezing, coughing, chest pain, hemoptysis

supplies. However, this is an inadequate method for weaponization of BWA, as methods used for water purification destroy a broad spectrum of BWAs (Klietmann and Ruoff, 2001; Crudy, 2001; Norris and Fowler, 1997; Linsel, 2001).

(iii) *Explosives (artillery, missiles, and detonated bombs)*: This is also an inefficient delivery system for BWAs. BWAs loses its virulence due to heat generated by the explosion (Klietmann and Ruoff, 2001; Crudy, 2001; Norris and Fowler, 1997; Linsel, 2001).

(iv) *Injection/absorption*: This causes the infection through percutaneous inoculation. This delivery method is also inadequate as most of the BWAs will be blocked by intact skin and the wounded skin allows the penetration of BWAs by absorption to the body (Klietmann and Ruoff, 2001; Crudy, 2001; Norris and Fowler, 1997; Linsel, 2001).

Why there is need of protection?

The protection against BWAs has become a worldwide security concern in light of the many recent international threats utilizing BWAs. In a BWA scenario, early protection will be highly beneficial to combat terrorism. The main aim of using a proper protection ensemble is not only to re-establish the ability of personnel to work in normal ways, but also to play a crucial role for the timely and effective treatment of casualties in the BWA scenario. In general, protection can be technically classified in to three categories, namely individual protection, collective protection, and medical protection. The individual protection deals with the protection of individuals by using over garments, mask, hood, overboots, protective gloves, etc. In battlefield conditions, the breathable air is very important for a combatant in order to combat terrorism and is one of the most challenging requirements for security monitoring (Prasad et al., 2008; Sidell et al., 1997a; Noll, 2005; Marzi, 1989; Katz, 1989; Boopathi et al., 2008; Singh et al., 2016). Regardless of the type, concentration, or method of attack, the best immediate protection against BWAs is the respiratory and body protective ensembles (Hinds, 1999). A number of variables are to be considered for the development of protective equipment which include weight, comfort, level of protection, and duration of protection required. It is the role of authorities to make judgments and to decide how to handle counterterrorism by choosing suitable protective measures for the combatants. The aim of personal protective gear is also to prevent the transfer of BWA from victims or the surrounding environment to rescue or medical workers. Different types of personal protective equipment (PPE) may be used, depending on the hazard present (Prasad et al., 2008; Sidell et al., 1997a; Noll, 2005; Marzi, 1989; Katz, 1989; Boopathi et al., 2008; Singh et al., 2016). The rapid pace of development in intelligent textiles and their compatibility with advanced microfabrication technology in conjunction with advanced materials in the field of protective equipment have culminated in the emergence of a new generation of protective ensembles.

Principles of body and respiratory protection

The first and foremost objective of body and respiratory (physical) protection is to limit the exposure of individuals or groups of individuals to BWAs, thus preventing or reducing the number of causalities by supplying proper medical aid in a protected zone. Whether the protection is individual or collective, the same twin aspects of physical protection holds good for both forms of protection: establishment of artificial barrier and supply of breathable air

Establishment of artificial barrier

In order to escape from any BWAs eventuality, the personnel should be separated from the contaminated zone by means of an artificial barrier to avoid exposure to BWAs. The ideal barrier has to provide protection against all the possible sources of BWAs, either in the form of liquid, aerosol, or gases, and should provide durable protection to individuals against the BWAs. Trustworthiness of physical protection largely depends on the effectiveness of the barrier, which depends mainly on its intrinsic properties, such as thickness, composition, temperature resistance, materials, etc. (Prasad et al., 2008; Boopathi et al., 2008; Singh et al., 2016; Hinds, 1999). In the case of IPE, the barrier normally employed is a rubber face piece, gloves, carbon coated fabric/polyurethane foam/activated charcoal cloth/spherical carbon coated fabric, etc., which resists the penetration of BWAs (Prasad et al., 2008; Boopathi et al., 2008; Singh et al., 2016; Hinds, 1999; Lakoff, 2008; Chemical and Biological Terrorism: Research and Development to Improve Civilian Medical Response, 1999; Smisek and Cerny, 1970; Gardner, 1998; Jansen et al., 2014; Sen, 2007). While in the case of collective protection, the barrier is usually a concrete wall, metal panels, polymer panels etc., which will offer the requisite protection provided that it is airtight (Prasad et al., 2008; Boopathi et al., 2008; Singh et al., 2016; Hinds, 1999; Lakoff, 2008; Chemical and Biological Terrorism: Research and Development to Improve Civilian Medical Response, 1999; Smisek and Cerny, 1970; Gardner, 1998; Jansen et al., 2014; Sen, 2007).

Supply of breathable air

Supply of breathable air is accomplished through various methods based on individual and collective protection. In the case of individual protection, this need is catered for by direct connection of a canister with a face mask or direct supply of fresh air through air gas cylinder (Prasad et al., 2008; Singh et al., 2016; Gardner, 1998). In the case of collective protection, contaminated air is supplied from an air blower through a filter comprising of HEPA/ULPA to make it breathable.

Individual protection

Individual protection involves an integrated approach, which requires all types of methods and equipment to protect the individual against any possible exposure of BWAs. Individual protection is based on twin aspects or in combination, namely respiratory protection and skin protection. Individual protection can be classified into two parts (Boopathi et al., 2008; Singh et al., 2016; Hinds, 1999; Lakoff, 2008; Chemical and Biological Terrorism: Research and Development to Improve Civilian Medical Response, 1999; Gardner, 1998): respiratory protection and body protection.

Respiratory protection

In a BWA scenario, respiratory protection is more important when compared to skin protection, as the most probable potential route of exposure is inhalation of aerosols. Respiratory protection is classified in two parts, based on supply of breathable air: air-purifying devices (gas masks), and air-supplying devices (self-contained breathing apparatus: SCBA).

Air-purifying devices

Gas masks, consisting of a face piece integrated with a respiratory cartridge (canister) come under this category (Prasad et al., 2008; Noll, 2005; Gardner, 1998). This device removes gases, vapors, and/or aerosols from the inhaled air and thus protects the respiratory tract. Gas masks are commercially available in variable sizes as they need to cater for different face sizes. It is important to understand that the mask only filters the inhaled air by physically retaining the contaminants; however, it cannot supplement oxygen deficiency. The canister attached to a respiratory mask contains two main components, i.e., high efficiency particulate aerosol (HEPA) or ultra-low penetration aerosol (ULPA), and a gas filter (Prasad et al., 2008; Noll, 2005; Gardner, 1998).

HEPA and ULPA: The canister contains a HEPA filter media for aerosol retention and is composed of a layer of glass fibers of 0.1–10μm diameter. The air space between the fibers is larger than 0.3μm (Hinds, 1999; Chemical and Biological Terrorism: Research and Development to Improve Civilian Medical Response, 1999; Gardner, 1998). HEPA and ULPA are very effective against BWA aerosols. The aerosol particles from the contaminated air are captured over the surface of the filter medium by Van der Waal's forces. HEPA can remove 0.3μm particle size with an efficiency of 99.7%, which gives protection against bacterial BWAs. Moreover, ULPA filters have an efficiency of 99.999% for particle size 0.12μm or larger, which is closer to virus size (Crudy, 2001; Norris and Fowler, 1997; Prasad et al., 2008; Gardner, 1998). These airborne particles are trapped by HEPA or ULPA through a combination of the following mechanisms:

(i) *Interception*: Particles follow air streamline, and come into contact with the fibers for capturing.
(ii) *Sedimentation*: Particles settle down due to their motion under gravity when they come into contact with the fiber.
(iii) *Impaction*: Larger particles do not follow air streamline because of inertia and impact on a fiber for capturing.
(iv) *Diffusion*: This mechanism operates for particles having size less than 0.1μm diameter. Particles diffuse through the fiber surface by Brownian movement, where they are captured.

In general, the efficiency of the aerosol retention depends on the fiber diameter of the medium, particle size of the aerosols, and rate of airflow through the filter.

Gas filter

A gas filter is a bed of impregnated carbon that removes the toxic gases from contaminated air by adsorption (Hinds, 1999; Lakoff, 2008; Chemical and Biological Terrorism: Research and Development to Improve Civilian Medical Response, 1999; Smisek and Cerny, 1970). Generally, high surface area materials are used as adsorbent material for the adsorption of various gases.

Air-supplying devices

SCBA is used to supply the combatant with fresh air from a cylinder (Eitzen and Takafuji, 1997; Gardner, 1998; Jansen et al., 2014; Sen, 2007). These systems are independent from ambient contaminated air and provide high protection and low breathing resistance when compared to canister. Oxygen cylinder SCBAs are more useful than fixed systems, but they provide air for a limited time only.

Body protection

Body protection is also very important and plays a crucial role in BWAs defense (Prasad et al., 2008; Sidell et al., 1997a; Noll, 2005; Marzi, 1989; Katz, 1989; Boopathi et al., 2008; Gardner, 1998; Sen, 2007; Truong and Wilusz, 2005). The protective suit is also called the "second skin of combatant," which protects from BWAs on the battlefield. The well-known nuclear, biological, and chemical (NBC) suit is primarily designed for protection against chemical warfare agents but also offers some level of protection against BWAs (Prasad et al., 2008; Sidell et al., 1997a; Noll, 2005; Marzi, 1989; Katz, 1989; Boopathi et al., 2008; Gardner, 1998; Sen, 2007; Truong and Wilusz, 2005). Bio-suits are primarily designed to protect against the BWA threat and they don't offer any chemical protection. However, the main advantage of a bio-suit is that it is lighter in weight as compared to an NBC suit. The average weight of bio-suit is approx. 500g, compared to the NBC suit of 3kg. In a BWA scenario, respiratory protection is more important than body protection, as most BWA (some BWA penetrate through mucosal membrane) do not penetrate unbroken skin (Gardner, 1998; Jansen et al., 2014). However, at the same time it is also important to always have body protection in a BWA scenario to eliminate exposure to BWAs. Body protection consists of the following, along with a respiratory mask:

(i) NBC suit
(ii) Integrated hood mask
(iii) Facelet mask
(iv) Impermeable overboots
(v) Impermeable gloves

NBC suit

NBC suits are classified in two categories: impermeable and permeable.

*Impermeable **NBC** suit:* This suit does not permit the passage of air and bacterial aerosol, and has a high resistance to penetration by BWA (Marzi, 1989; Katz, 1989; Boopathi et al., 2008; Gardner, 1998; Sen, 2007; Truong and Wilusz, 2005; Karkalić and Popović, 2004; Rivin and Kendrick, 1997). Since it is impermeable, it causes heat exhaustion within a short period of time. In the case of impermeable suits, the barrier generally used is butyl and halobutyl rubber, Viton, and polymers like polyterafluoroethylene, polyvinyl chloride, and neoprene (Marzi, 1989; Katz, 1989; Boopathi et al., 2008; Gardner, 1998; Sen, 2007; Truong and Wilusz, 2005; Karkalić and Popović, 2004; Rivin and Kendrick, 1997). In order to reduce heat stress in this suit, provision of microclimate cooling and SCBA can be given (Marzi, 1989; Boopathi et al., 2008; Gardner, 1998; Jansen et al., 2014; Sen, 2007; Truong and Wilusz, 2005; Karkalić and Popović, 2004; Rivin and Kendrick, 1997).

*Permeable **NBC** suit:* Compared to an impermeable suit, a permeable suit is more comfortable owing to air permeability and permits the removal of perspiration transported through filter material, resulting in release of heat stress. Permeable NBC suits are mostly based on three layers. The outer layer is oil and water repellent, fire retardant, and also has antistatic properties. The middle layer is filter fabric consisting of various adsorbents such as granular activated carbon (Marzi, 1989; Katz, 1989; Boopathi et al., 2008; Gardner, 1998; Jansen et al., 2014; Sen, 2007; Truong and Wilusz, 2005; Karkalić and Popović, 2004; Rivin and Kendrick, 1997; Gurudatt et al., 1997), activated carbon sphere (Marzi, 1989; Gardner, 1998; Jansen et al., 2014; Sen, 2007; Truong and Wilusz, 2005; Karkalić and Popović, 2004), activated carbon fabric (Rivin and Kendrick, 1997; Gurudatt et al., 1997; Lee et al., 2014; Sun et al., 2007; Zeng and Pan, 2008; Lin and Zhao, 2016; Frank et al., 2012), and perm selective membrane (Figueiredo et al., 2011; Viriyanbanthorn et al., 2006; Kissa, 1981; Karkalic, 2006; Ellingsen and Karlsen, 1983). The third layer or inner layer is placed next to the skin in order to provide comfort and at the same time to prevent the contact between the filter layer and the body (Marzi, 1989; Gardner, 1998; Jansen et al., 2014; Sen, 2007; Truong and Wilusz, 2005; Karkalić and Popović, 2004; Rivin and Kendrick, 1997; Gurudatt et al., 1997; Lee et al., 2014; Sun et al., 2007; Zeng and Pan, 2008; Lin and Zhao, 2016; Frank et al., 2012).

Recent developments in IPE: Currently, IPE are heavyweight, impermeable or permeable adsorptive protective garments, which cannot meet the critical demand of simultaneous high comfort and protection, and which provide a passive rather than active response to an environmental threat. Various research and development activities have been accomplished in the direction of multifunctional, intelligent, and advanced materials of a higher breathability as a filter fabric, in order to provide heat stress relief and also to add functionalities in the protective clothing. A high water vapor transport rate is important for reducing the heat stress potential of biological protective clothing.

Bui et al. (2016) made a flexible polymeric membrane based on two components: a highly breathable CNT-membrane that provides an effective barrier against biological threats; and a thin responsive functional layer coated on the membrane surface, which opens and closes depending on the requirements. Due to extra small size of CNT pores, it facilitates effective protection against BWA (Bui et al., 2016). Grandcolas et al. (Grandcolas et al., 2011) prepared a self-decontaminable textile based on WO_3 -modified titanate nanotubes for the photocatalytic destruction of weapons of mass destruction (WMD). Recent reports indicate the use of bioactive fibers in protective textiles for protecting combatants against risks caused by pathogenic bioaerosols, which will be helpful to reduce or eliminate epidemic diseases and infections (Abdou et al., 2008; Gupta et al., 2008; Kenawy et al., 2002; Shin et al., 1999; Tan et al., 2000). An antimicrobial textile coated with silver and metal oxide nanoparticle was fabricated for defense against BWA with antibacterial or self-decontamination ability (Gugliuzza and Drioli, 2013; Brewer, 2011a).

Fabrication of smart breathable material, which is triggered by environmental threat, will be benefited to combatants in terms of comfort and light weight (Schreuder-Gibson et al., 2003). In the past few decades, perm selective membrane of cellulose acetate, PTFE, polyvinyl alcohol, and polyallyl amine have been used in protective textiles, aiming to reduce the physiological burden and provide better protection against BWAs (Brewer, 2011b; Wu, 1994). Bio-inspired membranes are good heat exchangers, water condensers, and also barriers against BWAs, with self-cleaning capability, and some reports also successfully demonstrated that the interface of nanotechnology and material science can improve the combatant life in BWAs war scenario (Qui et al., 2004; Nicoletta et al., 2012; Kazuhiro et al., 2012; Bohringer et al., 2004). The electrospun nanoweb membrane has waterproof and water vapor permeability, which makes it a potential barrier against BWAs (Gibson et al., 1988). Schreuder-Gibson et al. (Schreuder-Gibson et al., 2002) have fabricated stretchable electrospun membrane of thermoplastic polyurethane elastomers (TPU), and these polymers have lower air permeability, higher water vapor diffusion, and aerosol filtration efficiency for BWAs protection.

Integrated hood mask
An integrated hood mask (IHM) provides protection against BWAs, and is specifically intended for soldiers with facial injuries (Prasad et al., 2008; Marzi, 1989; Katz, 1989; Gardner, 1998). It consists of a respiratory mask, canister, and hood. The hood is made of layered fabric. The front panel is provided with smugly fitting cutouts for anchoring into the mask.

Facelet mask
This is semi-protective mask, which is more comfortable compared to a normal respiratory mask, has very low breathing resistance, and a high level of speech transmission (Prasad et al., 2008; Marzi, 1989; Katz, 1989; Gardner, 1998). Hence, it is appropriate when full individual protection is not deemed necessary. The facelet comprises of an activated charcoal cloth filter for filtering low level concentrations of BWAs.

Impermeable overboots

The impermeable overboots are used for protection of the feet and generally worn over the top of any shoes. They are generally made of chemical resistant material such as neoprene, PVC, or butyl rubber. The overboot plays a critical role in BWA warfare scenarios (Norris and Fowler, 1997; Linsel, 2001; Katz, 1989; Gardner, 1998).

Impermeable gloves

Impermeable gloves are generally worn over cotton gloves and are made of butyl rubber (Norris and Fowler, 1997; Linsel, 2001; Katz, 1989; Gardner, 1998; Jansen et al., 2014; Sen, 2007). They are available in three sizes and the palm of the glove is serrated for better grip.

Testing and classification of commercially available BWA suits

Thermoplastic elastomers (TPE) polymers have ample potential in fabrication of suit right from conventional waterproof breathable fabric to barrier fabrics. Continuous or monolithic TPE films have a properties blend of permeability, abrasion resistant, strength and bio barrier. Most of the breathable BWA suits are fabricated from monolithic film (Schledjewski et al., 1997; Zapletalova et al., 2006). Protective clothing performance requirement against infective agents is defined in European standard EN 14216. The test standard deals with the protective potential of IPE against microorganisms in liquid, aerosol, or solid dust particles. Table 2 shows the types of Bio-suit along with the relevant standard (BS-EN-14126-2003, 2004; Ryan, 2016).

Collective protection

Collective protection involves the methods and equipment that is used to protect a group of individuals, food, water, and sensitive equipment from exposure to biological contamination (Eitzen and Takafuji, 1997; Crudy, 2001; Norris and Fowler, 1997; Linsel, 2001; Prasad et al., 2008; Gardner, 1998). The goal of collective protection is to provide personnel with an uncontaminated environment, which

Table 2 Types of protective clothing against infective agents

Type of protective suit	Description	Relevant standard
Type 1-A, 1-B, 1-C	Gas tight	EN 943-1
Type 2-B	Non gas tight	EN 943
Type 3-B	Protection against pressurized liquid chemicals	EN 466
Type 4-B	Protection against liquid aerosols	EN 465
Type 5-B	Protection against airborne solid particles	EN ISO 13982
Type 6-B	Protection against liquid chemicals	EN13034

allows the personnel to carry out tactical functions. The advantage of this is that individuals inside the protected area are not required to wear any IPE. Specifically, collective protection allows combatants to operate safely for extended periods at near-normal levels of effectiveness while under BWA threat. It also plays a vital role in medical application for treatment of casualties even in a contaminated environment. Indeed, this collective protection plays a very crucial role in real warfare scenarios, in order to provide the individual relief from the burden of individual protection equipment, thus allowing effective military and non-military operations even in contaminated environments (Eitzen and Takafuji, 1997; Crudy, 2001; Norris and Fowler, 1997; Linsel, 2001; Prasad et al., 2008; Gardner, 1998). There are two general categories of collective protection: standalone shelters and integrated systems.

Standalone shelters

This shelter is a separate building that is designed and constructed to withstand the range of BWA by providing barriers such as concrete walls, and is equipped with a filtration system for purified air (Eitzen and Takafuji, 1997; Crudy, 2001; Norris and Fowler, 1997; Linsel, 2001; Prasad et al., 2008; Gardner, 1998). In general, standalone shelters are built away from the potential debris hazard. The air supplied to shelters is purified by a combined system of HEPA/ULPA and activated carbon (Eitzen and Takafuji, 1997; Crudy, 2001; Norris and Fowler, 1997; Linsel, 2001; Prasad et al., 2008; Gardner, 1998). The air is drawn through an aerosol filter and a layer of active carbon by means of a blower.

Integrated system

This system is designed to provide protection against BWA through the use of filtered air under positive pressure to a variety of aircraft facilities, ships, vans, and vehicles. An integrated shelter is a specially designed and constructed room or area within or attached to a larger building (Eitzen and Takafuji, 1997; Crudy, 2001; Norris and Fowler, 1997; Linsel, 2001; Prasad et al., 2008; Gardner, 1998). Shelter types, and some commercially available collective protection shelters (Crudy, 2001; Norris and Fowler, 1997; Linsel, 2001; Prasad et al., 2008; Gardner, 1998), are given below.

Underground shelters

An underground shelter is an enclosed space specially designed to overpressure protected spaces. Examples include: French AMF 80, Barton APD shelters, American ASR-100-AV-NBC Safe Cell, American NBC air filtration system ASR-100-AV-NBC-COMP, etc.

Inflatable shelters

These shelters are portable and equipped with sealed NBC liner with easy deployment. The area inside the tent is over-pressured with filtered air. The entrance is fitted with an over-pressured air lock. Examples include: M28, Chemical biological protective shelter system, joint transportable collective protection system, collective protection system amphibious backfit, chemical/biological protective shelters, Temet NBC protection system LSS-88, etc.

Hardened shelters

This shelter is fixed and it is designed to withstand a wide range of threats. Examples include: glass reinforced polymers (GRP) polymer made shelters, metal stud blast walls, steel moment frame and braced frame buildings.

Level of protection

The amount and type of protection required in any BWA incident depends upon the hazard and the duration of exposure anticipated. There are four levels of personal protection, namely A, B, C, and D, for different types of respiratory protection (Gardner, 1998; Ryan, 2016; Chan et al., 2002).

- *Level A*: At this level, the highest level of respiratory and body protection is required due to high concentration of BWA in the atmosphere. This level includes a respiratory, skin, eye, and mucous membrane through BWA resistant suit, boots, and gloves, together with an SCBA.
- *Level B*: In this, full respiratory protection is required along with protection to skin. This equipment consists of a BWA-resistant suit, BWA resistant boots and gloves, and SCBA.
- *Level C*: At this level, protection should be selected when the types of airborne substance are known and concentration is measured. Air-purifying respirators are used rather than SCBA and skin or eye exposures are unlikely.
- *Level D*: This is limited to coveralls or other work clothes, boots, and gloves.

Performance requirement of protective clothing and facemask against BWA

The selection of protective clothing that provides protection against BWA depends on the proposed end use of the protective equipment. Some common end uses include protecting patients as well as medical professionals who are exposed to blood-borne and other pathogens. The majority of biohazard test methods and guidelines are focused on applications for military personal. With the current bio threat, biological protection is important for first responders, workers with the potential of exposure to hazardous materials, and the general public. Test methods for the operational performance of a typical bio protection suit are shown in Table 3 (Scott, 1999; Kendall et al., 2008).

Protection through vaccine and antibiotics

Although available protective equipment and devices can be provided to the combatant, still there is an important need for the development of vaccines to neutralize and eliminate the effects of a potential bio-terrorist attack. Research activities in the field of microbiology aiming to neutralize BWA have resulted in the development of new generation vaccines to combat and control the outbreak of disease (Chemical and Biological Terrorism: Research and Development to Improve Civilian Medical Response, 1999; Zajtchtchuk and Bellamy, 1997; Pomeratnsev et al., 1997). Continued research and development on new technology will be able to provide countermeasures to the potentially more vaccine resistant new bio-threat with the aid of genetic engineering techniques. Table 4 shows a list of some developed vaccines and antibiotics against BWA.

Table 3 Bio suit test methods and standards

S. no.	Test method	Standard
1	ASTM F1670-03	Used for protective clothing to test the penetration of synthetic blood
2	ASTM F1671-03	Used for protective clothing to test the penetration of blood born pathogen using bacteriophage
3	ASTM F1819-04	Used for protective clothing to test the penetration of synthetic blood using a mechanical pressure technique
4	ASTM F1862-00a	Used for face mask to test the penetration of synthetic blood
5.	ISO 16604	Clothing for protection against contact with blood and body fluids. Determination of resistance of protective clothing materials to penetration by blood-borne pathogens. Test method using Phi-X174 bacteriophage
6.	BS-EN-14126-2003	Protective clothing. Performance requirements and tests methods for protective clothing against infective agents.
7.	ISO 16603	Synthetic blood penetration test
8.	ISO 22610	Wet bacterial penetration test
9.	ISO 22612	Dry microbial penetration test

ASTM, American Society for Testing and Materials; ISO, International Organization for Standardization'; BS-EN, British Standard European Norm.

Table 4 Vaccine and antibiotics against BWA

Agent	Availability of vaccine/antibiotics	References
Anthrax	Available Used in some counties, FDA licensed Vaccine and antibiotics both available	Shlyakhov and Rubinstein (1994), Turnbull (1991), Centers for Disease Control and Prevention (1999)
Plague	Available, FDA licensed	Meyer et al. (1974), Centers for Disease Control and Prevention (1996)
Tularemia	Available, investigational new drug	Sandstrom (1994)
Q fever	Available, investigational new drug, Australia licensed	Marmion et al. (1984)
Smallpox	Available, FDA licensed Vaccine and antibiotics both available	Centers for Disease Control (1991)
Glanders	Not available	Gilligan (2002)
Botulism	Not available, investigational new drug, Antitoxin can arrest the process for some time	Anderson et al. (1981)

Materials used for the fabrication of IPEs

Different materials have been used for the fabrication of IPE, based on the concentration of agents and level of protection needs. Table 5 shows a list of elastomers or materials used for the fabrication of IPEs.

State-of-the-art products available for protection

Personal protective equipment are required in BWA scenarios to overcome the toxic effects of BWA. Recently, efforts have been directed towards the development of intelligent protective gear, which not only protect but also can sense, react, and adapt themselves to the existing environment conditions. The details of some existing and commercial protective equipment/products are shown in Table 6.

Outlook, prospects, and challenges

Vulnerabilities to potential terrorist use of BWA are expanding vastly, owing to openly available new research and development activities for novel production and delivery methodologies to terrorists as a result of technology proliferation. Hence, it is of the utmost importance that effective prevention and protection strategies are to be developed, based on multidisciplinary involvement. Significant progress has been made in recent years in the field of protection equipment. Such protective equipment and devices are certainly protecting the combatant around the globe. However, the use of even the latest Bio suit, respirator, and other protective gear is greatly restricted by heat stress, BWA concentration, exposure time, and relatively low protection factors. Current protective equipment still struggles to find a balance between comfort and protection, and tends to provide a passive rather than active response to BWA threat. Use of most of the protective equipment is still at risk due to the weight, size, and heat stress of the protective suits. Consequently, the field of protective equipment has many paramount challenges to address and technological gaps.

In spite of the extensive progress that has been made in the field of protective equipment, the integration of smart functionalities like detection, self-detoxification, and ability to communicate to the environment in a biothreat scenario remains a challenge. A healthy collaboration between researchers in the material science, flexible electronics, and intelligent textile domains is imperative in order to sustain the momentum that has been burgeoning in this field. With continued innovation and detailed attention to core challenges of protective equipment, on body wearable detectors, and smart textile material, it is expected that protective equipment/devices will play a pivotal role in the emergent new generation protective ensembles. One of the important aspects of protection from the combatant side is that the IPEs should be able to be used immediately as and when threat is detected, protecting combatants and enabling them to take quick countermeasures against any BWA threat. Indeed simultaneous

Table 5 Materials used for the fabrication of IPEs

S. no.	IPE name	Elastomers/materials	References
1	Face piece, facelet, escape mask, NBC overboots, NBC gloves	Natural rubber, nitrile rubber, butyl rubber, halogenated butyl rubber, polyurethane	Sidell et al. (1997b), Organization for the Prohibition of Chemical Weapons (n.d.)
2	Visor of face piece	Polycarbonate, polyurethane, laminated safety glass	Gardner (1998), Organization for the Prohibition of Chemical Weapons (n.d.)
3	Facelet	Outer layer: polyester/cotton Middle layer: polyurethane foam/non-woven polyester Inner layer: polyester	Gardner (1998), Chemical Defence (1988)
4	NBC suit: impermeable	Barrier or filter fabric: butyl and halobutyl rubber, viton, polytetrafluoroethyene, polyvinyl chloride, neoprene, thermally bonded fine HDPE, polyvinylidene chloride, polyamide or polyester with polyethylene layer	Marzi (1989), Katz (1989), Boopathi et al. (2008), Gardner (1998), Sen (2007), Truong and Wilusz (2005), Karkalić and Popović (2004), Rivin and Kendrick (1997)
5	NBC suit: permeable	Outer layer: plainer ripstop meta aramid, ripstop cotton, core spun polyester cotton, cotton polyester blend, polyester cotton laminated to mineral fabric, nylon, etc. Outer layer is also given a treatment of fluorocarbon or silicone to make it water and oil repellent and flame retardant, Meta aramid, para aramid, self-detoxifying outer layer with metal oxide and nanoparticle (TiO$_2$, Ag, MgO etc) equipped with wearable sensor. Middle layer or filter fabric: granular activated powder, granular activated carbon, activated carbon sphere, perm selective membrane (cellulose acetate, polyurethane, PTFE, polyvinyl alcohol, pollyallyl amine, nafion membrane etc.). Inner layer: Knitted or woven fabric, cellulose	Marzi (1989), Gardner (1998), Jansen et al. (2014), Sen (2007), Truong and Wilusz (2005), Karkalić and Popović (2004), Rivin and Kendrick (1997), Gurudatt et al. (1997), Lee et al. (2014), Sun et al. (2007), Zeng and Pan (2008), Lin and Zhao (2016), Frank et al. (2012), Figueiredo et al. (2011), Viriyanbanthorn et al. (2006), Kissa (1981), Karkalic (2006), Ellingsen and Karlsen (1983), Bui et al. (2016), Grandcolas et al. (2011), Abdou et al. (2008), Gupta et al. (2008), Kenawy et al. (2002), Shin et al. (1999), Tan et al. (2000), Gugliuzza and Drioli (2013), Brewer (2011a,b), Schreuder-Gibson et al. (2003), Wu (1994), Qui et al. (2004), Nicoletta et al. (2012), Kazuhiro et al. (2012), Bohringer et al. (2004), Gibson et al. (1988)
6	NBC Gloves	Butyl rubber, halogenated butyl rubber, neoprene	Linsel (2001), Katz (1989), Gardner (1998)
7	NBC boots	Neoprene molded sole, on which, butyl coated nylon fabric, Butyl rubber, halogenated butyl rubber, neoprene	Norris and Fowler (1997), Linsel (2001), Katz (1989), Gardner (1998), Jansen et al. (2014), Sen (2007)

Table 6 State-of-the-art products available for protection

S. no.	IPE name	IPE commercial name	Country/company developed
1	NBC suit/Bio suit	M-82 permeable protective suit, Saratoga suit, Swift Responder 3 (SR3) CBRN Protection Suit, Mk 1V Military CBRN Protection Suit, Phoenix Lightweight CBRN Over-Garment, Cougar and Panther - Military CBRN Protective Garment Systems, Kärcher Safeguard 3002-A1, Kärcher Safeguard 2002-HP, Kärcher Safeguard 3002-A1, Kärcher Safeguard 1001 Lightweight Decontamination Suit, Kärcher Heavy Safeguard 6002, PROTEC Max, M8 suit, TYVEK and TYCHEM bio suit, Microgard suit, EUROLITE NBC – Protection Suit – R110	China, Blucher GmbH (Germany), Remploy's, (UK), Kärcher Futuretech GmbH (Germany), Paul boye (France), British, UK, Dupont (United States), MICROGARD Limited (UK), Eurolite NBC protection, Ahlstrom (Finland)
2	NBC respirator/Bio respirator	Sfera (face protection), Selecta, 3M 6000 Series Full Face Masks, 3M 7000 Series Full Face Masks, Pro2000 Combination Filters, S10 respirator	D.P·I (Italy), Avon (USA), Scott Safety (UK)
3	NBC over boot	CBRN Overboot, Protective overboot, Acton NBC Chemical Protective Overboot	AirBoss Defence (USA), Eurolite NBC protection (Austria), Altama Botach, Avon (USA), Kärcher Futuretech GmbH (Germany)
4	NBC gloves	CBRN gloves, Protective gloves, Acton NBC Protective gloves	AirBoss Defense (USA), Eurolite NBC protection (Austria), Supergum Ltd., (Israel), Altama Botach, Avon (USA), Kärcher Futuretech GmbH (Germany)

work needs to be done on material, textile, wearable detection, and decontamination aspects; work is also vital on the seamless and flawless integration of these multifunctional properties. This is possible when engineers and scientists collaborate, along with intimate coordination and collaboration with combatants or users.

In future, one should think of smart and intelligent protective gear that not only protects, but also detects and displays the concentration of threat, along with biodegradable and self-decontamination capability, which can degrade or neutralize the biothreat. It is clear from the above that the protective gear/devices field offers exciting collaborative opportunities to achieve multifunctional properties, such as high air permeability, low heat stress, cooling provision, self-destroying BWA capability, and intelligent response to external stimulus present in the environment in the protective gears, so that the combatant gets success in terms of protection.

References

Abdou, E.S., Elkholy, S.S., Elsabee, M.Z., Mohamed, E., 2008. Improved antimicrobial activity of polypropylene and cotton nonwoven fabrics by surface treatment and modification with chitosan. J. Appl. Polym. Sci. 108, 2290–2296.

Anderson, J.H., Lewis, G.E., Lewis, G.E., 1981. Clinical evaluation of botulism toxoids. In: Biomedical Aspects of Botulism. New York Academic Press, pp. 233–246.

Atlas, R.M., 2002. Bioterrorism: From threat to reality. Annu. Rev. Microbiol. 56, 167–185.

Bohringer, B.R., Vande Ven, H.J.M., Spijkers, J.C.W., 2004. Nonporous, breathable membrane containing polyamide. US6706413.

Boopathi, M., Singh, B., Vijayaraghavan, R., 2008. A review on NBC body protective clothing. Open Text. J. 1, 1–8.

Brewer, S.A., 2011a. Recent advances in breathable barrier membranes for individual protective equipment. Recent Pat. Mater. Sci. 4, 1–14.

Brewer, S.A., 2011b. Recent advances in breathable membranes for individual protective equipment. Recent Pat. Mater. Sci. 4 (1), 11.

BS-EN-14126-2003, 2004. Protective clothing performance requirements and test methods for protective clothing against infective agents.

Bui, N., Meshot, E.R., Kim, S., Peña, J., Gibson, P.W., Jen Wu, K.J., Fornasiero, F., 2016. Ultrabreathable and protective membranes with sub-5 nm carbon nanotube pores. Adv. Mater. 28, 5871–5877.

Centers for Disease Control, 1991. Vaccine (smallpox) vaccine: recommendations of the immunization practices advisory committee (ACIP). MMWR Morb. Mortal. Wkly Rep. 40, 1–10.

Centers for Disease Control and Prevention, 1996. Prevention of plague: recommendations of the advisory committee on immunization practices. MMWR Morb. Mortal. Wkly Rep. 45, 1–15.

Centers for Disease Control and Prevention, 1999. Bioterrorism alleging use of anthrax and interim guidelines for management: United States, 1998. MMWR Morb. Mortal. Wkly Rep. 48, 69–74.

Chan, J.T., Yeung, R.S., Tang, S.Y., 2002. Hospital preparedness for chemical and biological incidents in Hong Kong. Hong Kong Med. J. 8, 440–446.

Chemical Defence, 1988. Chemistry in Britain. 24, 657–688.

Christopher, G.W., Cieslak, T.J., Pavlin, J.A., Eitzen, E.M., 1997. Biological warfare: a historical perspective. J. Am. Med. Assoc. 14, 364–381.

Christopher, G.W., Cieslak, T.J., Pavlin, J.A., 1999. EM Eitzen biological warfare: a historical perspective. In: Lederberg, J. (Ed.), Biological Weapons. Limiting the Threat. The MIT Press, Cambridge, MA, pp. 17–35.

Committee on R&D Needs for Improving Civilian Medical Response to Chemical and Biological Terrorism Incidents, Institute of Medicine, National Academy of Sciences, Chemical and Biological Terrorism, 1999. Research and Development to Improve Civilian Medical Response. National Academy Press, Washington, DC.

Crudy, E., 2001. Chemical and Biological Warfare. Copernicus Books, New York, 14–18.

Davis, C.J., 1999. Nuclear blindness: an overview of the biological of the biological weapons programs of the former Soviet Union and Iraq. Emerg. Infect. Dis. 5, 509–512.

Eitzen, E.M., Takafuji, E.T., 1997. Historical overview of biological warfare. In: Sidell, F.R., Takafuji, E.T., Franz, D.R. (Eds.), Medical Aspects of Chemical and Biological Warfare. Office of the Surgeon General, Borden Institute, Walter Reed Army Medical Center, Washington, DC, pp. 415–423.

Ellingsen, F., Karlsen, J., 1983. Transport mechanisms through a porous membrane and the subsequent effect on the protection when incorporated into multilayer clothing. In: Proc. Int. Symp. Protection Against Chemical Warfare Agents, Stockholm, Sweden.

Figueiredo, J.L., Mahata, N., Pereira, M.F.R., Montero, M.S., Montero, J., Salvador, F., 2011. Adsorption of phenol on supercritically activated carbon fibers: effect of texture and surface chemistry. J. Colloid Interface Sci. 357 (1), 210–214.

Frank, E., Hermanutz, F., Buchmeiser, M.R., 2012. Carbon fibers: precursors, manufacturing, and properties. Macromol. Mater. Eng. 297, 493–501.

Gardner, T.J., 1998. Jane's Protective Equipment. Jane's Information Group, UK, 21–221.

Gibson, P., Schreuder-Gibson, H., Pentheny, C., 1988. Electrospinning technology: direct application of tailorable ultrathin membrane. J. Coated Fabrics 28, 63–72.

Gilligan, P.H., 2002. Therapeutic challenges posed by bacterial bioterrorism threats. Curr. Opin. Microbiol. 5, 489–495.

Grandcolas, M., Sinault, L., Mosset, F., Louvet, A., Keller, N., Keller, V., 2011. Self-decontaminating layer-by-layer functionalized textiles based on WO_3-modified titanate nanotubes. Application to the solar photocatalytic removal of chemical warfare agents. Appl. Catal. A Gen. 391, 455–467.

Gugliuzza, A., Drioli, E., 2013. Review on membrane engineering for innovation in wearable fabrics and protective textiles. J. Membr. Sci. 446, 350–375.

Gupta, B., Jain, R., Singh, H., 2008. Preparation of antimicrobial sutures by pre irradiation grafting onto polypropylene monofilament. Polym. Adv. Technol. 19 (12), 1698–1703.

Gurudatt, K., Tripathi, V.S., Sen, A.K., 1997. Adsorbent carbon fabrics: new generation armour for toxic chemicals. Def. Sci. J. 47 (2), 239–250.

Hinds, W.C., 1999. Aerosol Technology: Properties, Behaviour, and Measurement of Airborne Particles, second ed. Wiley's, New York.

Chemical and Biological Terrorism: Research and Development to Improve Civilian Medical Response. 1999. National Academy Press, Washington, DC 34–43. https://doi.org/10.17226/6364.

Jansen, H.J., Breeveld, F.J., Stijnis, C., Grobusch, M.P., 2014 Jun. Biological warfare, bioterrorism, and biocrime. Clin. Microbiol. Infect. 20, 488–496.

Karkalic, R., 2006. Optimization of thin layered active charcoal sorption materials embedded into the NBC protective materials in the function of protective characteristics and physiologic compliance. Ph.D. thesis Military Academy, Belgrade, Serbia.

Karkalić, R., Popović, R., 2004. Complex performances of the contemporary textile materials covered with active charcoal. In: The 6th Yugoslav Materials Research Society Conference Yucomat 2004, Herceg Novi, September 13–17. pp. 106–108.

Katz, M.G., 1989. A new approach to heat stress relief in chemical protective clothing. In: Proceedings of the Third International Symposium on Protection Against Chemical Warfare Agents Stockholm, Sweden 11–16 June. Swedish Defence Research Establishment, UMEA, pp. 25–31.

Kaufmann, A.F., Meltzer, M.I., Schmid, G.P., 1997. The economic impact of a bioterrorist attack: Are prevention and post-attack intervention programs justifiable? Emerg. Infect. Dis. 3, 83–94.

Kazuhiro, M., Satoshi, Y., Takashi, T., 2012. Abutted or superimposed members for series flow integral or coated layers. USPC20120067812.

Kenawy, E.R., Abdel-Hay, F.I., El-Shanshoury, A.E.R.R., El-Newehy, M.H., 2002. Biologically active polymers. V. Synthesis and antimicrobial activity of modified poly(glycidyl methacrylate-co-2- hydroxyethyl methacrylate) derivatives with quaternary ammonium and phosphonium salts. J. Polym. Sci. A Polym. Chem. 40, 2384–2393.

Kendall, J.R., Presley, S.M., Austin, G.P., Smith, P.N., 2008. Advances in Biological and Chemical Terrorism Countermeasures. CRC Press1–243.

Kissa, E., 1981. Capillary sorption in fibrous assemblies. J. Colloid Interface Sci. 83, 265–272.

Klietmann, W.F., Ruoff, K.L., 2001. Bioterrorism: implications for the clinical microbiologist. Clin. Microbiol. Rev. 14, 364–381.

Lakoff, A., 2008. The generic biothreat, or, how we became unprepared. Cult. Anthropol. 23, 399–428.

Lederberg, J., 1999. Introduction. In: Lederberg, J. (Ed.), Biological Weapons. Limiting the Threat. The MIT Press, Cambridge, MA, pp. 3–5.

Lee, T., Ooi, C.H., Othman, R., Yeoh, F.Y., 2014. Activated carbon fiber-the hybrid of carbon fiber and activated carbon. Rev. Adv. Mater. Sci. 36 (2), 118–136.

Lin, J., Zhao, G., 2016. Preparation and characterization of high surface area activated carbon fibers from lignin. Polymers 8 (10), 369.

Linsel, G., 2001. "Bioaerosole—Entstehung und biologische Wirkungen" (Biologically Contaminated Aerosols—Origin and Biological Effects). The German Federal Institute for Occupational Safety and Health—BAuA, Berlin.

Marmion, B.P., Ormsbee, R.A., Kyrkou, M., 1984. Vaccine prophylaxis of abattoir-associated Q fever. Lancet 2, 1411–1414.

Marzi, W.B., 1989. Development of a new impermeable NBC protective suit for German civil defence. In: Proceedings of the Third International Symposium on Protection Against Chemical Warfare Agents, Stockholm, Sweden 11–16 June. Swedish Defence Research Establishment, UMEA, pp. 21–24.

Meselson, M., Guillemin, J., Hugh-Jones, M., Langmuir, A., Popova, I., Sherlokov, A., Yampolskaya, O., 1994. The Sverdlovsk anthrax outbreak of 1979. Science 266, 1202–1208.

Meyer, K.F., Cavanaugh, D.C., Bartelloni, P.J., Marshall, J.D., 1974. Plague immunization. I. Past and present trends. J. Infect. Dis. 29, 13–18.

Nicoletta, F.P., Cupelli, D., Formoso, P., DeFilpo, G., Colella, V., Gugliuzza, A., 2012. Light responsive polymer membranes: a review. Membranes 2, 134–197.

Noll, G.H., 2005. Hazardous Materials: Managing the Incident, third ed. Red Hat, Chester, MD.

Norris, J., Fowler, W., 1997. Nuclear Biological and Chemical Warfare on the Modern Battlefield. Brassey's Ltd., UK32–35.

Pomeratnsev, A.P., Startsin, N.A., Mockov, Y.V., Marnin, L.I., 1997. Expression of cereolysin AB genes in *Bacillus anthracis* vaccine strain ensures protection against experimental hemolytic anthrax infection. Vaccine 15, 1846–1850.

Poupard, J.A., Miller, L.A., 1992. History of biological warfare: catapults to capsomeres. Ann. N. Y. Acad. Sci. 666, 9–20.

Prasad, G.K., Singh, B., Vijayraghavan, R., 2008. Respiratory protection against chemical and biological warfare agents. Def. Sci. J. 58 (5), 686–697.

Organization for the Prohibition of Chemical Weapons, Protective Equipment: Information. n.d. www.opcw.org.

Qui, Z.M., Clark, J.C., Fan, W.W., 2004. Water and oil repellency imparting urethane oligomers comprising perfluoroalkyl moieties. US6803109.

Rivin, D., Kendrick, C., 1997. Adsorption properties of vapor-protective fabrics containing activated carbon. Carbon 35, 1295–1305.

Ryan J., Biosecurity and Bioterrorism: Containing and Preventing Biological Threats, Elsevier, 2nd ed., 2016, Cambridge, MA.

Sandstrom, G., 1994. The tularemia vaccine. J. Chem. Technol. Biotechnol. 59, 315–320.

Schledjewski, R., Schultze, D., Imbach, K., 1997. Breathable protective clothing with hydrophilic thermoplastic elastomer membrane films. J. Ind. Text 27, 105–114.

Schreuder-Gibson, H., Gibson, P., Senecal, K., Sennett, M., Walker, J., Walter, Y., Ziegler, P., 2002. Protective textile materials based on electrospun nanofibers. J. Adv. Mater. 34 (3), 44–55.

Schreuder-Gibson, H.L., Truong, Q., Walker, J.E., Owens, J.R., Wander, J.D., Jones, W.E., 2003. Chemical and biological protection and detection in fabrics for protective clothing. MRS Bull. 28, 574–578.

Scott, R.A., 1999. Chemical and biological terrorism: research and development to improve civilian medical response. Institute of Medicine (US) Committee on R&D Needs for Improving Civilian Medical Response to Chemical and Biological Terrorism Incidents In: Textiles for Protection. first ed.. National Academies Press (US), Washington, DC.

Sen, A.K., 2007. Coated Textiles: Principles and Applications, second ed. CRC Press, Boca Raton, FL181–188.

Shah, J., Wilkins, E., 2003. Electrochemical biosensors for detection of biological warfare agents. Electroanalysis 15, 157–167.

Shin, Y., Yoo, D.I., Min, K., 1999. Antimicrobial finishing of polypropylene nonwoven fabric by treatment with chitosan oligomer. J. Appl. Polym. Sci. 74 (12), 2911–2916.

Shlyakhov, E.N., Rubinstein, E., 1994. Human live anthrax vaccine in the former USSR. Vaccine 12, 727–730.

Warfare, weaponary, and the casualty. In: Sidell, F.R., Takafuji, E.T., Franz, D.R. (Eds.), Medical Aspects of Chemical and Biological Warfare. Text Book of Military Medicine Series, Part 11997a. Office of the Surgeon General, Department of Army, Washington, DC, pp. 361–393. (Chapter 16).

Warfare, weaponry, and the casualty. In: Sidell, F.R., Takafuji, E.T., Franz, D.R. (Eds.), Medical Aspects of Chemical and Biological Warfare. Text Book of Military Medicine Series, Part 11997b. Office of the Surgeon General, Dept. of Army, Washington, DC, pp. 361–393. (Chapter 16).

Singh, B., Singh, V.V., Boopathi, M., Shah, D., 2016. Pressure swing adsorption based air filtration/purification systems for NBC collective protection. Def. Life Sci. J. 01, 127–134.

Smisek, M., Cerny, S., 1970. Active carbon, manufacture, properties and applications. Elsevier Publishing Co, New York.

Stockholm International Peace Research Institute (SPIRI), 1971. The Rise of CB Weapons: The Problem of Chemical and Biological Warfare. Humanities Press, New York.

Sun, J., He, C., Zhu, S., Wang, Q., 2007. Effects of oxidation time on the structure and properties of polyacrylonitrile-based activated carbon hollow fiber. J. Appl. Polym. Sci. 106 (1), 470–474.

Tan, S., Li, G., Shen, J., Liu, Y., Zong, M., 2000. Study of modified polypropylene nonwoven cloth. II. Antibacterial activity of modified polypropylene nonwoven cloths. J. Appl. Polym. Sci. 77, 1869–1876.

Truong, Q., Wilusz, E., 2005. In: Scott, R.A. (Ed.), Chemical and Biological Protection in Textiles for Protection. CRC Press, Boca Raton, FL, pp. 562–567.

Turnbull, P.C.B., 1991. Anthrax vaccines: past, present and future. Vaccine 9, 533–539.

Viriyanbanthorn, N., Stacer, R.G., Mead, J.L., Sung, C., Schreuder-Gibson, H., Gibson, P., 2006. Breathable butyl rubber membranes formed by electrospinning. J. Adv. Mater. 38, 40–47.

Wheelis, M., 1991. Biological warfare before 1914. In: van Courtland, M.J.E. (Ed.), Biological and Toxin Weapons: Research, Development, and Use From the Middle Ages to 1945. Vol. 1. Stockholm International Peace Research Institute, Stockholm, Sweden, pp. 8–34.

Wu, H.S., 1994. Gas permeable coated porous membranes. US5286279.

Zajtchtchuk, R., Bellamy, R.F., 1997. Textbook of Military Medicine: Medical Aspects of Chemical and Biological Warfare. Office of the Surgeon General, Department of the Army, Washington, DC.

Zapletalova, T., Michielsen, S., Pourdeyhimi, B., 2006. Polyether based thermoplasticpolyure-thane melt blown nonwovens. J. Eng. Fibers Fab. 1, 62–72.

Zeng, F., Pan, D., 2008. The structural transitions of rayon under the promotion of a phosphate in the preparation of ACF. Cellulose 15 (1), 91–99.

Environmental sampling and bio-decontamination— Recent progress, challenges, and future direction

10

Vipin K. Rastogi[a], Lalena Wallace[b]

[a]U.S. Army Futures Command—Combat Capabilities Development Command, Chemical Biological Center, Edgewood, MD, United States, [b]DTRA, CB Research Center of Excellence Division, APG, Edgewood, MD, United States

Introduction and background

Since the anthrax attacks in 2001 via mailing of spores through USPS, an increased awareness of bioterrorism is evident not just in the United States, but across the globe (Craft et al., 2014; Dias et al., 2010; Erenler et al., 2018; Polyak et al., 2002; Wagar, 2016). There is a heightened awareness with respect to the nation's preparedness and capabilities necessary to minimize the impact of bioterrorism on human health, and to cleanup and recover from such incidents. Such bioterrorism incidents on a large- or wide-area scale have two principal consequences: (a) adverse impact on human and animal health; and (b) contamination of schools, hospitals, roadways, highways, parks, and building interiors and exterior surfaces. The focus of this review article is to summarize the consequence management with respect to recent progress and challenges that remain for environmental sampling and decontamination in the event of infrastructure contamination.

The dissemination of anthrax spores via seven letters sent through the USPS in 2001 resulted in the contamination of several building interiors, the death of five people, and sickness of 17 personnel. The extent of contamination was difficult to ascertain. In order to determine the extent of human exposure, over 100,000 nasal swabs were taken. In addition, 120,000 environmental samples from over 280 postal facility buildings were obtained to determine the extent of contamination (Franco and Bouri, 2010). Environmental samples included both dry and wet swabs or wipes, which placed enormous strain on laboratories processing and analyzing such samples. The cost of *Bacillus anthracis* spore remediation, restoration, and reoccupation varies, but overall cost is estimated to be well over 300 million dollars. A combination of decontamination techniques were selected to achieve this daunting task, and included removal of sensitive or important items to off-sites (spore inactivation achieved via nondestructive methods, i.e., ethylene dioxide, irradiation, or chlorine

Handbook on Biological Warfare Preparedness. https://doi.org/10.1016/B978-0-12-812026-2.00010-4

dioxide gas), use of sporicidal solutions such as bleach or liquid chlorine dioxide, and fumigation of highly contaminated buildings with chlorine dioxide gas and/or vaporous hydrogen peroxide. Additionally, spore removal was also achieved using HEPA (high efficiency particulate air) vacuuming from smaller areas.

A number of excellent reviews have summarized the cleanup efforts following Amerithrax—the Federal Bureau of Investigation (FBI) case name for the 2001 anthrax attacks, which lasted over several weeks after September 18, 2001 (Government Accounting Office, 2003; Schmitt and Zacchia, 2012; Canter et al., 2005). In this review we summarize the progress made in the past decade on three specific issues: environmental bio-sampling, bio-decontamination, and tools available to meet the current challenge of dealing with the consequences of infrastructure decontamination as a result of bioterrorism. We will discuss the current status and recent progress made with respect to environmental sampling of biological organisms, in particular *B. anthracis*. In addition, we will review the current status and recent studies addressing infrastructure decontamination. Briefly, we will also discuss approaches developed in the last 5 years for decontamination of sensitive platforms, like large-frame aircraft interior. Finally, we will discuss the current challenges with respect to bio-sampling and wide-area decontamination, along with future research needs in these areas to improve recovery and restoration following an intentional release of BW agent.

Environmental bio-sampling

Environmental sampling is a critical component of predecontamination and post-decontamination analysis during the recovery phase following contamination of building interiors. Following a wide-area release, rapid and effective qualitative and quantitative assessment of BW agent is acutely needed to declare the zones of contamination and the severity of contamination. Environmental sampling is at the very core of this assessment. Surface sampling is just as important an exercise after a cleanup effort in declaring a facility or contaminated zones ready for reoccupation. What follows is a review of progress made in environmental bio-sampling from abiotic surfaces.

In 2012, the Centers for Disease Control and Prevention (CDC) revised surface sampling procedures for *B. anthracis* sampled from smooth, nonporous surfaces (https://www.cdc.gov/niosh/topics/emres/surface-sampling-bacillus-anthracis.html). The utility of this method, which utilizes foam swabs, is limited to small areas of contamination. Prior to the release of this validated protocol, another study conducted by Rose and colleagues was published, which included a validation study of a cellulose sponge wipe-processing method by nine Laboratory Response Network-affiliated laboratories. The intent of this protocol was to address the issue of surface sampling from larger surfaces (Rose et al., 2011). One of the major issues highlighted in this study was the need to collect and analyze a large number of samples when assessing wide area contamination. Further work showed that this sampling protocol could be adapted using composite sampling to increase the amount of surface area sampled without increasing the laboratory processing time, labor, and consumables (Tufts et al., 2014). Composite sampling involves combining samples and testing pools of samples. This type of sampling has been proposed for other types of sampling as well. France et al. investigated the use of composite sampling approaches for spore contamination of soil (France et al., 2015).

Further modifications of the sponge wipe-processing method have been explored in order to reduce the burden of analysis by decreasing sample processing times, reducing waste, and improving recoveries at low concentrations. A Fast Analysis method involving filter plating of various volumes from the same sample has shown promise as an alternative protocol (Ahmed Abdel-Hady et al., 2019).

Since wide area contamination involves various surfaces, studies investigating sampling of numerous surfaces—biotic and abiotic—are necessary in order to assess the extent of contamination fully. Many studies published to date have examined sampling from hard smooth surfaces, such as stainless steel, glass, and vinyl tile (Piepel et al., 2011). Additional surfaces will need to be assessed in the future to fill this data/knowledge gap.

Past studies evaluating sampling methods have varied widely with respect to study design (Piepel et al., 2011). Some of the notable differences in these studies include the choice of BWA surrogate, method used for deposition of the contaminant, surface type tested, how samples were stored, method of sample collection, and method used for sample processing and reporting of recovery efficiencies. Standardization of methods and study design would benefit the field by providing data that could be compared regardless of which laboratory is providing the information.

Sampling, performed postdecontamination, is necessary to determine the effectiveness of the decontamination process. However, there are challenges involved with postanalysis sampling. In a study published in 2013, Calfee and colleagues at the Environmental Protection Agency filled a knowledge gap pertaining to the fate of spores co-collected with decontaminant residue, specifically pH-adjusted bleach (Calfee et al., 2013). Additional studies such as this one are needed to determine the impact of other decontaminant residue types co-collected in postdecontamination samples.

With each novel decontamination strategy that is proposed for wide area decontamination, new challenges arise with respect to sample analysis. Germination of spores prior to decontamination has been proposed in order to improve decontamination strategies, but this presents additional issues with respect to sampling. As such, Mott et al. conducted studies to develop an effective processing protocol that would recover both germinated and germinating spores from various surfaces. They determined the most efficient extraction was observed with an extraction buffer made up of PBS with 0.01% Tween 80 (Mott et al., 2017).

As they are introduced into the market, new technologies are being explored for use to improve sampling efficiencies. For example, cleaning robots have been investigated for use in sampling of spores from different indoor surfaces. It was found that some commercially available domestic cleaning robots were just as efficacious at surface sampling as the currently used surface sampling methods (Lee et al., 2013). It will be important to continue to explore such technologies in the future in order to improve sampling efficiencies and reduce labor and analysis costs associated with environmental sampling of biological materials.

Thus far, surface sampling is limited to two-dimensional and relatively smooth horizontal surfaces. Protocols are not available for sampling from complex three-dimensional surfaces. In our laboratory, the use of hydrogel as a biological sampler was investigated. Hydrogel is a water-based gel that is applied as a thick viscous material to a contaminated surface and dries into a thin film within a few hours (Fig. 1). The dried film is then peeled off the surface, hydrated for 1–2 h, and then analyzed for encapsulated spores. It was determined that encapsulated spores were present in the hydrogel, proving that it is possible to use hydrogel as a biological sampler (bio-hydrogel). To determine the effectiveness of a hydrogel as a bio-sampler, four diverse surface types—painted steel, pinewood, polycarbonate, and small screws (after intentional contamination)—were included. The results showed 50%–100% recovery with the use of bio-hydrogel (Smith et al., 2014).

One of the problems plaguing the field of bio-sampling is the assays that follow sample collection. False negative results are always a concern associated with sample analysis. Efforts have been made to reduce the rate of false negatives from surface sampling. Cultivation methods, while being the gold standard technique for detection, do not provide the specificity that results from molecular techniques. It was found that modified rapid viability-polymerase chain reaction (RV-PCR) gave lower

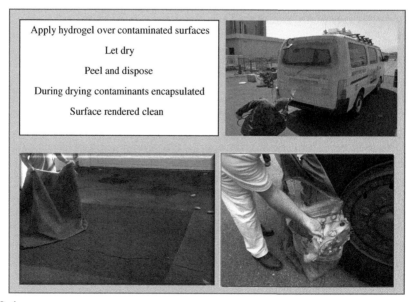

Apply hydrogel over contaminated surfaces

Let dry

Peel and dispose

During drying contaminants encapsulated

Surface rendered clean

FIG. 1

Hydrogel concept and applicability.

false negative rates than cultivation methods (Hutchison et al., 2015). Further investigations of such assays should be conducted in the future to reduce and/or eliminate the level of false negatives and hence improve bio-sampling data applicability and utility.

Infrastructure decontamination

Significant progress has been made since the cleanup effort of building interiors following the Amerithrax incident. In general, for cleanup of air-suspended spores in HVAC (heating, ventilation, and air conditioning) ducts and complex large building interiors, fumigants, such as chlorine dioxide gas (CD), vaporous hydrogen peroxide (VHP), and/or methyl bromide, have been used (Rastogi et al., 2009, 2010a; Pottage et al., 2012; Serre et al., 2016; U.S. EPA, 2015). In contrast, for surface cleanup on a smaller scale, i.e., room or small building, alternative approaches for spore cleanup include the use of liquid disinfectants, such as pH-adjusted bleach and peroxide-based commercial products, e.g., Spor-Klenz®, for surface disinfection (see Table 1 for a complete list of commercial sporicidal chemical products and physical approaches) (Edmonds et al., 2014). The dosage of CD gas required for a thorough cleanup ranges between 7000 and 9000 ppm-h, and for VHP, 450–600 ppm-h.

The use of biological indicators (BIs) provides confirmation on process parameters. Recently, a multiagency led study was conducted under the Bio-response Operational Testing and Evaluation (BOTE) project (Environmental Protection Agency, 2013). This study was spearheaded by the Office of Research and Development, National Homeland Security Research Center, U.S. Environmental Protection Agency (EPA). The study was supported by the Department of Homeland Security (DHS), Department of Defense (DOD), Centers for Disease Control (CDC), Federal Bureau of Investigation (FBI), and Department of Energy National Labs (DOE). In this study, a remediation exercise using three common sporicidal technologies was completed after a building was contaminated with aerosolized spores of *Bacillus globigii*, a simulant for the BWA, *B. anthracis* (Ames). This was one of the most exhaustive studies of its kind. In this building-scale study, it was interesting to learn that none of the three decontamination options resulted in complete spore kill, even though the spore density was 4–6 logs on the first floor and 2 logs on the second floor. Of the three technologies, VHP performed the worst, with almost one-third of the environmental samples detecting positive hits. Both, CD gas (9000-ppmv-h dose) and pH-adjusted bleach (6000-ppm) with at least a 10 min contact time, resulted in maximal spore kill on BIs and on surface, since minimal number of positive hits were observed for environmental samples and BIs. Curiously, more positive hits were recorded from the second floor, even though the spore deposition was at least two orders of magnitude less. This could possibly be due to less than optimal fumigant concentrations achieved on the second floor, due in part to poor mixing and/or distribution of the fumigants.

Table 1 Effective sporicidal approaches

General approach	General chemistry and commercial products	Use condition	Efficacy details	Comments and reference
Chemical/oxidant disinfectants	20%–28% H_2O_2 and 1%–6% peroxyacetic acid (PAA). Peridox®, Contec, Inc.; Steriplex Ultra, BioMed; Minncare Cold Sterilant, Minntech; Oxonia Active, Ecolab, Inc.; Spor-Klenz RTU, Steris Corp	Ready-to-use	6-Log reduction on glass and steel surface with 30 min contact time	Registered sporicidal decontaminant for *Bacillus anthracis* spore inactivation on dry, hard, nonporous surfaces
	Decon Green—developed by the U.S. Army—ECBC; 35% H_2O_2	Ready-to-use	High efficacy on steel, glass, aluminum, porcelain, granite, brick, and butyl rubber. Much lower efficacy on wood, concrete, and asphalt	Technology Evaluation Report. Washington, DC: U.S. Environmental Protection Agency. EPA/600/R-10/087, 2010
	EasyDECON200, EFT Holdings, Inc. 8% H_2O_2	Ready-to-use	High efficacy on steel, glass, aluminum, porcelain, granite, brick, butyl rubber, and concrete.	U.S. Environmental Protection Agency. EPA/600/R-10/087, 2010
	CASCAD Surface Decon Foam; Allen-Vanguard Corp.; Hypochlorite and hypochlorous acid	A specialized sprayer required for foam generation. 30 and 60 min contact times recommended for nonporous and porous surfaces	High efficacy on steel, glass, aluminum, porcelain, granite, brick, butyl rubber, and concrete.	U.S. Environmental Protection Agency. EPA/600/R-10/087, 2010
	pH-adjusted Bleach; EPA developed with no one commercial vendor	Must be used within 3 h after preparation; 10–30 min contact times	High efficacy on steel, glass, aluminum, porcelain, granite, brick, butyl rubber, asphalt, and concrete.	U.S. Environmental Protection Agency. EPA/600/R-10/087, 2010

Category	Agent/Method	Conditions	Efficacy	Reference
Fumigant	CD Gas; Sabre Oxidation technologies and ClorDiSys, Inc.	Fumigant and must be generated at the point-of-use. Requires 75% RH, 75F, 3000-ppmv for 3h contact	High efficacy on steel, glass, concrete, wallboard, carpet, and ceiling tile.	Rastogi et al. (2009). AEM
	Vapor hydrogen peroxide (VHP); Steris Corp.	Vapor phase; requires RH <35%, 150–200-ppmv with a dwell time of 3–4h	High efficacy on steel, glass, butyl rubber, porcelain, carpet, ceiling tile. Much lower efficacy on concrete	Rastogi et al. (2009). AEM
	Methyl bromide; Matheson Tri-gas; 99.5% methyl bromide gas	Fumigant; 212mg/L gas with a dwell time 36–48h contact	High sporicidal efficacy on limited surfaces	U.S. EPA. Washington, DC: U.S. Environmental Protection Agency, EPA/600/R-13/110, 2013
Physical	UV-C; germicidal 254-nm; 20,000J/m^2	Radiation; contaminated objects must be in direct path of radiation	<4-Log spore kill against *Bacillus globigii* spores	Kesavan et al. (2014)
	Hot humid air; 75–80°C, 80%–90% RH, and a dwell time of 7 days	Hot humid air; must be pumped in continually	>7-Log spore kill of *B. thuringiensis* and *B. anthracis* spores	Rastogi et al. (2010b) and Buhr et al. (2012, 2016)

Thus far, discussion has largely focused on facility and infrastructure decontamination, and the use of chemical approaches with aggressive oxidant have largely been successful in achieving the expected spore kill. In many cases, adverse effects on surfaces have been observed (like corrosion and malfunctioning of electric circuits) with the use of aggressive chemistries. Building may also contain sensitive equipment, like electrical items, computers, phones, and other similar items. Due to incompatibility of sensitive equipment, alternative approaches must be explored. In addition, extensive international air travel involving large-frame commercial or freight aircraft, raises the possibility of an aircraft contamination following an intentional terrorist event. For sensitive equipment and platforms, such as an aircraft interior, such chemical approaches are not applicable. Two physical approaches, i.e., hot humid air and UVC irradiation, are appropriate alternatives. In the past 8 years, hot humid air has been explored as an alternative approach for decontaminating large-frame aircraft interior surfaces (Rastogi et al., 2010b; Buhr et al., 2012, 2016). Sustainment of 7 days of hot air at 75–80°C, with 80%–90% RH was found to be an effective sporicidal approach for 7-log reduction in spore viability on surfaces relevant to aircraft interior. It is not clear if such a sustainment of hot humid air parameters is compatible with electronics and avionics within an aircraft interior (Dr. Shawn Park, Boeing Corporation, Seattle, WA; personal communication).

The second physical approach is exposure to ultraviolet (UV) radiation. Ultraviolet (UV) light has been used for disinfection since the mid-20th century. It is used for drinking water and wastewater treatment, air disinfection, and commodity treatment (Cundith et al., 2002). UV light exists within the spectrum of light between 10 and 400 nanometers (nm). The germicidal range of UV is within the 100 and 280 nm wavelengths (UV-C), with the peak for germicidal activity being between 253 and 265 nm. UV systems using low-pressure, mercury-arc germicidal lamps produce the highest amounts of UV radiation with 90% at 254 nm. Lethal dose$_{90}$ (LD$_{90}$) for *B. subtilis* spores has been reported to be between 245 and 326 Joules/meter^2 (J/m^2) at 254 nm (Nicholson and Galeano, 2003; Moeller et al., 2008). In those studies, a suspension of spores was exposed to UV-C. In a more recent study, Kesavan et al. reported LD$_{90}$ values of 138, 725, and 1128 J/m^2 for single, 2.8-μm, and 4.4-μm clusters of *B. globigii* (BG), respectively (Kesavan et al., 2014). In that study, only 3–4-log kill of BG spores on dry filter was reported even after a fluence of >20,000 J/m^2. Thus far, no study has been conducted demonstrating a 6-log kill of *B. globigii* spores. Currently, our laboratory, in collaboration with the U.S. EPA National Homeland Security Research Center, is investigating UVC efficacy and kill kinetics against *B. anthracis* Ames spores (unpublished work, Rastogi and Wood et al., 2019).

Two critical issues with current decontamination options are the logistical burden of transport and storage at the point of use, and the shelf life once disinfectants are prepared, and/or the bottle with concentrated stock is opened. The active components are a relatively small fraction compared to the rest of the bulk water. Shelf life becomes an issue for disinfectants such as diluted bleach, which must be used within 3 h of preparation, and must be kept away from sun and high temperatures. Special transport measures are needed to deal with metal incompatibility and possible corrosion resulting from chlorine-based products. Are there any alternatives to stocking liquid-based disinfectants?

Ideally, the active components should be in solid powder form. In more recent years, the Bioxy series of powders, manufactured by Atomes Inc. (Quebec, Canada), offer an alternative decontamination option. When needed, a solution can be prepared in any available water. A 2%–5% solution generates peracetic acid and peroxide at neutral pH. Such a disinfectant preparation is noncorrosive, nonhazardous, and is biodegradable. In a recent limited scope study, the sporicidal potential of Bioxy solution was demonstrated against spores of *B. anthracis* Sterne strain (Rastogi et al., 2017). In the future, it will be of interest to explore broad-scale applicability of Bioxy solution and investigate material incompatibility. Some of the unique advantages of Bioxy-type products include: (a) multiyear shelf life due to powder formulation; (b) noncorrosive properties; (c) biodegradability; (d) nonhazardous classification; (e) safe for the user and surfaces with no adverse environmental impact; and (f) significantly reduced logistical burden for first responders and combatants.

Another novel approach for bio-decontamination is the use of a sporicidal hydrogel, such as DeconGel, which has been tested in the Rastogi laboratory. The base material is a hydrogel polymer, which is 30% aqueous. The polymer hydrogel is applied as a paint over contaminated surfaces, or could even be sprayed using an airless paint sprayer (Rastogi et al., 2017; Dagher et al., 2017). The polymer dries into a thin film within 12–18 h after surface application. During the drying process, the gel encapsulates contaminants, such as radioactive material, oil, and biological material. The film can be peeled off the surface, rendering the surfaces clean. With respect to spores, the base gel simply encapsulates, and live spores can be retrieved after gel hydration for 2 h in water at 37°C. However, reformulation of the gel using a sporicidal chemistry resulted in inactivation of the spores over the surface and within the gel (Fig. 2). This work was performed in the Rastogi laboratory, in collaboration with CBI Polymers, Inc. The patented gel technology was demonstrated to clean surfaces of oil/grease and radioactive materials. More testing will need to be performed with the gel technology with respect to the diverse nature of surfaces, type of BW agents, and sporicidal additives, before this type of technology could be considered a viable option. Three specific advantages of the hydrogel technology are: (a) minimal waste generation (dried peeled film and no liquid waste); (b) minimal re-aerosolization (the spores are immediately locked in); and (c) the hydrogel technology allows enhanced contact and resident time for the active ingredient to act on the spore surface, permitting the use of reduced concentrations of sporicidal active ingredients.

Challenges

Evidently, the challenges in the wake of a large-scale bio-release are enormous with respect to detection, recovery, cleanup, and restoration. Much of the scientific research in both environmental sampling and surface decontamination has been done on smooth surfaces, such as steel, glass, and a few other building interior surfaces.

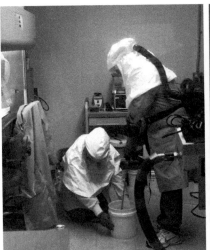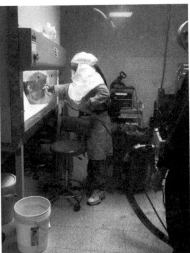

FIG. 2

DeconGel application.

A wide area bio-release on a large scale is expected to contaminate various infrastructure exterior and outdoor surfaces. Such surfaces include vegetation, grass, roads, asphalt, concrete, vehicles, wood and metal sheds, and animal housings. There is a paucity of information available on how to sample from many of these surfaces, and how to decontaminate such surfaces. One of the most challenging tasks in the future will be to determine how to demarcate and quarantine contaminated zones. Re-aerosolization of spores from contaminated areas and their unintentional transport to clean zones poses another challenge that must be addressed. Limited studies have been performed on outdoor materials, so this remains an area in this field of research that has been largely unaddressed (U.S. EPA, 2015). It is also clear that no one decontamination technology option will be suited for all surface types.

Another serious concern is the lack of acceptable risk-based criteria for re-occupancy following cleanup and restoration. Currently, environmental sampling with zero positive hits is the only acceptable criteria for declaring a facility or building "clean" for re-occupancy. In the case of a wide area release, such a rigid criterion may simply be unachievable.

Future research directions

Since 2001, impressive technological and scientific progress has been made in the areas of detection, surface decontamination protocols, array of decontamination options, and standardized test method development to assess sampling and lab-scale disinfectant efficacy. Meeting the consequences of a large-scale wide-area release of

BW agent still poses a daunting challenge with respect to cleanup efforts. We recommend several specific areas of research focus to address the gaps in knowledge and robust set of data.

1. Cleanup guidelines—When is a facility safe and ready for reoccupation? What are the acceptable standards and protocols for cleanup? What is an acceptable risk level after facilities have been cleaned up? Can physical data alone answer this question? How clean is safe? There is no documented threshold for cleanup of weaponized form of anthrax causing spores, below which there is no risk to human health. Persistence and LD_{50} of spores present considerable difficulties in defining the uncertainties surrounding the cleanup adequacy. Federal guidelines are needed to define the level of acceptable risk for facility reoccupation following a cleanup effort. Such guidelines must be supported by scientific data on long-range persistence, re-aerosolization, and microbial risk analysis. Future research, therefore, in the area of microbial risk analysis, fate of spores in building interiors, and potential for re-aerosolization would be extremely helpful in developing cleanup standards and guidelines.

2. Environmental sampling—In the event of a large-scale release, the shear number of samples will be a very daunting exercise. Future research into alternative rapid and quantitative approaches to detecting viable spores is highly desirable. Sampling from complex three-dimensional surfaces is another area of research, where no, or very limited, technical approaches are currently available. Adequate sampling methods and protocols are also needed for sampling from exterior surfaces, such as concrete, asphalt, vegetation, and brick walls. Furthermore, even though the focus of this review has been sampling and decontamination of spores, which are known to be highly persistent, the range of BW threats also includes Gram-negative bacterial cells, such as *Francisella tularensis, Yersinia pestis, Burkholderia mallei*, and *Brucella melitensis*. Loss of viability of vegetative cells is well-documented when subjected to drying or desiccation (Rastogi et al., 2017; Kramer et al., 2006; Sinclair et al., 2008). In order for the first responders and response planners to draw a threat assessment, it is imperative that viable cells are retrieved upon sampling. Future research, therefore, should be focused in the area of viability extension following environmental sampling until analysis and assessments are drawn. Improved air sampling devices and technologies are highly desirable.

3. Environmental bio-decontamination—While great strides have been made in decontamination of building interior surfaces, very limited study data is currently available for the effectiveness of commercially available technologies for outside surfaces and materials commonly encountered in the context of large-area release. Future research must be focused on acceptable cleanup standards for exterior surfaces, vegetation, and varied surface ground structures. Research into methodologies and protocols need to be pursued for efficient and cost-effective technologies to decontaminate vegetation, exterior structures, and varied surfaces. Another area of research is effective decontamination of vertical

surfaces. Foams or gels could be developed and tested to generate data in support of effective decontamination. Effective protocols and technologies are also needed for decontaminating vertical and angular surfaces within or outside infrastructure.

Future research coordination with aircraft manufacturers could also focus on investigating compatibility of hot humid air parameters with electronics and avionics to ensure suitability of airworthiness of an aircraft after cleanup.

Conclusions and closing remarks

We have come a long way from 2001, with respect to environmental sampling and decontamination of infrastructure, due in large part to multiagency coordination, including Department of Defense, U.S. EPA's National Homeland Security Research Center, Centers for Disease Control, Department of Homeland Security, and a host of other investigative agencies. Wide-area BWA release over urban populated areas poses a very daunting challenge for local, state, and federal agencies engaged in contamination characterization, cleanup, and restoration efforts. *Bacillus anthracis* spores top the list of Category-A Bio-warfare agents, due to their long-term persistence and resistance to common disinfection practices. Consequence management following a bioterrorist attack will require a very well-coordinated effort among first responders, subject matter experts, and policy makers. A significant effort will be required in educating the public on the risks associated with microbial risk uncertainties. Needless to say, continued efforts are needed to develop technologies, methodologies, and protocols to address efficient quantitative sampling from complex biotic and abiotic surfaces. Many of the current approaches will need significant modifications, if and when large areas are to be decontaminated. The hazardous waste generated in soil and water bodies resulting from rains must be addressed with respect to removal and cleanup of BWA. The monumental task of cleanup, restoration, and re-occupation will undoubtedly engage self-help (from the public), public education, risk-based cleanup standards, and extensive coordination among all local, state, and federal partners.

References

Ahmed Abdel-Hady, M.W.C., Aslett, D., Lee, S.D., Wyrzykowska-Ceradini, B., Robbins Delafield, F., May, K., Touati, A., 2019. Alternative fast analysis method for cellulose sponge surface sampling wipes with low concentrations of Bacillus spores. J. Microbiol. Methods 156, 5–8.

Buhr, T.L., et al., 2012. Test method development to evaluate hot, humid air decontamination of materials contaminated with *Bacillus anthracis* Sterne and *B. thuringiensis* Al Hakam spores. J. Appl. Microbiol. 113 (5), 1037–1051.

Buhr, T.L., et al., 2016. Hot, humid air decontamination of a C-130 aircraft contaminated with spores of two acrystalliferous *Bacillus thuringiensis* strains, surrogates for *Bacillus anthracis*. J. Appl. Microbiol. 120 (4), 1074–1084.

Calfee, M.W., Ryan, S.P., Griffin-Gatchalian, N., Clayton, M., Touati, A., Slone, C., McSweeney, N., 2013. The effects of decontaminant residue on the viability of Bacillus spores during wipe sample storage. Biosafety (S1)https://doi.org/10.4172/2167-0331.S1-001.

Canter, D.A., et al., 2005. Remediation of *Bacillus anthracis* contamination in the U.S. Department of Justice mail facility. Biosecur. Bioterror. 3 (2), 119–127.

Craft, D.W., Lee, P.A., Rowlinson, M.C., 2014. Bioterrorism: a laboratory who does it? J. Clin. Microbiol. 52 (7), 2290–2298.

Cundith, C.J., Kerth, C.R., Jones, W.R., Mccaskey, T., 2002. Air-cleaning system effectiveness for control of airborne microbes in a meat-processing plant. J. Food Sci. 67 (3), 1170–1174.

Dagher, D., et al., 2017. The wide spectrum high biocidal potency of bioxy formulation when dissolved in water at different concentrations. PLoS One 12 (2), e0172224.

Dias, M.B., et al., 2010. Effects of the USA PATRIOT Act and the 2002 Bioterrorism Preparedness Act on select agent research in the United States. Proc. Natl. Acad. Sci. U. S. A. 107 (21), 9556–9561.

Edmonds, J.M., Sabol, J.P., Rastogi, V.K., 2014. Decontamination efficacy of three commercial-off-the-shelf (COTS) sporicidal disinfectants on medium-sized panels contaminated with surrogate spores of *Bacillus anthracis*. PLoS One 9 (6), e99827.

Environmental Protection Agency, 2013. Bio-Response Operational Testing and Evaluation (BOTE) Project.

Erenler, A.K., Guzel, M., Baydin, A., 2018. How prepared are we for possible bioterrorist attacks: an approach from emergency medicine perspective. ScientificWorldJournal 2018, 7849863.

France, B., et al., 2015. Composite sampling approaches for *Bacillus anthracis* surrogate extracted from soil. PLoS One 10 (12), e0145799.

Franco, C., Bouri, N., 2010. Environmental decontamination following a large-scale bioterrorism attack: federal progress and remaining gaps. Biosecur. Bioterror. 8 (2), 107–117.

Government Accounting Office, 2003. Capitol Hill Anthrax Incident EPA's Cleanup Was Successful; Opportunities Exist to Enhance Contract Oversight.

Hutchison, J.R., Piepel, G.F., Amidan, B.G., Sydor, M.A., Deatherage Kaiser, B.L., 2015. False negative rates of a macrofoam-swab sampling method with low surface concentrations of two *Bacillus anthracis* surrogates via real-time PCR. https://www.pnnl.gov/main/publications/external/technical_reports/PNNL-24204Rev1.pdf.

Kesavan, J., Schepers, D., Bottiger, J., Edmonds, J., 2014. UV-C decontamination of aerosolized and surface-bound single spores and bioclusters. Aerosol Sci. Technol. 48, 450–457.

Kramer, A., Schwebke, I., Kampf, G., 2006. How long do nosocomial pathogens persist on inanimate surfaces? A systematic review. BMC Infect. Dis. 6, 130.

Lee, S.D., et al., 2013. Evaluation of surface sampling for Bacillus spores using commercially available cleaning robots. Environ. Sci. Technol. 47 (6), 2595–2601.

Moeller, R., et al., 2008. Roles of the major, small, acid-soluble spore proteins and spore-specific and universal DNA repair mechanisms in resistance of *Bacillus subtilis* spores to ionizing radiation from X rays and high-energy charged-particle bombardment. J. Bacteriol. 190 (3), 1134–1140.

Mott, T.M., et al., 2017. Comparison of sampling methods to recover germinated *Bacillus anthracis* and *Bacillus thuringiensis* endospores from surface coupons. J. Appl. Microbiol. 122 (5), 1219–1232.

Nicholson, W.L., Galeano, B., 2003. UV resistance of *Bacillus anthracis* spores revisited: validation of *Bacillus subtilis* spores as UV surrogates for spores of *B. anthracis* Sterne. Appl. Environ. Microbiol. 69 (2), 1327–1330.

Piepel, G.F., Amidan, B.G., Hu, R., 2011. Laboratory studies on surface sampling of *Bacillus anthracis* contamination: summary, gaps, and recommendations. https://www.pnnl.gov/main/publications/external/technical_reports/PNNL-20910.pdf.

Polyak, C.S., et al., 2002. Bioterrorism-related anthrax: international response by the Centers for Disease Control and Prevention. Emerg. Infect. Dis. 8 (10), 1056–1059.

Pottage, T., et al., 2012. Low-temperature decontamination with hydrogen peroxide or chlorine dioxide for space applications. Appl. Environ. Microbiol. 78 (12), 4169–4174.

Rastogi, V.K., et al., 2009. Quantitative method to determine sporicidal decontamination of building surfaces by gaseous fumigants, and issues related to laboratory-scale studies. Appl. Environ. Microbiol. 75 (11), 3688–3694.

Rastogi, V.K., et al., 2010a. Systematic evaluation of the efficacy of chlorine dioxide in decontamination of building interior surfaces contaminated with anthrax spores. Appl. Environ. Microbiol. 76 (10), 3343–3351.

Rastogi, V.K., Wallace, L., Smith, L.S., Shah, S.S., Foster, R., 2010b. Laboratory-Scale Demonstration of Hot Moist Air as a Bio-Decon Technology for Large-Frame Aircraft Interior Surfaces. . ECBC-TR-831.

Rastogi, V.K., Smith, L.S., Edgington, G., Dagher, M., Dagher, D., Dagher, F., 2017. The sporicidal potency of bioxy formulations in decontaminating bio-warfare agents. Clin. Microbiol. Infect. Dis. 2, 1–4.

Rose, L.J., et al., 2011. National validation study of a cellulose sponge wipe-processing method for use after sampling *Bacillus anthracis* spores from surfaces. Appl. Environ. Microbiol. 77 (23), 8355–8359.

Schmitt, K., Zacchia, N.A., 2012. Total decontamination cost of anthrax letter attacks. Biosecur. Bioterror. 10, .

Serre, S., et al., 2016. Whole-building decontamination of *Bacillus anthracis* Sterne spores by methyl bromide fumigation. J. Appl. Microbiol. 120 (1), 80–89.

Sinclair, R., et al., 2008. Persistence of category A select agents in the environment. Appl. Environ. Microbiol. 74 (3), 555–563.

Smith, L.S., Rastogi, V.K., Burton, L., Rastogi, P.R., Parman, K., 2014. A Novel Hydrogel-Based Bio-Sampling Approach. . ECBC-TR-1328.

Tufts, J.A., et al., 2014. Composite sampling of a *Bacillus anthracis* surrogate with cellulose sponge surface samplers from a nonporous surface. PLoS One 9 (12), e114082.

U.S. EPA, 2015. Surface Decontamination Methodologies for a Wide-Area *Bacillus anthracis* Incident.

Wagar, E., 2016. Bioterrorism and the role of the clinical microbiology laboratory. Clin. Microbiol. Rev. 29 (1), 175–189.

Biological and toxin warfare convention: Current status and future prospects

11

Chacha D. Mangu

National Institute for Medical Research, Mbeya Medical Research Center, Mbeya, Tanzania

The imminent danger

While the advancement in chemical and biological sciences and technology has contributed toward the improvement of quality of life of mankind, it has also brought calamity with it. Technology has allowed pathogens to be replicated in laboratory, mainly for research purposes; however, similar technologies may be used for harm. Global political instability increases threats of war and terror with government armies, terrorists, and revolutionary groups using chemicals, microorganisms, and toxins as weapons of mass destruction. A record of 259 intra- and inter-state conflicts including wars, revolutions, and *coup d'état* have been documented to have occurred between 1946 and 2014 and about 40 in 2014 alone (Pettersson and Wallensteen, 2015). This draws attention to the imminent danger for both the armed forces and civilians if a biological weapon might be considered as an option in combat.

Since the discovery of microbes as a cause of ill-health to mankind, evidence of infectious diseases making their appearances contemporaneously in battle fronts causing more deaths than combat cannot go unnoticed (Connolly and Deadly, 2002). Despite the fact that epidemics have followed wars throughout as natural disasters (Short, 2010), history has recorded overtime the intentional use of disease agents to intimidate and weaken the adversary armies since 1000 BC (Riedel, 2004; Frischknecht, 2008). As this phenomenon became increasingly eminent, in circumstances of no apparent account, it is theoretically unclear and difficult to deduce whether infectious diseases among combatants are the result of natural processes or hostilely introduced to cause harm. The only significant difference, therefore, between a naturally occurring epidemic and one that arises as a result of biological warfare is "motive."

The UN defines biological weapons as complex systems that disseminate disease-causing organisms or toxins to harm or kill humans, animals, or plants (United Nations, 2018). Holding the same view, Joshua Lederberg, in his address regarding biological warfare to the U.S. committee on foreign relations, in 2001, defined biological warfare as use of agents of disease for hostile purposes, encompassing attacks on human health and survival and extending to plant and animal crops (Lederberg, 2001).

Handbook on Biological Warfare Preparedness. https://doi.org/10.1016/B978-0-12-812026-2.00011-6

In contrast to ancient times, where human copses and animal carcasses containing infectious pathogens were used to introduced diseases in enemy camps, in the modern era, through advancing laboratory technologies, disease agents have been grown, new pathogens discovered and produced, while those already naturally occurring have been enhanced for mass production, safe storage, and effective diffusion as weapons. The possibilities for genetic engineering, for example, suggest leeway for a biological weapon that could arise from a novel recombinant virus created purely for such a purpose, with an enhanced virulence that can cause nearly 100% fatalities (United Nations, 2018). It is such availability of disease agents and the ease with which they can be obtained through affordable technology that increases the emerging pathogens' as candidates for biological weaponry and the likelihood of making use of such weapons. Ideal biological weapons would be specially produced and safely contained in a media package that is attached to a special delivery mechanism for expulsion only under a definite command over an intended target with a purpose to cause harm and mass destruction. An international conference to set guiding rules to prohibit the development, production, and stockpiling of biological and toxin weapons, which later formed the BWC, transpired behind such pending and yet evidently inevitable danger.

The convention

The Biological Weapons Convention (BWC) is the multilateral disarmament treaty banning the development, production, and stockpiling of an entire category of weapons of mass destruction (United Nations, 2007). After discussion and agreement of the State Parties, it was opened for signature on 10 April 1972, and entered into force on 26 March 1975. In 2016, more than 40 years since its inception, the BWC confirmed the membership of 178 State Parties.

Previous efforts to prevent the use of weapons of mass destruction, including biological weapons, have been noted in different international declarations and disarmament treaties since the 1907 Hague Convention (IV), which proposed respecting the laws and customs of wars on land (ICRC, 1907); and later the 1925 Geneva Protocol, which banned the use of biological and chemical weapons on the battlefield (United Nations Office for Disarmament Affairs, 1925). Such disarmament treaties, despite being implemented in different stages, only condemned the use but did not specifically ban development, production, or stockpiling of weapons of mass destruction and in many cases discussions on such remained inconclusive. In 1969, the United Kingdom (UK) got the ball rolling toward BWC by presenting a draft convention at the Eighteen Nation Disarmament Conference (ENDC) calling for separate treatment of biological weapons and setting higher priority over the chemical weapons. Although opposed by many states, the agenda gained attention after the United States (US), under President Richard Nixon, renounced every form of offensive biological weapons program including research, development, production, and stockpiling of biological agents applicable in warfare (Tucker and Mahan, 2009). Later, in March 1971, the Soviet Union and her allies introduced a revised version of the draft convention addressing only biological weapons, opening room for further negotiations toward BWC. As the negotiations proceeded, on 05 August 1971, US and the Soviet Union each presented draft conventions separately, yet with identical propositions, to the Conference of the Committee on Disarmament (CCD), which was then forwarded to the UN General Assembly on 28 September 1971. The conventions were then adopted on 16 December 1972 as the BWC, and opened for signature on April 10, 1972 (United Nations, 2017).

The BWC is therefore the first disarmament treaty to ban completely all processes involving weapons made of biological agents from development to deployment. The BWC has undergone eight reviews, conducted every five years during the Review Conference for the purpose of obtaining consensus and enhancing its implementation. There are fifteen articles to the convention; importantly, member State Parties are required to comply with the following key agreements in regard to BWC articles (United Nations, 2017):

- Never under any circumstances develop, produce, stockpile, acquire, or retain biological weapons.
- To destroy or divert to peaceful purposes biological agents, toxins, weapons, equipment, and means of delivery prior to joining.
- Not to transfer, or in any way assist, encourage or induce anyone else to acquire or retain biological weapons.
- To take any national measures necessary for implementation of the BWC.
- To request the UN Security Council to investigate alleged breaches of the BWC and to comply with its subsequent decisions.
- To assist states that have been exposed to danger as a result of violation of the BWC.
- To facilitate the fullest possible exchange of equipment, materials, and information for peaceful purposes.

The BWC, its expeditions, plights, and current status

The BWC itself, as a result of prolonged efforts by the international community to establish new guiding principles that would supplement the Geneva Convention, is so far the best treaty that could keep the world free from BW. However, several deficiencies thwart effective implementation of the convention. Below is an account of some of these.

Lack of agreement

Despite the fact that the BWC is the second treaty to obtain higher universality, there has been daunting lack of agreement on major articles in the convention as a result of persisting divergent views and positions on various key issues. Understanding that key components with any convention are agreement and transparency among its members, confidence-building measures (CBMs) were introduced in 1986, during the second review, "in order to prevent or reduce the occurrence of ambiguities, doubts, and suspicions, and in order to improve international co-operation" (United Nations Office for Disarmament Affairs, 2015). However, in the end, lack of trust and transparency are still echoed among the members. The Ad Hoc Group of the States Parties to the BWC was formed in 1994 to negotiate and develop a legally binding verification regime for the convention, envisioning states declaring the facilities and activities involving research and development of munitions-prone biological agents relevant to the treaty and routine on-site visits to declared facilities, as well as challenge inspections of suspect facilities and activities. The chair of the Ad Hoc Group, ambassador Tibor Toth of Hungary, pointed out that some of the concessions would be difficult for different sides while giving a hint that the propositions were "give-and-take" (United Nations, 2000). However, in 2001, proceedings of the fifth review were adjourned following a failure to conclude the negotiations on the draft protocol to make it a legal instrument. In 2016, the eighth review conference was described as disappointing, since one member State Party blocked the consensus on a broadly acceptable compromise text (Pearson and Sims, n.d.). Despite the fact that there have been, to date, eight reviews, with industrious efforts to find concession among the members, evidently, even after over four decades of the convention, finding consensus on some key issue that would have made the convention more effective has proven elusive. The International Committee of the Red Cross (ICRC), in its statement echoes this:

> Over the past five years of annual meetings, a great deal of information has been shared and many proposals have been made on how to implement the treaty and improve its effectiveness. Disappointingly, however, there has been little collective agreement.

> **International Committee of the Red Cross (2016)**

A bumpy road is anticipated as a result of further compromise that would be required of the State Parties to obtain consensus for improving effectiveness in implementation of the convention.

Unachieved universality

Despite the agreed fact by nearly all the states globally on the increased urgency to address the threat of biological weapons (BW), State Parties have failed to respond collectively. There is a lack of universality in the convention, as only 178 State Parties have pledged their membership, and 6 more as signatories, as of May 2017 (United Nations, 2017). The 12 states that are not members can easily, either by themselves or being used by states with ill intention, execute activities in developing, producing, and stockpiling weapons of mass destruction.

Lack of compliance and related verification

The BWC, unlike the Nuclear Nonproliferation Treaty and the Chemical Weapons Convention, does not have an international organization to verify if the State Parties are complying with the treaty. A group of governmental experts (VEREX) was established at the Third Review Conference (1991) to identify and examine potential verification measures from a scientific and technical standpoint which led to formation of an Ad Hoc Group in 1994 to negotiate and develop a legally binding verification regime for the convention. The blocking of the draft protocol in 2001 presented by the Ad Hoc Group, which tabled propositions for verification of compliance with the BWC in regard to biodefense programs, vaccine production facilities, and scientific facilities including BSL-3 and BSL-4 laboratories through voluntary invited visits and inspections under the patronage of international BWC organization, was a major setback.

Noncompliance with the BWC has been recorded from member State Parties even after signing the convention. Events such as the "Sverdlovsk case," where anthrax spores were suspected to have accidentally leaked from the Soviet secretive Compound 19 military research facility on the southern edge of the city of Sverdlovsk (now Yekaterinburg) on April 2, 1979, and the "Yellow Rain" case of 1981, where the Soviet Union was accused of being involved in the production, transfer, and use of trichothecene mycotoxins in Laos, Kampuchea, and Afghanistan, raised the alarm for BW activities, and hence, noncompliance of the Soviet Union (Moodie, 2001; Harris, 1987). Iraq withheld cooperation from the UN Security Council special body, UNSCOM, and later, the UN Monitoring, Verification, and Inspection Commission (UNMOVIC) during their inspection processes in 1998 and 2002, respectively (Zanders et al., 2003). In 2006, the US presented allegations citing Iran, North Korea, and Syria for possible noncompliance (United Nations, 2006).

Article IV of the convention requires each State Party to take measures under its jurisdiction and constitutional processes to comply with the convention (Harris, 1987). However, while a state might declare it has shut down production and carried out a complete destruction of its BW stockpiles, there is no legal mandate to open the facilities for verification, and if it does so, it will be purely voluntarily. This lack of verification lowers a convention down to another political declaration that lacks comprehensive implementation mechanisms.

Implementation challenges

Due to challenges in implementation of the convention by the member State Parties, the sixth review conference (2006) agreed to establish an Implementation Support Unit (ISU) to support State Parties in the administration and comprehensive implementation of the convention, promotion of universalization, and exchange of the CBMs. In its 2016 report of activities, the ISU raised concern over the constraint in executing its activities as a result of being understaffed and underfunded, and hence, its failure to effectively respond to individual State Parties' calls for assistance (Anon, 2012–2016). Such constraint leaves most of the implementation dilemmas among the State Parties unresolved.

The future of the convention

For a sustainable future, some teeth need to be put into the convention to invigorate the treaty and affirm the comprehensive prohibition of all types of misuse of biological agents and toxins, including their development and production with hostile incidents. One of the most important assignments is to clear all the existing gray areas, to improve consensus among the member State Parties, and move disputed articles into enforcement. Members must agree to compromise to enable mechanisms for enhancing verification for compliance to improve trust among State Parties.

On the other hand, the BWC has to succeed in the era of advancement of biotechnology such as genetic recombinant engineering, cell culture, and microbial discovery. Such technologies are becoming affordable and can be implemented by portable equipment in real time. A few personnel can accomplish tasks in a very short time that a few years back needed quite a large workforce. The current technology, therefore, can facilitate in-house production of BW with strictly limited facilities, and hence improved secrecy. Since such technologies are becoming affordable, they may fall into the hands of malicious individuals who may use them for terrorist attacks. Therefore, control of biotechnological activities for peaceful purposes has to be improved with inclusive interstate collaboration. Such control must not leave out handling and transportation of hazardous biological and toxic agents between states and handling facilities such as laboratories and research institutions.

The "US Anthrax" case, where envelopes of anthrax spores were mailed to US senators in 2001 reminds us that acts of terrorism cannot be contained by a convention. In the surge of non-state players in conflicts, acquisition and deployment of BW for acts of terror cannot be ignored. Improving universality of the BWC to acquire 100% State Parties membership is one of the essential steps toward preventing proliferation and infiltration of BW that might happen through the non-member State Parties.

In conclusion, measures that are being agreed within the framework of the convention will have limited scope to bring about a new dawn of the convention unless there is a guiding and executing structure. Compliance verification teams and implementation organization under supervision of the United Nations have been proposed to ensure that there is an official structure that has a mandate to execute the requirements to assist the State Parties. The value of BWC is embedded within its ability to improve global biosafety and biosecurity. The convention will soon lack value if the articles are not observed by the intended State Parties and fail to bring a positive change toward a world free of BW of mass destruction. However, with the increasing commitment of the member States Parties, the future holds hope of a safer world now and for the generations to come.

References

Anon, 2012–2016. Report of the Implementation Support Unit on Its Activities to Implement Its Mandate. (Report number BWC/CONF.VIII/PC/7).
Connolly, M.A., Deadly, H.D.L., 2002. Comrades: war and infectious diseases. LANCET Suppl. 360, 23–24.

Frischknecht, F., 2008. The history of biological warfare. In: Richardt, A., Blum, M. (Eds.), Decontamination of Warfare Agents. WILEY-VCH Verlag GmbH & Co. KGaA, Weinheim, German, pp. p1–10.

Harris, E.D., 1987. Sverdlovsk and Yellow Rain: two cases of the Soviet noncompliance? Int. Secur. 11 (4), 41–95.

ICRC, 1907. Convention (IV) Respecting the Laws and Customs of War on Land and its Annex: Regulations concerning the Laws and Customs of War on Land. The Hague. (Online). Available at: https://ihl-databases.icrc.org/ihl/INTRO/195.

International Committee of the Red Cross, 2016. Biological Weapons Review Conference: ICRC Statement. International Committee of the Red Cross, Geneva, Switzerland. (Online). Available from: www.icrc.org.

Lederberg, J., 2001. Biological warfare. Emerg. Infect. Dis. 7 (6), 1070–1071. (Online) Available from: https://doi.org/10.3201/eid0706.010636.

Moodie, M., 2001. The Soviet Union, Russia, and the biological and toxin weapons convention. Nonproliferation Rev. 59–69. (Online). Available from: https://www.nonproliferation.org/wp-content/uploads/npr/81moodie.pdf.

Pearson, G.S., Sims, N.A., n.d. Report From Geneva: The BTWC Eighth Review Conference: A Disappointing Outcome (Havard Susex Program Occasional Paper 2017, Review No. 16) (Online). Available from: www.unog.ch/80256EDD006B8954/(httpAssets)/96E73A407E3 6F9D0C12580ED00354AB3/$file/REPORT_FROM_GENEVA_46+E.pdf.

Pettersson, T., Wallensteen, P., 2015. Armed conflicts, 1946-2014. J. Peace Res. 52 (4), 536–550.

Riedel, S., 2004. Biological warfare and bioterrorism: a historical review. BUMC Proc. 17, 400–406.

Short, A.V.M.B., 2010. War and disease: war epidemics in the nineteenth and twentieth centuries. ADF Health 11 (1), 15–17.

Tucker, J.B., Mahan, E.R., 2009. Presidents Nixon's Decision to Renounce the U.S. Offensive Biological Weapons Program. Case Study Series. National Defense University Press, Washington, DC.

United Nations, 2000. Highlights of Press Conference By Tibor Toth, Chairman of the Ad HOC Groups of States Parties to the Biological Weapon Convention, Held at the Palais des Nations on 4 August 2000. United Nations Information Services, Geneva, Switzerland.

United Nations, 2006. Confronting Noncompliance With the Biological Weapons Convention—Submitted by the United States of America Geneva, Switzerland. (BWC/CONF.VI/WP.27).

United Nations, 2007. Biological Warfare Convention: An Introduction. United Nations Publication, Geneva, Switzerland.

United Nations, 2017. The Biological Weapons Convention: An Introduction. United Nation, Geneva, Switzerland. (Online). Available from: http://www.un.org/disarmament.

United Nations, 2018. What are Biological and Toxin Weapons? United Nations, Geneva, Switzerland. (Online). Available from: https://www.unog.ch/80256EE600585943/%28http Pages%29/29B727532FECBE96C12571860035A6DB?OpenDocument.

United Nations Office for Disarmament Affairs, 1925. 1925 Geneva Protocol. Protocol for the Prohibition of the Use in War of Asphyxiating, Poisonous or Other Gases, and of Bacteriological Methods of Warfare. The League of Nations, Geneva, Switzerland. (Online). Available from: https://www.un.org/disarmament/wmd/bio/1925-geneva-protocol/.

United Nations Office for Disarmament Affairs, 2015. Guide to Participating in the Confidence-Building Measure of the Biological Weapon Convention, revised ed. United Nations, Geneva, Switzerland.

Zanders, J.P., Hart, J., Kuhlau, F., Guthrie, R., 2003. Non Compliance With the Chemical Weapons Convention. Lessons From and for Iraq (SIPRI Policy Paper No. 5).

Next generation agents (synthetic agents): Emerging threats and challenges in detection, protection, and decontamination

12

Anshula Sharma[a], Gaganjot Gupta[a], Tawseef Ahmad[a], Kewal Krishan[b], Baljinder Kaur[a]

[a]*Department of Biotechnology, Punjabi University, Patiala, India*
[b]*Department of Defence and Strategic Studies, Punjabi University, Patiala, India*

Potential biological weapons and warfare agents

Biological weapons are refers to those which contain replicating infectious and lethal forms of life including bacteria, viruses, fungi, protozoa, prions, or poisonous chemical toxins produced by living organisms (Rogers et al., 1999). Biological warfare agents (BWAs) as bioweapons have been widely used in wars because of their easy availability, low production costs, easy transportation and dispersal, and nondetection by basic security systems. Biowarfare agents are responsible for the spread of human diseases associated with high morbidity and mortality rates. Further, these agents can multiply in the host organism and get transmitted to others individuals, causing erratic consequences, which lead to mass geographical spread. Due to low production costs and easy cultivation, any developed or under developed country can afford their manufacturing and maintenance. These are available in liquid as well as in dry forms with extensive storage life. People who have not previously encountered these biowarfare agents usually do not have any natural immunity in their body against these agents, thus are highly prone to infections. Moreover, in comparison to common human diseases, the etiological agents causing these deadly diseases have highly hostile animal reservoirs (called zoonotic in nature) and are difficult to diagnose and cure. The nature, properties, and lethal effects caused by many zoonotic biological agents that have been used for biowarfare purposes over the years and have led to serious epidemiological outbreaks, are summarized in Table 1.

Handbook on Biological Warfare Preparedness. https://doi.org/10.1016/B978-0-12-812026-2.00012-8

Table 1 Epidemiological spread of human disease causing pathogens (*likely to be used as biological weapons*)

Disease	Causal agent	Carriers	Effects on humans	Countries affected	References
Bacterial diseases					
Anthrax	*Bacillus anthracis*	White deer, biting flies, sheep, camels, antelopes, cattle, humans	Sore throat, mild fever, fatigue and muscle aches, mild chest discomfort, shortness of breath, nausea, coughing up blood, painful swallowing	United States, Europe, Asia, Africa, Caribbean, Middle East	Dutta et al. (2011), Jansen et al. (2014)
Brucellosis	*Brucella* sp.	Goats, sheep, reindeer, pigs, caribou, humans	Fever, back pain, body aches, poor appetite and weight loss, headache, night sweats, weakness, abdominal pain	Europe, Africa, Asia, Latin America, Arctic and sub-arctic parts of North America	Thavaselvam and Vijayaraghavan (2010)
Botulism	*Clostridium botulinum*	Fish, birds, snails, earthworms, maggots, nematodes, humans	Difficulty in swallowing or speaking, dry mouth, facial weakness, blurred or double vision, drooping eyelids, trouble breathing, nausea, vomiting and abdominal cramps, paralysis	South Africa, United States	Sobel (2005), Jansen et al. (2014)
Enterotoximea	Staphylococcal enterotoxin B	Sheep, goats, worms	Loss of appetite, abdominal discomfort, bloody diarrhea	Middle Asia, Kazakhstan	Spencer and Scardaville (1999), Horn (2003)
Glanders	*Burkholeria mallei*	Horses, donkeys, mules, humans	Fever, muscle aches, chest pain, muscle tightness, headache	Asian, Middle East	Van Zandt et al. (2013), Go and Sansthan (2014)
Melioidosis	*Burkholderia psuedomallei*	Rodents, humans	Cough, chest pain during breathing, high fever, headache, muscle soreness, weight loss	Southeast Asia, Australia; Rarely in Tropic and subtropic areas of world	Jansen et al. (2014), Madad (2014)
Plague	*Yersinia pestis*	Rodents, fleas, humans	Sudden onset of fever, headache, chills, weakness, swollen, tender, and painful lymph nodes	Europe, North Africa	Christie (1982), Jansen et al. (2014)
Q fever	*Coxiella burnetii*	Birds, ticks, sheep, cattle, goats, cats, rabbits, humans	High fever, chills or sweats, cough, chest pain, headache, clay-colored stools, diarrhea, nausea	Worldwide except New Zealand	Maurin and Raoult (1999)

Disease	Agent	Host/Vector	Symptoms	Location	References
Tularemia	*Francisella tularensis*	Arthropods, aquatic rodents, rabbits, humans	Skin ulcer, swollen lymph nodes, severe headaches, fever, chills, fatigue	United States	Gürcan (2014)
Toxicosis	Ricin	Humans	Weight loss, anxiety, intolerance to heat, fatigue, hair loss, weakness, hyperactivity, irritability, apathy, depression, sweating	United States	Jansen et al. (2014)
Viral diseases					
Dengue	*Flavivirus*	Mosquitos	High fever, severe headaches, pain behind the eyes, severe joint and muscle pain, fatigue, nausea, vomiting	America, China, Europe, Southeast Asia	Murray et al. (2013)
Ebola	*Ebolavirus/ filovirus*	Bats, monkeys, gorillas, chimpanzees, humans	Fever, headache, joint and muscle aches, weakness, diarrhea, vomiting, stomach pain, loss of appetite	Africa, Europe	Jansen et al. (2014), Madad (2014)
Hepatitis	*Viruses*	Humans	Fatigue, dark urine, pale stool, abdominal pain, loss of appetite, weight loss	Africa and Asia	Lemoine et al. (2013)
HIV	*Lentivirus*	Humans	Headache, diarrhea, nausea and vomiting, fatigue, aching muscles, sore throat, swollen lymph nodes	Asia, United States	Fettig et al. (2014), Maartens et al. (2014)
Influenza Type A (Spanish Flu H1N1 and Swine Flu)	*Influenza virus*	Humans	Fever, cough, sore throat, runny or stuffy nose, muscle or body aches, headaches, fatigue	Asia, Europe	Taubenberger and Morens (2008), Zimmer and Burke (2009)
Lassa virus	*Arenavirus*	Rodents, humans	Fever, general weakness, headache, sore throat, muscle pain, chest pain, nausea, vomiting, diarrhea, cough	Africa	Raabe and Koehler (2017)

Continued

Table 1 Epidemiological spread of human disease causing pathogens (*likely to be used as biological weapons*)—cont'd

Disease	Causal agent	Carriers	Effects on humans	Countries affected	References
Measles	*Morbillivirus*	Humans	Fever, dry cough, runny nose, sore throat, inflamed eyes	Africa, Asia, Europe, United States	Abad and Safdar (2015)
Rabies	*Lyssaviruses rabies virus* and *Australian bat lyssavirus*		Fever, headache, muscle aches, loss of appetite, nausea, fatigue		Yousaf et al. (2012)
Severe acute respiratory syndrome (SARS)	*Coronavirus*	Animals, humans	Fever, headache, loss of appetite, diarrhea, dry cough, fatigue, breathing problems	China, Canada, United Kingdom	Vijayanand et al. (2004)
Smallpox	*Variola virus*	Humans	Skin rash, severe headache, backache, abdominal pain, vomiting, diarrhea.	Europe, North Africa, United States	Henderson et al. (1999)
Venezuelan equine encephalitis	*Alphavirus*	Rodents, bats, birds, mosquitoes, horses, humans	High fevers and headaches, central nervous system disorders	United States, Canada, Argentina	Weaver et al. (2004)
Yellow fever	*Flavivirus*	Mosquitoes	Fever, headache, nausea and vomiting. Serious cases may cause fatal heart, liver, and kidney conditions. Fever, chills, loss of appetite, nausea, muscle pains, particularly in the back, headaches	Africa, Asia, South America	Gardner and Ryman (2010), Monath and Vasconcelos (2015)

The utilization of these BWAs in previous bioterror incidences leading to adverse consequences has been well documented in various review articles (Jansen et al., 2014; Madad, 2014; Krishan et al., 2017). Following the historical pattern of biological terror attacks and disease outbreaks, it can be concluded that biotechnology has accidently unleashed a new threat to mankind in the form of virulent bioweapons to inflict mass causalities and devastation (Lesser et al., 1999). So, highly efficient and cost-effective medical countermeasures based on ethnic specificity of the biological agent need to be developed with the purpose of bio-preparedness and biosecurity (Horn, 2003; Jansen et al., 2014; Pal et al., 2016). Associated risks with human pathogens must be evaluated on the basis of rate of mortality, availability of treatment and prophylactic measures, need for hospitalization, public perception, and epidemiological spread (Rotz et al., 2002).

A brief history of biological warfare

Any documentation related to history of biological warfare is difficult to extract from the literature since all the allegations are based solely on eye-witness accounts and circumstantial evidence after the event. Moreover, the allegations in most of the cases have been denied by the accused parties. In 1155 AD, at a battle in Tortona, Italy, Barbarossa broadened the scope of biological warfare by using the dead bodies of soldiers as well as animals to pollute wells. During war with the French in 18th century, British forces under the direction of Sir Jettray Ahmest gave blankets that had been used by smallpox and yellow fever victims to the native Americans in order to spread disease (Frischknecht, 2003). During the US civil war, in 1863 AD, General Johnson used dead bodies of sheep and pigs to pollute drinking water at Videsburg, Mississippi. During World War I, in 1915, the first allegation was made against the Germans that they had attempted to employ the biological agents of Cholera against Italy and Plague against Britain. Later, in 1916, they were accused of using Anthrax at Bucharest, Romania (Metcalfe, 2002). The first incidence of biological warfare was documented during World War I when millions of deaths were recorded due to pandemic outbreak of The 1918 Spanish Influenza virus (H1N1) that triggered naturally. This was the most striking example of indirect offense where a disease causing biological agent was used to deteriorate combat capabilities of enemy forces at the war front. This perhaps led to the establishment of the concept of biowarfare agents and bioterrorism. The period from 1940 to 1969 can be considered as the golden age of biological warfare research and development. In the last few decades, several incidences of bioterrorism/biological warfare were recorded (Ainscough, 2002).

The history of biological warfare programs in the United States and Former Soviet Union (FSU) is extensively documented in the literature. As a rapidly evolving super power, the FSU initiated its biological warfare program during the mid-1960s to 1970s, when they started showing interest in genetics and genetic manipulations of potential human pathogens. Despite being a signatory member of the 1972 Biological and Toxin Weapon Convention (BTWC), the FSU developed highly potent and deadly chimeric biological warfare agents. Extensive research was carried out to weaponize the developed agents to powder or aerosol formulations for direct loading into munitions such as spray tanks and cluster bombs. The first genetically engineered vaccine tolerant pathogen *Francisella tularensis*, causing tularemia, was established under the "Enzyme" program during the 1950s to 1960s, which modernized the concept of biological warfare. Until 1992, they had a repository of 52 highly contagious strains that could overcome all the barriers of immune systems and current medical treatments. Anthrax strain 836, Pasechnik's superplague strain, glanders strain, myelin toxin forming *Yersinia pestis*, tularemia (Schu S-4), and viruses like Ebola, Marburg and influenza are only a few to name them. They also tried their hands at the production of chimeric viruses by introducing genetic elements from Venezuelan equine encephalitis (VEEV), Ebola (EBOV), and Marburg (MARV) into native smallpox virus (Ainscough, 2002).

In 1997, Russian scientists published a research paper in the journal *Vaccines* where they suggested a method of introducing *Bacillus cereus* genes into *B. anthracis* for making it resistant to Russian anthrax vaccine. Introduction of antibiotic resistant genes in pathogenic strains can significantly enhance lethality of disease by reducing treatment options (Athamna et al., 2004). Similarly, the prophylactic effects can be circumvented by suppression of the immune system through the expression of immune modifier genes using viral vectors, e.g., expression of mouse interlukin-4 in recombinant Ectromelia virus suppresses immune functions of the host and overcomes genetic resistance to mouse pox. In 1998, a DNA sequence based investigation on the preserved samples of 11 victims revealed simultaneous occurrence of 4 distinct virulent variants of *B. anthracis*, showing the continuation of a biological warfare program, which was reportedly denied from time to time by FSU. Surprisingly, genetic manipulation procedures were adopted to enhance resistance of the existing etiological agents to high temperature and the available range of therapeutic antibiotics and prophylactic measures, and to make novel immune-suppressive agents that could be easily weaponized when needed. These facts indicate the violation of 1972 BTWC by FSU, as they were continuously engaged in developing designer lethal agents, which they might have integrated into their special war plans (Ainscough, 2002).

Russian biowarfare program came to light in October 1989 when a top ranking microbiologist and Director of "The Institute for Ultra Pure Biological Preparations" in Biopreparat, Dr. Vladimir Pasechnik was defected to the UK. Later, a lower level bench scientist in Pasechnik's lab, referred by the code name "Temple Fortune," also defected to the United Kingdom, 3 years after the defection of Dr. Pasechnik, where he replicated his previous account of a biowarfare program and then disclosed it to the British government. In late 1992, Dr. Kanatjan Alibekov (now known as Ken Alibek) became the third defector from the Russian biowarfare program (Mangold and Goldberg, 1999; Tucker, 1999; Ainscough, 2002). In 1999, Alibek's article on "Biohazard" narrated his first-hand experience as a member of the FSU program and emphasized the extensive research and development of genetically manipulated biological warfare agents, large scale production facilities, weaponization of BW agents, defensive measures, and future BW goals of the Russian Government

(Alibek, 2008). Former USSR president Mikhail Gorbachev (1990–91) and, later on, Russian president Boris Yeltsen (1991–99) had announced the termination of their biowarfare programs in the early 1990s. But many intelligent analysts suspected the execution of a biowarfare program in a very secretive mode and its substitution in the Russian military doctrine.

Earlier, in 1997, the Office of the Secretary of Defence, United States, also released a special report on *Proliferation: Threats and Responses* indicating then circulating trends in biological warfare capabilities (Cohen, 1997). There is also a thriller novel written by Richard Preston in 1998 titled *The Cobra Event*, which describes fictional bioterror attacks on the United States in 1997, where a genetically engineered virus designated as "Cobra" was used to spread a "Designer Disease called Brain Pox," which symptomatically resembles smallpox, Lesch-Nyhan syndrome, and the common cold (O'Toole, 1999). Soon after the September 11 terror attacks on the World Trade Centre and the Pentagon in 2001, anthrax-laced letters were sent to national legislators of the United States, which resulted in the spread of terror among civilians and triggered the adoption of prophylactic measures by communities on a mass scale. These two events were sufficient to sensitize President Bill Clinton to extract deficiencies in the US national security and resulted in the establishment of The Homeland Security Council to co-ordinate efforts of the existing national agencies and organizations to overcome future security challenges (Ainscough, 2002).

In 2003, Aken and Hammond provided three major evidences that indicate violation of the 1972 BTWC treaty by the US Govt., as recorded in one of their scientific writings on "Genetic Engineering and Biological Weapons" published in the journal *EMBO Reports*. The US military has repeatedly discussed the possible use of biotechnology for upgrading offensive warfare potential by developing material degrading microbes to destroy biofuels, construction materials, and stealth paints. The first evidence emerged in 1998, when J. Campbell at the Naval Research Laboratory in Washington DC described the possible application of genetically modified fungi to destroy military paints in 72 h. The second evidence came to light in 1990, when the United States started conducting field trials on pathogenic *Pleospora papaveracea* strains against the drug producing crop, opium poppy. Potential risks were successively tested by evaluating crop destruction in 2001 in Tashkant, Uzbekistan. Similarly, pathogenic *Fusarium oxysporum* strains were developed in the United States to destroy coca plants with field test scheduled to be held in Columbia, 2001. However, worldwide protest against field trials on pathogenic destruction of drug producing (cocaine, benzoylecgonine, ecgonine) cash crops produced strong public opposition leading to termination of the project. The third evidence came from the use of psychoactive substances (sleeping gas- BZ) as biological weapons in the Moscow hostage crisis in 2002, which caused death of more than 170 people. The US Marine Corps also investigated the military usefulness of benzodiazepines and alpha-2 adrenoreceptor agonists as potential weapons. Other BTWC signature states have developed potential biowarfare agents through extensive research and development, but as far as their delivery is concerned, it is still in its infancy as compared to weaponization of biowarfare agents by FSU (Van Aken and Hammond, 2003).

Emergence of next generation biological weapons

With the advancement in genetic engineering and synthetic biology techniques, complex genetic manipulations have become possible for creation of "tailor-made" microorganisms. Harmless bacteria or viruses can be made pathogenic or infectious by genetic manipulation mediated via multiple gene transfers and through construction of synthetic or chimeric microorganisms. Moreover, genetically engineered biological agents have the ability to resist the existing treatment therapies and may potentially be used as biowarfare agents. Biological agents with novel/altered pathogenic characteristics, such as enhanced survivability, infectivity, virulence, and drug resistance are referred to as "next generation bioweapons." Decoding of the human genome and recent breakthroughs in genetic engineering, gene therapy, and drug delivery approaches will eventually enhance the chances of use of potentially pathogenic microorganisms as next generation bioweapons (Ainscough, 2002).

The JASON advisory group has been used to provide technical advice to the US Dept. of Defence, briefing on near term future threats due to development of genetically engineered bioweapons (Ainscough, 2002). Steven M. Block, a member of the JASON group, has raised several concerns over the potential bioterrorist activities in the country using next generation bioweapons (Block, 2001). They classified biowarfare agents into six major groups, as described below.

Binary biological weapons

Russian scientists were masters of binary biological weapons technique, which was used to enhance virulence of several human pathogens causing anthrax, dysentery, and plague. It includes a dual component system consisting of a pathogenic host strain and virulence genes bearing plasmids, which could be individually propagated at a large scale. Just before their deployment into a bioweapon, these components would have been mixed together and subsequent biotransformation would have taken place within the munition acting as a bioreactor.

Designer genes

The decoding and availability of whole genome sequence data has provided ample opportunity to biotechnologists for designing and reconstruction of virulent genes, which introduces desired virulence characteristics in the existing repertoire of pathogenic microorganisms. Advanced synthetic biology techniques and genetic engineering techniques have led to the feasible construction of designer genes, which can further be used for the creation of genetically modified human pathogens.

Designer diseases

Recent breakthroughs in molecular and cellular biology have equipped biologists to develop designer diseases by creating designer pathogens as etiological agents to achieve desired symptoms of progress of a hypothetical disease. Their potential targets include somatic or germ cells of the body, in which tissue destruction is induced through apoptosis, enhanced cell proliferation causing cancerous effects on important tissue or organ systems, or through immunosuppressive effects, which are difficult to reverse, e.g., "Brain Pox" disease—the fictional disease described in the novel *The Cobra Event*, written by R. Preston.

Gene therapy based bioweapons

Gene therapy based treatment changes genetic composition of a patient through programmed repairing or replacement of faulty genes, and has huge potential in treating diseases causing high mortality in human populations. It can be accomplished in germ cells or somatic cells of the body based upon severity of the disease and to prevent its inheritance by future generations. It has been successfully tested in animal models, e.g., the vaccinia virus has been used as a vector to insert genes in mammalian cells, but the technology is still in its infancy as it is totally unethical to enroll human volunteers for introducing and predicting genetic effects of gene therapies. Retroviruses can be utilized as delivery vehicles as they can easily integrate themselves into the human genome and can overcome all barriers of the natural defense system of the human body.

Host swapping diseases

In the case of zoonotic diseases, where a pathogenic virus has a natural animal reservoir to reside and multiply with little or no effect observed in the carrier species, e.g., chimpanzee for HIV, fruit bats and monkeys for Ebola and Marburg, pigs for swine flu, etc., they can be readily transmitted to the humans through carrier animal species that are in close contact with the human population. Further, animal viruses may be genetically modified to utilize preferential human codons, thereby eliminating the chance of codon biasing, and such generated humanized viral agents would have serious implications in future biowarfare programs.

Stealth viruses

These consist of cryptic viral agents bearing potential human oncogenes that can be illicitly or secretly transferred to human genomes. Usually, they remain dormant for many years but exposure to a single natural stimulus can activate oncogenic determinants present on the stealth viruses and could cause vast destruction in the human population. For example, human herpes virus can cause oral and genital lesions after induction. Similarly, people who have contracted chicken pox previously present a natural reservoir of varicella virus that sometimes rejuvenates in the form of herpes zoster virus causing shingles disease in some people.

Synthetic biology assisted whole genome synthesis of bacterial clones and bacteriophages

Synthetic biology combines science and engineering approaches to design and construct novel pathways, devices, and living systems, as well as to re-design natural biological processes. An exponential increase in whole genome sequence data in the last three decades has provided synthetic biologists a virtual platform for designing and reconstruction of virulent effector elements to introduce necessary changes in the existing repertoire of pathogens through genetic manipulations or reference template assisted assembly of synthetic whole genome sequence. With the recent advancements in this field, it is now possible to artificially synthesize gene constructs with requisite amounts of pathogenic loci, which can be stitched together to create infectious dwarfed genomes or even whole genomes resembling natural human pathogens. Surprisingly, the artificial bacteria and viruses can be constructed using natural genetic segments isolated from extreme environments such as dead animals, fecal samples, or preserved tissues of viral victims buried in permafrost (Table 2). The following paragraphs emphasize important historical developments in the field of synthetic biology leading to development of synthetic native or chimeric (designer) bacterial and viral agents.

Synthesis of bacteriophage φX174

The first artificial bacteriophage, φX174, was constructed to understand the structure and functions of viral genomes infecting important bacterial strains of human relevance. Smith and coworkers in 2003 described synthesis of 5386 bp genome of φX174 by stitching together synthetic DNA fragments using polymerase cycle assembly techniques. Researchers at the Institute for Biological Energy Alternatives (IBEA) at Rockville, Maryland, wanted to use this technology for construction of artificial bacterial chromosomes consisting of a few million base pairs of DNA. Artificial bacterial chromosomes with important genes may be used to generate synthetic microbial factories to produce biofuels such as hydrogen and to cut down carbon emissions from coal degasification units (Smith et al., 2003).

Synthesis bacteriophage T7 genome by refactoring process

Chan and colleagues used refactoring methods to redesign T7 bacteriophage genome (39,937 bp) with the aim to study its important gene functions (Chan et al., 2005). They generated three chimeric bacteriophages, namely α-WT, WT-β-WT, and α-β-WT by removing overlapping genetic segments and replacing a 11,515 bp stretch of wild type (WT) genome with 12,179 bp synthetic α- and β-cassettes using recombination supporting *E. coli* BL21. Overlapping segments conserved but not needed for viral replication can be removed/replaced while maintaining its viability. This study revealed the potential of dwarfed genome to perform all replicative and functional activities and paved the way to the development of the first synthetic bacterial clone, as described below.

Synthesis of *M. genitalium* and *M. mycoides* clones using minimal genome content

Using systematic mutagenesis approach, researchers at The Institute for Genome Research (TIGR), Rockville, Maryland, identified 265–350 genes of urethritis

Table 2 Synthetic biology assisted construction of infectious agents

Year	Viral construct	Nature of genome	Construction	Test model	References
Synthetic viruses					
2002	Polio virus	ssRNA	DNA driven ssRNA synthesis and in vitro phage packaging	HeLa Cell Lines & CD155tg Mice	Cello et al. (2002)
2003	Phi X-174	dsDNA	PCA assisted assembly of synthetic genome	E. coli	Smith et al. (2003)
2005	The 1918 Spanish flu virus	ssRNA	Sequencing &RT-PCR assisted assembly of eight viral RNA fragments from preserved tissues of victims	Mice	Neumann et al. (1999), Fodor et al. (1999), Hoffmann et al. (2000), Taubenberger et al. (1997), Taubenberger et al. (2005)
2005	Bacteriophage T7	dsDNA	Removing overlapping sequences and replacing >30% of viral genome with synthetic constructs α and β	E. coli	Chan et al. (2005)
2006–07	Human endogenous retrovirus	RNA	(1) Whole genome synthesis of HERV-K$_{CON}$ (2) Site directed mutagenesis assisted chemically synthesized consensus sequence of HERV-K(HML-2) named "Phoenix"	(1) HEK 293T cell lines (2) HEK 293T, BHK21, G355.5, SH-SY5Y, HeLa, WOP cell lines	(1) Lee and Bieniasz (2007)) (2) Dewannieux et al. (2006)
2006–07	HIVcpz	RNA	Chemical synthesis of consensus viral string (RNA templates isolated from fecal samples)	Chimpanzee Pan troglodytes	Keele et al. (2006), Takehisa et al. (2007)
2008	SARS-like Coronavirus	RNA	Rationale design and synthetic viral cDNA assisted viral genome assembly	Murine Vero and DBT cell lines; HAE human cell lines and BALB/c mice	Li et al. (2005), Becker et al. (2008))
Bacteria					
2008	Mycoplasma genitalium syn-2.0	dsDNA	First synthetic dwarf genome (582,970 bp) consisting of 485 protein coding and 43 RNA coding genes. Segments joined by in vitro recombination (Work still continued)	E. coli	Gibson et al. (2008)
2010	Mycoplasma mycoides JCV-1-syn1.0	dsDNA	First synthetic bacterial cell consisting chemically synthesized genome with only 400 protein coding and 43 RNA coding genes	S. cerevisiae	Glass (2012)

causing *Mycoplasma genitalium*, essential for maintaining cell viability and supporting cell replication. It was the first attempt to construct a synthetic bacterial clone
using artificial assembled genetic constructs and it was initiated in 1999. Results of
the study were published in *Science*, indicating minimal genome content required for
DNA replication and repair, gene expression, cellular transport, and metabolism and
energy generation in a living prokaryotic cell. This study further led to the successful
synthesis of the first dwarfed (582,970 bp) genome of *M. genitalium* (Gibson et al.,
2008). Preliminary investigations on the genomic transplants in *M. laboratorium* and
M. capricolum have revealed the possibility of development of synthetic species using artificially constructed bacterial genomes. The slow growing *M. genitalium* was
replaced with more prolific strain *M. mycoides* to synthesize the first synthetic bacterial cell, named *M. mycoides* JCV-1-syn1.0, which was successfully booted to life in
2010 (Gibson et al., 2010; Sleator, 2010; Glass, 2012). Work is still under progress
to construct an entirely new designer strain, called *M. genitalium* syn2.0, containing
minimum set of essential genes required for life, to study the potential of synthetic
agents in bioremediation and biomedicine.

Synthetic biology assisted whole genome synthesis of native or chimeric viruses

Synthetic virology, an important branch of synthetic biology, has undoubtedly addressed many diseases inherited from ancestors, and epidemiology and pathogenicity of next generation agents as possible emerging threats in context to biowarfare
(Table 2 and Fig. 1). Synthesis of chimeric viral genomes with designer elements,

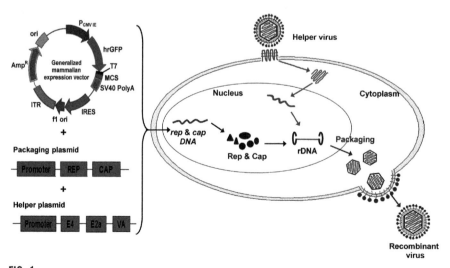

FIG. 1

Schematic representation showing construction of synthetic viruses.

construction of artificial viruses through in vitro phage assembly, and development of delivery systems to confer efficient transmission of designer agents among humans are the most favored trends in the field of synthetic virology these days.

Synthesis of the 1918 Spanish flu virus

The historical influenza pandemic, causing death of more than 50 million persons worldwide during 1918–19, remained undiscovered until 1995. Taubenberger and coworkers, in 1997, initiated efforts to recover viral RNA segments from lung tissue autopsy samples of a 21 year soldier and frozen tissue of an Inuit women buried in the permafrost, both were among the 1918 pandemic victims (Taubenberger et al., 1997). They reconstructed the genomes of the 1918 "Spanish Flu" virus from eight viral RNA segments using the techniques of gene sequencing and RT-PCR and, later, the authors successfully assembled artificial virus responsible for the Spanish Flu pandemic. The technique of "reverse genetics" allowed the construction of the first synthetic virus in October 2005, at the Armed Forces Institute of Pathology in Rockville over a span of 10 years (Neumann et al., 1999; Fodor et al., 1999; Hoffmann et al., 2000; Taubenberger et al., 2005, 2007). Out of a total of eight genes, hemeagglutinin (HA), neuraminidase (NA), and polymerase B1 (PB1) were considered important virulent factors contributing to the severity of the disease. Sixteen different variants were reported for HA antigens, mainly aglycoproteins in nature that help in attachment of virus to the host cell, while nine subtypes of NA, antigenic glycoproteins, were reported in humans and animals, with N1 and N2 linked to epidemics in man and others, specifically, for ducks and chickens. HA type-5 (HA5) and NA type-1 (N1) are important components of viral capsids required for assembly of infectious viral particles. Neuraminidase activity is important to release replicated viruses from the infected host cell. The Spanish Flu virus probably originated in birds, and evolved to cause the 1918 epidemic.

Synthesis of poliovirus

In 2002, Cello and coworkers artificially constructed poliovirus using cDNA driven synthesis of viral RNA genome in absence of its natural template and then, using a mixture of biologicals, the resulting RNA genomes were subsequently enveloped to generate artificial poliovirus (Cello et al., 2002). Synthetic virology approach was followed to investigate the functional features of the viral genome and the underlying mechanism of pathogenicity associated with its important virulence factors. Twenty-five mutations were introduced intentionally in the synthetic cDNA construct that served as genetic markers to assess properties of viral genome architecture and associated functional genetic loci. Subsequently, artificial viruses were tested for infection using HeLa cell lines and CD155tg mice models and confirmed as infectious. However, the rate of infection was considerably lesser than the wild type strain. Chemical synthesis of this first lytic animal RNA virus confirmed the accuracy of the deduced viral genome sequence and assigned oncological features to defined loci in the viral genome.

Synthesis of human endogenous retrovirus

Human endogenous retrovirus (HER) includes a class of degenerate human retroviruses including Human Mouse Mammary Tumor Virus-like 2 provirus (HML-2) of the Human Endogenous Retrovirus K provirus (HERV-K (HML-2)). Synthetic consensus sequence and site-directed mutagenesis were used to generate infectious proviral particles of (HERV-K (HML-2) called "Phoenix" (Dewannieux et al., 2006). Another proviral clone HERV-K$_{CON}$, a close relative of progenetor HERV-K (HML-2) variant that infested human genome few million years ago and inherited since then with the human genome in a Mendelian fashion, was generated using whole genome synthesis (Lee and Bieniasz, 2007). Thorough investigations were carried out in human cell lines, namely HEK293T, HeLa, SH-SY5Y, Baby Hamster Kidney BHK21, Feline G355.5, and Murine WOP cell lines. Out of these, no infection was reported in the case of HeLa and WOP cell lines by Phoenix. Ancestors of these proviral strains may be much less infectious, but these studies may provide valuable information related to ancient proviral genomic repertoires that have contributed immensely to human evolution and physiology.

Synthesis of HIVcpz

Viruses causing zoonotic infections, such as human immunodeficiency virus (HIV-1) and simian immunodeficiency virus (SIVcpz) have natural reservoirs in wild chimpanzees such as Pan Troglodytes troglodytes. A research group isolated viral nucleic acid strings from fecal samples from wild Pan Troglodytes troglodytes. They derived a consensus viral sequence that was synthesized artificially and used to produce infectious molecular clones of SIVcpz (Keele et al., 2006; Takehisa et al., 2007). Synthesis of in vitro particles might facilitate investigation of important viral elements that determine cross species transmission of these retroviral elements and mechanism of host adaptive responses to viral infections.

Synthesis of SARS-like coronavirus

From 2002 to 2003, an unknown infectious agent infected 8427 persons in China, of whom 813 died due to unknown cause and lack of appropriate treatment measures. The causal agent was soon identified as a new species of coronavirus named "severe acute respiratory syndrome virus coronavirus (SARS-CoV)" by the World Health Organization. SARS-CoV disappeared in July 2003, as rapidly as it had emerged in 2002. Bats are natural reservoirs of SARS-CoV and none of the human in vitro culture systems supports viral replication. The possibility of human adaptation of bat SARS-CoV was studied using synthetic SARS-CoV viral cDNA. Authors created artificial clones by exchanging the receptor binding domain (RBD) with that of human SARS-CoV capable of infecting VeroE6 cell, DBT-hACE2, DBT-cACE2 (murine cell lines), HAE human cell lines, and BALB/c mice. This was the example of the largest retroviral genome (approx. 30 kb) capable of replicating and infecting human cells. It is suspected that SARS-CoV may re-emerge again, and might be even more deadly than the previous form, due to the possibility of cross species transmission of virulent characters to existing repertoire of coronavirus infecting other mammals like civets (Li et al., 2005; Becker et al., 2008).

The authors also fear that replication of the acquired knowledge related to virulent genetic loci and assembly of designer pathogens is possible and might be utilized by bioweaponeers for construction of more deadly viruses and designer pathogens that confer efficient transmission among humans.

In vitro packaging of viral genomes

Rapid advancement in the field of DNA synthesis and sequencing is heading towards the deliberate, large-scale genetic manipulation of organisms that may be further extended to whole-genome synthesis of viruses. The main aim of this new approach is to understand an organism's pathogenic properties in relation to humans and to protect or treat human viral disease, if any. It is, however, necessary to understand the detailed mechanism of packaging viral genomes, which enhances infectivity of the synthesized host specific chimeric constructs. Different models have been proposed for encapsulation of viral genomes, as discussed below. For genome packaging, viruses must make a distinction between viral and host nucleic acid, which is assisted by an outer membrane capsid protein with receptor binding domain (RBD) that helps in recognition and subsequent binding to the target genome. In case of L-A virus of yeast, a stem-loop secondary structure and a site-specific sequence at 5′-end of the genome is recognized by polymerase-group antigen (pol-gag) fusion protein prior to packaging (Fujimura et al., 1992). Although only a few applications of virus synthesis have been described as yet, key recent findings have been the resurgence of influenza virus and polioviruses (Wimmer et al., 2009). Various methods of in vitro packaging of viral genes for the assembly of infectious viruses are described in subsequent paragraphs.

Mechanism for dsRNA viral genome packaging

In φ12 viruses containing dsRNA genome, the ssRNA acts as a substrate for the packaging motor protein, i.e., P4 ATPase. Inside the capsid, the positive-sense strand of the virus genome is packaged and replicated to form dsRNA. P4 ATPase comprises a hexameric multifunctional unit (lined with a helix α6 and loops L1 and L2) that plays an important role in procapsid assembly and ssRNA packaging. During infection, it acts as a passive channel for the extrusion of newly synthesized mRNA molecules from the virus. The genome of φ12 encloses three segmented dsRNAs, each comprising a positive strand sequentially recognized by the procapsid, which undergoes conformational changes to accommodate all the RNA segments. The loops L1 and L2 are vital for RNA binding and translocation. The phosphate-binding P-loop upon ATP hydrolysis changes its confirmation from "down" to "up" form, accompanied by similar transition of α6 and loop L2 from "up" to "down" position, signifying the driving force for RNA translocation. The arginine finger of hexameric ATPase motors plays an important role in sequential ATP hydrolysis, inducing conformational changes that direct the *trans* arginine finger to active center of ATPase (neighboring subunit), thus triggering the successive ATP hydrolysis.

Mechanism for dsDNA viral genome packaging

In bacteriophages containing crystalline dsDNA genome, at the end of the packaging process packaging motors are required to generate enough force to offset the pressure inside the viral capsid. The most influential packaging motor is the bacteriophage T4 genome packaging motor, which helps in packaging DNA by generating a force.

Mechanism of linear motor assisted viral genome packaging

An electrostatic interactions based mechanism has been proposed for viral genome packaging in bacteriophage T4. The packaging motor protein has two domains, i.e., N-terminal ATPase domain and C-terminal nuclease domain, which are linked together by small amino acids to provide the flexibility to the motor to ensure packaging. ATP hydrolysis is triggered upon dsDNA binding to C-terminal domain (gp17 subunit) and thereby positioning the *cis* "arginine" finger of N-terminal domain into the ATPase active center that further results in subsequent conformation changes. These changes align opposite charges of both N- and C-terminal domains that pull C-terminal domain towards N-terminal domain by the action of electrostatic forces, leading to packaging two base pairs of dsDNA (Sun et al., 2010).

Mechanism of rotary motor assisted viral genome packaging

The main component of DNA packaging machine is portal protein gp10 (dodecameric) of φ29 having a central α-helical channel lined with negative charges for easy passage of DNA. It also comprises a wider end inside the capsid and a narrower end protruding from the capsid; and energy from ATP hydrolysis is used for rotating the portal in order to drive the DNA into the procapsid (Hugel et al., 2007).

Examples of in vitro packaged viral genomes

Adeno-associated virus (AAV) are nonenveloped icosahedral parvoviruses containing ssDNA with intact inverted terminal repeats (ITR), resulting in the formation of important secondary structures in viral genome. Further, co-infection with helper virus (like adenovirus or herpes simplex virus, HSV) assists its replication with host cell polymerase (Ni et al., 1994). An in vitro course of action for the packaging of AAV was studied by Zhou and Muzyczka (1998). Synthesis of an infectious AAV particle was done by using its replicative-form of DNA as a substrate, AAV Rep, and capsid proteins in order to transfer recombinant gene to mammalian cells. Two types of products formed, both were heat-resistant, indicating an appropriate ratio of protein-to-DNA. In addition, products also share structural resemblance of mature AAV particles. An efficiently synthesized particle enclosing intact terminal repeats always shows chloroform, DNase I, and heat-resistance, pertaining to an authentic AAV particle. Resistance is known to be a vital property for packaging purposes. However, nonpathogenic and persistent nature, in combination with its broad range of infection, as reported by Wright and coworkers, marked this virus as an imperative nominee for a therapeutic gene transfer vector (Wright et al., 2003).

An in vitro study was conducted by Cashion et al. (2005) to investigate the application of gene therapy for treatment of neurological diseases by using human polyoma virus JCV derived virus-like particles (VLP). Here, VLP act as a delivery vector for the central nervous system (CNS) because JCV preferentially infects both oligodendrocytes and astrocytes. The construction of JCV-derived recombinant VP1 in insect cells and respective packaging strategies for purified VP1 and plasmid DNA were optimized. In order to illustrate tropism and species specificity of VP1-VLP containing plasmid DNA expressing EGFP in vitro, transduction of VP1-VLP was done in human and rodent brain-derived and nonbrain derived cells. Significant transduction was observed in human prostate cell line (PC-3); thus, assigning VP1-VLP as an efficient and selective delivery system for therapeutic genes to target specific cells in the brain (Cashion et al., 2005).

Techniques of synthetic biology foster current challenges in agriculture and industry, biological defense, environmental, and medical sciences and provide important breakthroughs in improving global scenarios of human and animal health. It is, however, difficult to rule out the possibility of replicating existing knowledge domains of synthetic biology for unlawful activities and spreading economic and physiological distress at a global scale among vulnerable human populations.

Biowarfare agent detection: Methods and challenges

Biowarfare agent threat is the foremost national and world security concern, attributable to their potential economic, psychological, and social impact. Effective protection against BWAs is quite difficult because of their intricate detection and expensive protection measures. To counter this threat, several countries have established their own biodefence programs to strengthen the strategies for detection, protection, and decontamination of biowarfare agents (Pal et al., 2016). Early detection and identification of BWAs is essential to initiate corrective emergency responses for management of such incidents. Efforts are being made across the globe for development of efficient technologies and systems for detection and identification of BWAs. Many advanced molecular and microbiological sensing techniques such as antibody-based immunoassays, cellular fatty acid profiling, flow cytometry, nucleic acid based detection, mass spectrometry, microbiological culturing, and genomic analysis have been used for primary identification of biological agents. These techniques are highly reliable, sensitive, and selective, and have been successfully applied for detection of potential biowarfare agents. However, despite the handiness of available techniques and tools, no foolproof system is available for the complete detection of hazardous biowarfare agents. Regardless of being highly efficient, these detection methods possess various drawbacks such as complicated and laborious isolation and purification procedures, low detection limits, contrasting etiology and pathology, and different physiochemical and structural attributes of bioagents, which ultimately affect the detection efficacy (Suter, 2003; Sapsford et al., 2008; Das and Kataria, 2010; Madad, 2014).

Microbiological culturing

Microbiological culturing is the conventional method used for the isolation and identification of biological agents such as bacteria, fungi, and viruses. Microbes have the ability to propagate in selective culture media, which allow only the targeted microorganism to grow. Selective culturing offers an additional benefit of long term viability and enrichment of the concerned microbe for further characterization. Various tests have been employed for the morphological identification and biochemical characterization of a particular biological agent. Microbiological culturing is highly reliable and specific, but is laborious and time consuming, which limits its efficacy (Pal et al., 2016).

Flow cytometry

Flow cytometry involves scattering of laser light and emission of fluorescence by excitation of dyes linked with bacterial cells. Cell size and cell count in the case of liquid suspension are estimated by laser light scattering. Monoclonal antibodies that are fluorescently labeled can also be used for detection of various pathogens. Biowarfare agents such as *B. anthracis*, *B. melitensis*, botulinum toxin, *F. tularensis*, and *Y. pestis* can be easily identified using flow cytometry techniques (McBride et al., 2003; Hindson et al., 2005).

Cellular fatty acid based profiling

In 1963, two separate reports, by Abel and coworkers and Kaneda, described bacterial identification methods based on cellular fatty acid profiling. Bacterial strains can be easily distinguished in terms of variability of their fatty acids structures and profiles. Firstly, conversion of cellular fatty acids to fatty acid methyl esters takes place, followed by analysis by gas liquid chromatography. GC chromatograms generate important fatty acid fingerprints that have been successfully employed for identification and characterization of various biological agents, viz. *B. anthracis*, *B. mallei*, *Brucella*, *B. pseudomallei*, *F. tularensis*, and *Y. pestis* (Abel and Peterson, 1963; Kaneda, 1963; Pal et al., 2016).

PCR based detection

Molecular biology techniques offer specific and rapid identification of biowarfare agents, as compared to conventional microbiological techniques. Polymerase chain reaction (PCR) based assays identify an organism on the basis of presence of specific DNA sequence(s) in the organism. Quantitative real-time PCR (Q-PCR) based on specific and nonspecific detection is also used for amplification and simultaneous detection of targets. PCR-based identification has been reported in the case of various biowarfare agents such as arenaviruses, *B. anthracis*, *C. burnetii*, filoviruses, *F. tularensis*, and *Y. pestis*. Recombinase polymerase amplification (RPA), an alternative form of DNA amplification technique, has been used to scan the presence of double stranded DNA templates for homologous sequences. RPA is rapid and highly sensitive technique, as it can detect even a single copy of the target in <20 min. RPA assays and reverse transcriptase RPA (RT-RPA) assays have been successfully used for detection of BWAs such as *B. anthracis*, *Brucella* sp., Ebola virus, *F. tularensis*, Marburg virus, Rift Valley fever virus, Sudan virus, variola virus, and *Y. pestis*. RT-PCR has also been used for the detection of chimeric viruses such as Zika virus, yellow fever virus, Ebola virus, and Mengla virus (Alfson et al., 2017; Kum et al., 2018; Yang et al., 2019). Disadvantages of nucleic acid based detection techniques include their inability to detect proteins such as toxins (Janse et al., 2010; Trombley et al., 2010; De Bruin et al., 2011).

Immunological methods

Immunoassays based on antigen-antibody interactions have been widely exploited for identification of potential biowarfare agents. Antibodies bind to specific antigens present on the surface of the cell and form a colored or detectable complex, which ultimately marks the presence/detection of a bioagent in the sample. Enzyme linked immunosorbant assay (ELISA) has been mainly used for quantitative detection of antigen following the basic principle of immunoassays. ELISA has been widely used for diagnosis of several diseases and simultaneous screening of large number of samples. Thus, the technique is highly efficient, economical, and reliable. To date, ELISAs have been successfully employed for the detection of biowarfare agents such as *B. anthracis*, *B. pseudomallei*, *B. mallei*, *Brucella abortus*, Ebola virus, *F. tularensis*, Marburg virus, toxins, and *Y. pestis*. Besides these, fluorescent microscopy has also been used for biowarfare agent detection, where a fluorescent labeled antibody attached to antigenic receptors present on the surface of the microbial cells aids in its detection. Immuno-histochemical based methods have also enabled detection of some viruses such as Alphaviruses and Chikanguniya viruses (Wang et al., 2008). Other immunoassay, namely, hand-held immuno-chromatographic assays (HHIAs), are also used for detection of biowarfare agents such as *B. anthracis, B. abortus, B. pseudomallei*, botulinum, *F. tularensis*, smallpox virus, Ricin toxin, variola virus, and *Y. pestis*. HHIAs are cost-effective, simple, rapid, and are performed on nitrocellulose or nylon membranes and are based on lateral flow immunoassays. Even having several advantages, these assays are less sensitive and specific as compared to other immunological methods (Gomes-Solecki et al., 2005; Wang et al., 2009; Ghosh and Goel, 2012; Sharma et al., 2013; Pal et al., 2016).

Next generation sequencing

DNA sequencing techniques have been used for the unambiguous identification of biological warfare agents. Next-generation sequencing (NGS) technologies have radically changed the traditional ways of DNA sequencing, and thus, have opened new vistas in the field of identification of bacterial and viral bio-threats from clinical and environmental samples. NGS involves simultaneous sequencing of multiple DNA fragments for determination of the desired sequence. In recent years, NGS technologies have gained much importance and validation as an effective biodefense strategy due to their highly specific and rapid detection capabilities. NGS techniques have been applied for *B. anthracis* detection in air and soil samples. Strain-specific polymorphism has also been identified by NGS in the case of *B. anthracis* and *Y. pestis*. *F. tularensis* was detected in human abscess samples of unknown etiology by next generation direct DNA sequencing technique. NGS technologies have been extensively used in medical diagnostics, mainly for the identification of novel infectious biological agents for which diagnostics and therapeutics are currently unavailable (Cummings et al., 2010; Kuroda et al., 2012; Lefterova et al., 2015).

Bio-sensors

Bio-sensors are analytical devices that generate response in the form of an electrical signal by interacting with the analyte in a biological component (biological warfare agent). The biological response produced is then converted to a detectable form by the transducer, which marks the presence of any biowarfare agent in the sample. Biosensors offer significant advantages in terms of high specificity and selectivity in comparison to conventional detection techniques. Thus, these are being widely used for biological detection. Bio-sensors have been categorized into different types according to the type of transducer and bioreceptor used.

Nanomaterials have also been used for the development of highly efficient and specific electrochemical bio-sensors for easy detection of biowarfare agents. A highly specific electrochemical immuno-biosensor consisting of bismuth nanoparticles (BiNPs) has been developed for anthrax PA toxin detection in a particular sample (Sharma et al., 2015). Another electrochemical immunosensor, consisting of gold and palladium bimetallic nanoparticles, has been developed for detection of *B. anthracis* with the 1 pg/mL detection limit (Sharma et al., 2016). *B. anthracis* was also identified using electrochemical genosensor loaded with gold nanoparticles by detecting its PCR amplicons with detection limit of 1.0 pM (Das et al., 2015). Botulinum neurotoxin type-E was also identified by an electrochemical immunosensor assembled with gold nanoparticles and graphene transducer (Narayanan et al., 2015). Wu and coworkers reported the identification of *B. melitensis* using an impedometric immunosensor loaded with gold nanoparticles and carbon electrodes (Wu et al., 2013).

Surface plasmon resonance (SPR) technique has also been used for label-free detection of various biological agents. Label free detection offers significant advantages over other methods, which require secondary labeled reagents for the detection purposes. Using SPR technique, rapid and specific detection of bioagents has been reported, such as of *B. anthracis*, botulinum neurotoxin, *Brucella*, *Staphylococcus* enterotoxin A (SEA) and B (SEB), and *Y. pestis*. However, piezoelectric bio-sensors using quartz crystal microbalances (QCM) have been considered as better alternatives to SPR, hence have also been extensively used for biological agent detection. A piezoelectric immunosensor with detection limit of 5×10^6 cells has been developed for detection of *F. tularensis*. An immunosensor assembled with QCM detection has been developed for detection of staphylococcal enterotoxin A in milk samples (Salmain et al., 2012; Ghosh et al., 2013).

Biophysical detector systems

Generally, biological detectors only detect the presence of any biowarfare agent in a particular environment without identifying the nature and type of that bioagent. However, if these detectors are attached to an identifier, then these become capable of identifying the nature of the particular biological agent. There are separate and independent units assembled in a single system for different purposes. For sample collection, various types of samplers/collectors are being used, such as cyclone samplers, viable particle size samplers, and virtual impactors. Whereas for detection/identification purposes, different types of detectors, such as fluorescence-based detectors and particle size-based detectors, are being used. Nowadays, biological detectors are also widely used for various biological agent detection (Pal et al., 2016).

Protection against next generation biological agents: Methods and challenges

For decades, humans have solely relied on vaccines for protection against infectious viruses. However, to date it has not become completely feasible to develop vaccines against all the viral infections; moreover vaccines are not believed to be completely effective against all the infections (Henderson et al., 2003; Quinn et al., 2008). So, challenges associated with development, efficacy, and safety of vaccines have eventually led to emergence of better alternatives for prevention of viral diseases. These alternatives are more efficient than vaccines as they directly target the particular virus and interrupt its life cycle at molecular level by use of antibodies, specific proteins, and oligonucleotides. Such strategies, referred to as "biochemical prevention and treatment," have been considered more successful than vaccines/chemical drugs for protection of humans against some pathogenic viruses such as hepatitis C virus (HCV), HIV, and human rhinovirus (HRV).

Biochemical prevention and treatment strategies generate immediate response and protection against a particular infection, whereas in the case of vaccines, it takes a longer time and booster doses for the generation of an immune response. These strategies work by following two mechanisms; one by blocking viral entry via use of host cell receptor blockers or by protein-based specific antiviral molecules, and secondly by targeting the viral mRNA and inhibiting viral replication by use of antisense oligonucleotides, ribozymes, and RNA interference (Le Calvez et al., 2004). Chimeric proteinaceous toxins have proved to be effective therapeutics for providing protection against HIV-1 infection. Two chimeric toxins, namely CD4-PE40 and 3B3 (Fv)-PE38, were designed to target the HIV envelope (Env), which selectively kill the infected cells. These chimeric toxins were further tested against mice models to investigate their potential therapeutic efficacy against HIV and markedly suppressed acute HIV-1 infection (Goldstein et al., 2000). However, peptide-based drugs generally face issues of low potency and unfavorable pharmacokinetics.

Monoclonal antibodies (MAbs) have been widely used as biochemical therapeutics. Development of chimeric, humanized, and antiviral antibodies has been very beneficial for treatment of many viral infections. An FDA approved monoclonal antibody known as Synagis was developed and proved successful in prevention and treatment of respiratory syncytial virus (RSV). This antibody inhibits viral replication by binding specifically to the RSV surface glycoprotein and has been considered as the primary medical means of providing protection against RSV (Cohen, 2000). Host cell receptor blockers have also proven to be efficient in inhibiting viral infections by blocking virus entry into the cell. Receptor-blocking has been commonly carried out via application of monoclonal antibodies that bind to specific epitopes present on the receptor molecules. A MAb was generated against ICAM-1 (adhesion molecule responsible for viral entry and attachment), which was capable of providing protection against infections caused by human rhinoviruses (HRV) (Marlin et al., 1990). However, the efficacy of MAbs is limited to an extent because of their low functional affinity for adhesion molecules in comparison to multivalent viral particles (Casasnovas and Springer, 1995). To overcome this challenge, avidity or functional affinity of antiviral antibodies was improved by the generation of recombinant antibodies. A tetravalent recombinant antibody, CFY196, with improved avidity was developed against ICAM-1 and was capable of preventing HRV infections (Charles et al., 2003).

Antisense-oligonucleotides (AS-ONs) are short synthetic oligonucleotides that inhibit viral protein production by blocking viral mRNA translation have also been explored for providing protection against viral infections. Vitravene, the first AS-ON based drug, is a potent antiviral agent for cytomegalovirus retinitis (a herpes-like eye disease). Vitravene binds complementarily to the viral messenger RNA and inhibits its translation and hence prevents the infection caused by human cytomegalovirus (Orr, 2001). Antisense phosphorodiamidate morpholino oligomers (PMOs) have also been used for providing protection against viral infections mainly caused by filoviruses (Iversen et al., 2012; Nan and Zhang, 2018).

Ribozymes are catalytically active oligonucleotides, which selectively bind and cleave target RNAs. Ribozymes have been considered as better alternatives to AS-ONs. Successful animal and cell based trials have been carried out, which confirms the use of ribozymes as potent antiviral agents. Ribozymes have been efficient viral inhibitors for infections such as influenza, hepatitis B and C, HIV, etc. (Yu et al.,

1993; Tang et al., 1994; Welch et al., 1996, 1997). HEPTAZYME, a modified ribozyme, cleaves target entry site of the hepatitis C virus and hence inhibits the infection. However, further research on use of ribozymes as biotherapeutics has been hampered by low potency and inefficient in vivo intracellular delivery. In contrast to this, RNA interference has significantly enhanced potency in comparison to other technologies. Therefore, only low levels of RNAi based antiviral drugs are sufficient to generate an effective immune response. Synthetic siRNAs have potential applications as potent antiviral agents. RNAi technology has been successfully applied for inhibition of replication of several pathogenic viruses such as filoviruses, influenza virus, HIV-1, poliovirus, and RSV (Jacque et al., 2002; Novina et al., 2003; Ge et al., 2003; Ursic-Bedoya et al., 2013).

Chimeric or designer viruses as candidates to study disease pathogenesis

EBOV and MARV viruses are highly lethal bat-borne filoviruses that cause severe hemorrhagic fever disease in humans (Sarwar et al., 2011). An EBOV outbreak in West Africa claimed at least 11,000 lives and caused a huge economic loss to the country during 2013–16 (Bausch, 2017). Baize and coworkers suggested that a single spill over from animal reservoir is sufficient to initiate a fresh outbreak of the EBOV (Baize et al., 2014). Egyptian fruit bats *Rousettus aegyptiacus* have been identified as likely reservoirs of MARV and EBOV without developing symptoms (Jones et al., 2015; Paweska et al., 2016). Many studies have reported essential involvement of Niemann-pick C1 glycoprotein for entry into bat and human cells. However, the magnitude of infection may be species dependent in these filoviruses (Carette et al., 2011; Côté et al., 2011; Hoffmann et al., 2016; Yang et al., 2019). Two recent studies have reported synthesis of chimeric viruses using EBOV and MARV leader and trailer sequences, as discussed below (Fig. 2).

Synthesis of chimeric LLOV-(EBOV/MARV/RESTV)

A filovirus called Lloviu virus (LLOV) was discovered for the first time in Spain, causing high mortality (Negredo et al., 2011). Recently, LLOV emerged in Northeast Hungary, causing increased mortality in *Miniopterus schrei bersii* bats (Kemenesi et al., 2018). Manhart et al. (2018) obtained partial genome sequence of Lloviu virus (LLOV) with no known pathogenicity showing it a close relative of EBOV. Both share the same replication strategy, as LLOV polymerase also binds to 3′ terminal nucleotides for recognition of promoter region. Authors also reported that human cells support replication and transcription of LLOV. Chimeric LLOV mini-genome using EBOV, MARV, and RESTV (Reston virus) leader and trailer regions was constructed to study sequence of events that occur during transcription and replication of infected human BRT7/5, HEK293T cells. Thus, mini-genome strategy proved to be significant enough to rescue infectious LLOV clones and for characterization of novel mini-genome filovirus (Manhart et al., 2018).

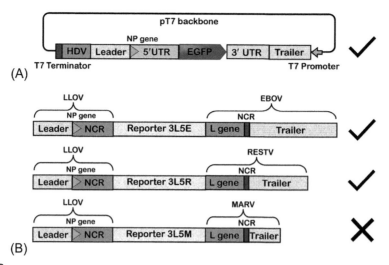

FIG. 2

Chimeric constructs showing efficient replication and expression of reporter genes where (A) schematic diagram of chimeric MLAV minigenome comprising T7 promoter and terminator region; HDV; leader and trailer sequences; enhanced green fluorescent reporter gene (EGFP); NP gene and 5'- and 3' UTRs; marks represent successful replication and expression of reporter genes and mark represents noncoding constructs (adapted from Yang, X.-L., Tan, C.W., Anderson, D.E., Jiang, R.-D., Li, B., Zhang, W., Zhu, Y., Lim, X.F., Zhou, P., Liu, X.-L., 2019. Characterization of a filovirus (Měnglà virus) from Rousettus bats in China. Nat. Microbiol. 1) and (B) chimeric LLOV minigenome containing reporter genes flanked by 3' leader and noncoding region (NCR) of LLOV NP gene and 5' NCR of L gene and trailer of EBOV, RESTV, or MARV (adapted from Manhart, W.A., Pacheco, J.R., Hume, A.J., Cressey, T.N., Deflubé, L.R., Mühlberger, E., 2018. A chimeric Lloviu virus minigenome system reveals that the bat-derived filovirus replicates more similarly to Ebolaviruses than Marburgviruses. Cell Rep. 24, 2573–2580. e4).

Synthesis of chimeric MLAV-(EBOV/MARV)

Yang and coworkers reported the characterization of another phylogenetically distinct relative, named Měnglà virus (MLAV), from *Rousettus* bats in China, which has 32%–54% genome sequence identity with known filoviruses. As per pairwise sequence comparison (PASC), analysis of genome designates MLAV as a new genus, i.e., Dianlovirus (family: filoviridae). Chimeric MLAV mini-genomes with EBOV or MARV leader and trailer sequences were constructed to study replication-competence and interspecies spillover transmission by transducing cell lines derived from bats, dogs, hamsters, humans, and monkeys. However, the assessment of risks involved in interspecies transmission needs to be evaluated in vivo to study pathogenesis of MLAV (Yang et al., 2019).

Chimeric viruses as important vaccines candidates

Despite the emerging threats of viral epidemics and their potential utilization for development of BWAs, there is no licensed vaccine available against chikungunya virus (CHIKV) (Wang et al., 2008; Darwin et al., 2011; Kaptein and Neyts, 2016) or West Nile NY99 virus (Huang et al., 2005). Chimeric viruses are affordable candidates for development of vaccines against contagious viruses, as shown in Fig. 3 (Wang et al., 2008).

Chimeric Zika virus

Zika virus (ZIKV) is a single-stranded RNA flavivirus transmitted by *Aedes* spp. mosquitoes and is associated with various congenital neurological complications. The 10.8 kb genome of ZIKV encodes a single polyprotein that, upon action of host and viral proteases, forms three structural proteins (C, PrM, and E) and seven non-structural proteins (NS1–5 NS2A, NS2B, NS3, NS4A, NS4B, and NS5) (Ye et al., 2016). Numerous approaches are being employed to develop live attenuated virus (Shan et al., 2017), inactivated whole virus (Sumathy et al., 2017), subunit DNA/RNA vaccines (Pardi et al., 2017; To et al., 2018), virus like particles (Espinosa et al., 2018; Salvo et al., 2018), over virus vectored (Xu et al., 2018), and chimeric ZIKV (Kum et al., 2018; Li et al., 2018) vaccines pertaining to several benefits including economical and long-term immunity. But only a few of them have generated appropriate prophylactic responses in tested in vivo models. Recently, a ZIKV vaccine candidate based on JEV, a licensed live attenuated SA14-14.2 flavivirus vaccine as a backbone was reported (Li et al., 2018).

Kum et al. (2018) prepared a chimeric virus vaccine construct (YF-ZIKprM/E) by swapping antigenic surface glycoproteins (prM/E) and capsid anchor (Canch) of yellow fever virus-17D (YFV-17D) with corresponding sequence of pre-epidemic Asian ZIKV isolate. Several tissue culture adaptive mutations were also introduced in chimeric virus to ensure efficient replication and extracellular viral release. In mosquito cells, YF-ZIKprM/E, in comparison to YFV-17D, replicates inadequately and has proven to be avirulent in AG129 mice and BALB/c pups. In addition, it also induces a protective immune response in immunocompetent C57BL/6 and NMRI mice models (Kum et al., 2018).

Another chimeric virus, CH-17-D/ZIKV, comprising prM/E proteins of ZIKV strain integrated into yellow fever virus 17-D attenuated backbone was constructed by Touret et al. (2018). Using Infectious Subgenomic Amplicons (ISA) reverse genetics methods, cleavage site between prepeptide and prM protein was modified. In Vero-E6 cells, study confers the chimeric strain to be fitter than parental one and is in close relation to 17-D vaccine strain in HEK-293T cells. Furthermore, preimmunized mice were protected against neuro-invasive disease following challenge with a heterologous ZIKV strain. Researchers also reported a live attenuated vaccine against YFV currently being commercialized (Chin and Torresi, 2013; Scott, 2016).

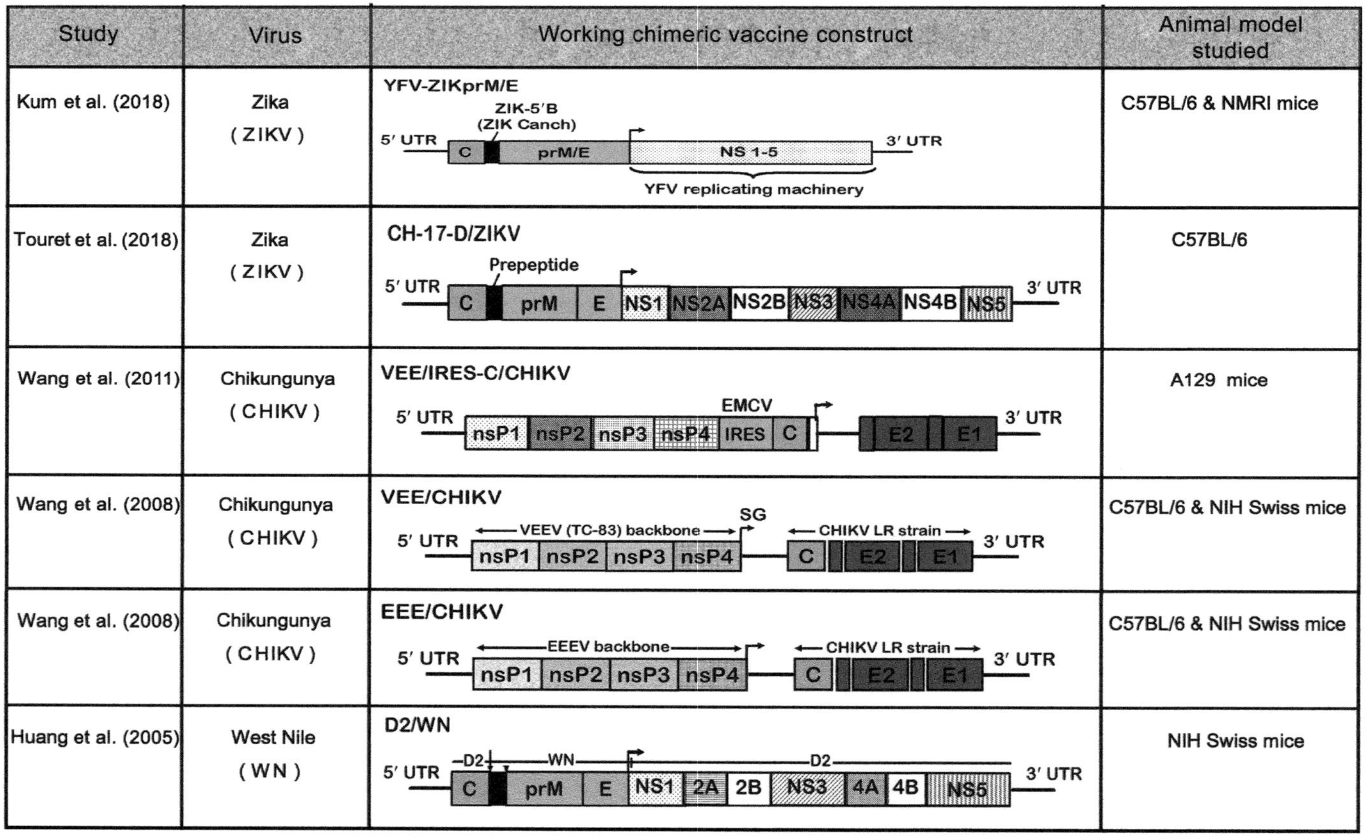

FIG. 3 Working constructs of chimeric viruses as potential vaccine candidates.

Chimeric West Nile virus

West Nile virus (WNV) is a flavivirus that causes infection in blood samples of vertebrates and frequently relies on the plaque reduction neutralization test (PRNT), being considered as the most specific and sensitive antibody detection test in case of arboviruses (Komar et al., 2009).

A chimeric virus, Dengue serotype 2/West Nile (D2/WN), was prepared by Huang et al. (2005) by co-expressing prM/E of WN NY99 virus and two D2 PDK-53 vaccine constructs, namely PDK53-E and PDK53-V. The integrated prM/E specific signal sequence from WN virus is an important determinant of chimeric viability. In addition, two mutations were introduced in chimeric cDNA clones at M-58 and E-191 positions to improve their viability. The feature of phenotypic markers of attenuation of PDK-53 vaccine was retained by D2/WN-E2 and -V2 chimeras; and reported to be immunogenic and protect mice from a high-level dosage challenge with wild-type WN NY99 virus. Furthermore, study favors D2 PDK-53 virus to be a carrier for development of chimeric flavivirus vaccines (live-attenuated) and chimeric D2/WN vaccine virus against WN disease (Huang et al., 2005).

Furthermore, a study was conducted by Komar et al. (2009) to evaluate the effectiveness of chimeric construct of YF-17D and WNV against wild-type WNV in PRNT test. Premembrane and envelope protein of WNV (strain New York 1999) were inserted in gene sequences of attenuated YF-17D strain resulting in infectious chimeric YF/WN virus. Since YF/WN was found to be more attenuated than the wild-type WNV and YF-17D strain, it was recommended as a surrogate diagnostic reagent in place of WNV for PRNT assays with the warning of reduced sensitivity for detecting low levels of antibodies. With chimeric construct, the titres of neutralizing antibody were found to be reduced by more than twofolds (Komar et al., 2009).

Chimeric CHIKV

CHIKV is an emerging alphavirus, first isolated from febrile humans in Tanzania in 1953, causing severely incapacitating disease characterized by fever, rash, and joint pains, which persists for months (Karabatsos, 1985). Earlier, a live attenuated and highly immunogenic CHIKV vaccine was developed using 181/clone 25 strain derived from a wild type Thai strain (Levitt et al., 1986). However, a small group of vaccinated humans developed symptoms of arthritis when clinical investigation of the developed vaccine was in Phase II safety trials (Edelman et al., 2000). Live attenuated vaccines are preferred over other prophylactic countermeasures involving use of dead organisms as they offer a quick and long lasting immunogenic response even after single immunization. However, there is very high probability of natural reversion and dissemination reactogenicity of virulence characters loading to development of viral diseases or viremia. In contrast to this, chimeric vaccines have very light or no chance of producing viremia, as tested in animal models (Wang et al., 2008).

Thus, a chimeric alphavirus/CHIKV vaccine candidate was constructed using three recombinant alphaviruses as backbone, i.e., sindbis virus (SINV)-AR339, TC-83 vaccine strain of Venezuelan equine encephalitis virus (VEEV), and eastern equine encephalitis virus (EEEV) strain expressing CHIKV structural protein genes. In

BHK-21, all chimeras replicated efficiently and were highly attenuated in C57BL/6 mice models producing robust neutralizing antibody response. Interestingly, TC-83 and EEEV backbones present ED greater immunogenicity, and vaccinated C57BL/6 mice were fully protected against disease after CHIKV challenge (Wang et al., 2008). Later on, Wang et al. (2011) provided convincingly evidence supporting production of appropriate immune responses by the developed chimeric viruses in both immuno-competent and immunocompromised (A129) mice (Wang et al., 2011).

More chimeric vaccine candidates, namely TC-83/CHIKV and EEE/CHIKV, were prepared using structural genes of CHIKV and nonstructural protein genes of VEEV (TC-83-attenuated vaccine strain) or EEEV. The potential of two constructs to infect the CHIKV vectors, *Aedes aegypti* and *Ae. Albopictus*, was assessed in comparison to parental wild strains and found to be poorly infectious and have lower dissemination rates that might be mediated by midgut infection barriers. Hence, both TC-83/CHIKV and EEE/CHIKV were adequately attenuated for mosquito infection to affirm their development as human vaccine for prevention of CHIKV (Darwin et al., 2011).

Chimeric enterovirus

Enterovirus 71 (EV71) causes "hand, foot, and mouth disease" of bovine animals and humans. It comprises 11 subgenotypes (A, B1 to B5, and C1 to C5). Hence, an EV71 vaccine is desirable for protecting in opposition to all 11 subgenotypes. In 2014, Ye and team reported the construction of two chimeras, HBcSP55 and HBcSP70, prepared by fusion of hepatitis B core antigen (HBc) with epitopes SP55 or SP70 of EV71, respectively, to study their potential and mechanism of action against EV71. Chimeras can be prepared and self-assembled into virus like particles (VLPs) due to the presence of epitopes displayed on the surface. Carrier- and epitope-specific antibody response was induced upon immunization with chimeric constructs in mice models against lethal EV71 infections. Interestingly, in comparison to anti-HBcSP70, anti-HBcSP55 serum was not able to hinder EV71 attachment to vulnerable cells; whereas in vitro at postattachment stage both sera counteract EV71 infection. Hence, chimeras exhibiting SP55 and SP70 epitopes proved to be a promising candidates for a broad-spectrum EV71 vaccine (Ye et al., 2014).

Chimera viruses have also been exploited for treatment of other diseases such as human herpes virus infection induced cancers in infected mice models. A chimera virus was developed (a mouse virus with a human viral gene) that inhibits human LANA protein (essential for maintaining infection and causing cancer) for treatment of human herpes virus infection and its associated cancers. Such strategies can also be applied for the generation of chimera viruses, which can be effective against some other lethal viruses such as the Epstein-Barr virus or the human papilloma virus responsible for cervical cancers (Habison et al., 2017). Very recently, a chimeric antigen receptor T (CAR-T) cell has been developed for the treatment of relapsed or refractory acute lymphoblastic leukemia. CAR-T cells specifically target and kill tumor cells expressing the tumor antigen. CAR-T cell therapies have also been employed for protection against hematologic malignancies, ovarian cancer, pancreatic cancer, and prostate cancer (Jhaveri and Rosner, 2018).

Decontamination procedures: Methods and challenges

Effective decontamination systems are required to combat the threat of bioterror attacks and to minimize adverse effects caused by hazardous biological agents. The traditional methods for decontamination of biowarfare agents involve the use of bleaches and decontamination solutions, which are generally referred as "wet" solutions. Spread of infectious agents is not limited to a particular environment or space, as they easily transmit form one place to another by means of their spores; hence, it is necessary to decontaminate concerned surfaces and buildings also. Localized small-scale remediation has usually been done by treating contaminated surfaces with liquid formulations of decontaminant solutions such as hydrogen peroxide, chlorine dioxide gas dissolved in water, phenolics, sodium hypochlorite, and quaternary ammonium compounds, or decontamination foams. Large-scale remediation can be done by fumigating with chlorine dioxide gas in specific locations.

Other tested decontamination agents include ethylene oxide, glutaraldehyde, hydrogen peroxide vapor, peracetic acid, ortho-phthalaldehyde, ozone, and paraformaldehyde. These chemical decontaminants are known to have potential effectiveness against *B. anthracis* spores, but chlorine dioxide gas is considered as one of the best decontamination alternative for the fumigation of heavily contaminated/infected areas. However, it is time-consuming, as it takes long contact time for such chemical solutions to disinfect viruses and spores before removing with fresh water. Washing with soap and water is also a general personal decontamination procedure. Washing with hot soapy water removes most of the biological contaminates from emergency responders who have been exposed to biological agents. Alcohol solutions are also considered effective for decontamination of hard nonporous surfaces. Generally, 70% alcohol solutions are used for the decontamination of most of the biological contaminates. But being highly flammable, use of alcohol solutions is restricted to a particular level.

Autoclaving, dry heat, thermal washer disinfection, ultrasonication, and sterilization are other commonly used decontamination procedures. These methods are effective against most of the biological agents but disposal of decontamination reagents and contaminated waste water is still challenging. Hence, the use of such perilous chemicals is limited to an extent because of requirement of special biosafety approvals for their storage, transport, and disposal. Also, some risks have been associated with use of wet chemicals as they often lead to the corrosion of materials such as leather, plastics, paints, metals, rubber, and skin. So, the use of these hazardous chemicals on sensitive equipment and materials is not recommended. Moreover, these chemicals are nonspecific in nature and, when released into the environment, lead to toxification and degradation of our natural resources. Thus, the existing decontamination systems are not thoroughly effective (Hawley and Eitzen, 2001; Raber et al., 2001; Kumar et al., 2010).

There is a strong need of ideal and eco-friendly decontamination technologies that focus on selective and effective disinfection of biowarfare agents. Decontaminants are required that are generally present in dry forms, can be easily transported with no mass storage requirement, and are fast working. Hence, alternative decontamination methods have been developed that include the use of ionizing and nonionizing radiations, thermal energy, and reactive gases produced by plasmas. Ionizing gamma radiations were also used for the decontamination of biological agents but somehow led to the destruction of sensitive equipment. Nonionizing ultraviolet (UV) radiation has also been tested for the destruction of some biological agents, but the success of this technique was limited because of resistance of dried spores to UV radiation. Thermal energy methods have also been tried, but their efficiency is limited by the temperature constraint; also, it leads to the damage of surfaces or equipment and the method is relatively time consuming.

A portable arc-seeded microwave plasma torch was developed and applied for the decontamination of biological warfare agents. Emission spectroscopy of the plasma torch revealed the production of ample amounts of reactive atomic oxygen that effectively oxidized the biological agents. Moreover, plasma or gas in a highly energized state is highly reactive, which is capable of destruction of all kinds of organic contaminants by means of a nonthermal method. For decontamination purposes, *B. cereus* was selected simulant of *B. anthracis* spores. The results revealed that all spores were killed in <8s at 3cm distance, 12s at 4cm distance, and 16s at 5cm distance away from the nozzle of the torch. Thus, plasma torch can be also used as an alternative decontamination technique (Lai et al., 2005). A research and development project was also initiated for providing novel decontamination systems against biowarfare agents with key focus on effective, economical, fast, nontoxic, and specific decontamination. The project also aimed for the development of adsorptive elimination technologies or deactivation technologies using photocatalysis technique (Seto, 2009).

Vacuum cleaning with HEPA filtration is also considered to be an effective decontamination method that reduces the particulate load to allow effective remediation. HEPA vacuuming is reported to be more efficient than fumigation procedures. In the present scenario, the existing decontamination systems are not so effective and specific, thus their efficiency is limited to an extent. Thus, the authors propose a strong need of development of more effective, specific, and eco-friendly decontamination technologies with key focus on selective and effective disinfection of biowarfare agents (Raber et al., 2001; Fitch et al., 2003; Weis et al., 2002).

Conclusions

Biological warfare can be used with impunity under camouflage of natural outbreaks of diseases to decimate human populations and to destroy livestock and crops of economic significance. With the rapid evolution in synthetic biology techniques, the construction of synthetic biological agents and their further use as next generation bioweapons has been rapidly increasing, which eventually has enhanced the risk of biological warfare compared to the past. Many draft and whole genome sequences of important pathogenic bacteria and viruses infecting humans have been decoded to date and are accessible through nucleic acid sequence databases, such as Genbank, EMBL, DDBJ, GDB: The Human Genome DB, Microbial Genome DB for Comparative Analysis (MBGD), Virulence Factors of Bacterial Pathogens (VFDB), The National Microbial Pathogen DB Resource (NMPDR), Virus Pathogen Resource (ViPR), Integrated DB for Viral Genomics (viruSITE), Barcode of Life Data Systems, CTD (Comparative Toxicogenomics DB), etc. So, there exists a virtual platform in the form of essential genes, virulence factors, or synthetic constructs with humanized infectious elements that provide huge scope to the bioweaponeers to develop next generation bioweapons based on designer genes or designer diseases models for instigating serious consequences in future bioterror attacks. A simple alteration in genetic compliment may make a pathogen more deadly than the existing natural forms. The emergence of next generation bioweapons including chimeric agents previously unknown to man can be even more dangerous and challenging than natural agents as they can cross all the barriers of pathogenicity. Also, due to current limitations in the methods of detection, protection, and decontamination, there exists a huge knowledge gap that needs attention for developing appropriate defensive strategies against biowarfare agents.

Historical evidence has clearly predicted an asymmetric correlation between offensive and defensive biowarfare strategies. Discontinuation of biowarfare programs can have a serious limitation on the nation's ability to develop appropriate defensive tools such as antibiotics, vaccines, and other therapeutics. Deployment of biowarfare programs in the military doctrine of a nation without endangering military alliances is always advantageous for national security and protecting the vulnerable civilian population. Additionally, domestic laws against use of bioweapons should be enacted. The Biological and Toxin Weapon Convention should be strengthened through a legal binding instrument. The authors highly recommend the use of physical protective and prophylactic measures to eliminate natural spread of existing contagious biowarfare agents at a mass scale, especially among the most vulnerable populations such as children with poor immunity and armed forces involved in direct combat at the war front.

References

Abad, C. & Safdar, N., 2015. The reemergence of measles. Curr. Infect. Dis. Rep. 17, 51.

Abel, K., Peterson, J., 1963. Classification of microorganisms by analysis of chemical composition I: feasibility of utilizing gas chromatography. J. Bacteriol. 85, 1039–1044.

Ainscough, M.J., 2002. Next Generation Bioweapons: The Technology of Genetic Engineering Applied to Biowarfare and Bioterrorism. AIR UNIV MAXWELL AFB AL.

Alfson, K., Avena, L., Worwa, G., Carrion, R., Griffiths, A., 2017. Development of a lethal intranasal exposure model of Ebola virus in the cynomolgus macaque. Viruses 9, 319.

Alibek, K., 2008. Biohazard. Random House.

Athamna, A., Athamna, M., Abu-Rashed, N., Medlej, B., Bast, D., Rubinstein, E., 2004. Selection of *Bacillus anthracis* isolates resistant to antibiotics. J. Antimicrob. Chemother. 54, 424–428.

Baize, S., Pannetier, D., Oestereich, L., Rieger, T., Koivogui, L., Magassouba, N.F., Soropogui, B., Sow, M.S., KeÏTA, S., De Clerck, H., 2014. Emergence of Zaire Ebola virus disease in Guinea. N. Engl. J. Med. 371, 1418–1425.

Bausch, D.G., 2017. West Africa 2013 Ebola: From Virus Outbreak to Humanitarian Crisis. In: Marburg-and Ebolaviruses. Springer.

Becker, M.M., Graham, R.L., Donaldson, E.F., Rockx, B., Sims, A.C., Sheahan, T., Pickles, R.J., Corti, D., Johnston, R.E., Baric, R.S., 2008. Synthetic recombinant bat SARS-like coronavirus is infectious in cultured cells and in mice. Proc. Natl Acad. Sci. 105, 19944–19949.

Block, S.M., 2001. The growing threat of biological weapons: the terrorist threat is very real, and it's about to get worse. Scientists should concern themselves before it's too late. Am. Sci. 89, 28–37.

Carette, J.E., Raaben, M., Wong, A.C., Herbert, A.S., Obernosterer, G., Mulherkar, N., Kuehne, A.I., Kranzusch, P.J., Griffin, A.M., Ruthel, G., 2011. Ebola virus entry requires the cholesterol transporter Niemann–Pick C1. Nature 477, 340.

Casasnovas, J.M., Springer, T.A., 1995. Kinetics and thermodynamics of virus binding to receptor. Studies with rhinovirus, intercellular adhesion molecule-1 (ICAM-1), and surface plasmon resonance. J. Biol. Chem. 270, 13216–13224.

Cashion, L., Ast, O., Citkowicz, A., Harvey, S., Mitrovic, B., Masikat, M.R., Kauser, K., Larsen, B., Rubanyi, G.M., Harkins, R.N., 2005. 170. In vitro transduction of cells to determine tropism using viral-like particles derived from JC virus VP1. Mol. Ther. 68, S68.

Cello, J., Paul, A.V., Wimmer, E., 2002. Chemical synthesis of poliovirus cdna: generation of infectious virus in the absence of natural template. Science 297, 1016–1018.

Chan, L.Y., Kosuri, S., Endy, D., 2005. Refactoring bacteriophage T7. Mol. Syst. Biol. 1.

Charles, C.H., Luo, G.X., Kohlstaedt, L.A., Morantte, I.G., Gorfain, E., Cao, L., Williams, J.H., Fang, F., 2003. Prevention of human rhinovirus infection by multivalent fab molecules directed against ICAM-1. Antimicrob. Agents Chemother. 47, 1503–1508.

Chin, R., Torresi, J., 2013. Japanese B encephalitis: an overview of the disease and use of Chimerivax-JE as a preventative vaccine. Infect. Dis. Ther. 2, 145–158.

Christie, A., 1982. Plague: review of ecology. Ecol. Dis. 1, 111–115.

Cohen, W.S., 1997. Proliferation: Threat and Response. DIANE Publishing.

Cohen, A., 2000. Effectiveness of palivizumab for preventing serious RSV disease. J. Resp. Dis. 2, S30–S32.

Côté, M., Misasi, J., Ren, T., Bruchez, A., Lee, K., Filone, C.M., Hensley, L., Li, Q., Ory, D., Chandran, K., 2011. Small molecule inhibitors reveal Niemann–Pick C1 is essential for Ebola virus infection. Nature 477, 344.

Cummings, C.A., Chung, C.A.B., Fang, R., Barker, M., Brzoska, P., Williamson, P.C., Beaudry, J., Matthews, M., Schupp, J., Wagner, D.M., 2010. Accurate, rapid and high-throughput detection of strain-specific polymorphisms in *Bacillus anthracis* and *Yersinia pestis* by next-generation sequencing. Investig. Genet. 1, 5.

Darwin, J.R., Kenney, J.L., Weaver, S.C., 2011. Transmission potential of two chimeric Chikungunya vaccine candidates in the urban mosquito vectors, *Aedes aegypti* and *Ae. albopictus*. Am. J. Trop. Med. Hyg. 84, 1012–1015.

Das, S., Kataria, V.K., 2010. Bioterrorism: a public health perspective. Med. J. Armed Forces India 66, 255–260.

Das, R., Goel, A.K., Sharma, M.K., Upadhyay, S., 2015. Electrochemical DNA sensor for anthrax toxin activator gene atxA-detection of PCR amplicons. Biosens. Bioelectron. 74, 939–946.

De Bruin, A., De Groot, A., De Heer, L., Bok, J., Wielinga, P., Hamans, M., Van Rotterdam, B., Janse, I., 2011. Detection of Coxiella burnetii in complex matrices by using multiplex quantitative PCR during a major Q fever outbreak in The Netherlands. Appl. Environ. Microbiol. 77, 6516–6523.

Dewannieux, M., Harper, F., Richaud, A., Letzelter, C., Ribet, D., Pierron, G., Heidmann, T., 2006. Identification of an infectious progenitor for the multiple-copy HERV-K human endogenous retroelements. Genome Res. 16, 1548–1556.

Dutta, T., Sujatha, S., Sahoo, R., 2011. Anthrax—update on diagnosis and management. J. Assoc. Physicians India 59, 573–578.

Edelman, R., Tacket, C., Wasserman, S., Bodison, S., Perry, J., Mangiafico, J., 2000. Phase II safety and immunogenicity study of live chikungunya virus vaccine TSI-GSD-218. Am. J. Trop. Med. Hyg. 62, 681–685.

Espinosa, D., Mendy, J., Manayani, D., Vang, L., Wang, C., Richard, T., Guenther, B., Aruri, J., Avanzini, J., Garduno, F., 2018. Passive transfer of immune sera induced by a Zika virus-like particle vaccine protects AG129 mice against lethal Zika virus challenge. EBioMedicine 27, 61–70.

Fettig, J., Swaminathan, M., Murrill, C.S., Kaplan, J.E., 2014. Global epidemiology of HIV. Infect. Dis. Clin. 28, 323–337.

Fitch, J.P., Raber, E., Imbro, D.R., 2003. Technology challenges in responding to biological or chemical attacks in the civilian sector. Science 302, 1350–1354.

Fodor, E., Devenish, L., Engelhardt, O.G., Palese, P., Brownlee, G.G., GarcÍA-Sastre, A., 1999. Rescue of influenza A virus from recombinant DNA. J. Virol. 73, 9679–9682.

Frischknecht, F., 2003. The history of biological warfare: human experimentation, modern nightmares and lone madmen in the twentieth century. EMBO Rep. 4, S47–S52.

Fujimura, T., Ribas, J.C., Makhov, A.M., Wickner, R.B., 1992. Pol of gag–pol fusion protein required for encapsidation of viral RNA of yeast LA virus. Nature 359, 746.

Gardner, C.L., Ryman, K.D., 2010. Yellow fever: a reemerging threat. Clin. Lab. Med. 30, 237–260.

Ge, Q., McManus, M.T., Nguyen, T., Shen, C.-H., Sharp, P.A., Eisen, H.N., Chen, J., 2003. RNA interference of influenza virus production by directly targeting mRNA for degradation and indirectly inhibiting all viral RNA transcription. Proc. Natl Acad. Sci. 100, 2718–2723.

Ghosh, N., Goel, A., 2012. Anti-protective antigen IgG enzyme-linked immunosorbent assay for diagnosis of cutaneous anthrax in India. Clin. Vaccine Immunol. 19, 1238–1242.

Ghosh, N., Tomar, I., Lukka, H., Goel, A., 2013. Serodiagnosis of human cutaneous anthrax in India using an indirect anti-lethal factor IgG enzyme-linked immunosorbent assay. Clin. Vaccine Immunol. 20, 282–286.

Gibson, D.G., Benders, G.A., Andrews-Pfannkoch, C., Denisova, E.A., Baden-Tillson, H., Zaveri, J., Stockwell, T.B., Brownley, A., Thomas, D.W., Algire, M.A., 2008. Complete chemical synthesis, assembly, and cloning of a Mycoplasma genitalium genome. Science 319, 1215–1220.

Gibson, D.G., Glass, J.I., Lartigue, C., Noskov, V.N., Chuang, R.-Y., Algire, M.A., Benders, G.A., Montague, M.G., Ma, L., Moodie, M.M., 2010. Creation of a bacterial cell controlled by a chemically synthesized genome. Science 329, 52–56.

Glass, J.I., 2012. Synthetic genomics and the construction of a synthetic bacterial cell. Perspect. Biol. Med. 55, 473–489.

Go, P.C.V.V.E., Sansthan, A., 2014. Glanders-A re-emerging zoonotic disease: a review. J. Biol. Sci. 14, 38–51.

Goldstein, H., Pettoello-Mantovani, M., Bera, T.K., Pastan, I.H., Berger, E.A., 2000. Chimeric toxins targeted to the human immunodeficiency virus type 1 envelope glycoprotein augment the in vivo activity of combination antiretroviral therapy in thy/liv-SCID-Hu mice. J. Infect. Dis. 181, 921–926.

Gomes-Solecki, M.J., Savitt, A.G., Rowehl, R., Glass, J.D., Bliska, J.B., Dattwyler, R.J., 2005. LcrV capture enzyme-linked immunosorbent assay for detection of Yersinia pestis from human samples. Clin. Diagn. Lab. Immunol. 12, 339–346.

Gürcan, Ş., 2014. Epidemiology of tularemia. Balkan Med. J. 31, 3–10.

Habison, A.C., De Miranda, M.P., Beauchemin, C., Tan, M., Cerqueira, S.A., Correia, B., Ponnusamy, R., Usherwood, E.J., McVey, C.E., Simas, J.P., 2017. Cross-species conservation of episome maintenance provides a basis for in vivo investigation of Kaposi's sarcoma herpesvirus LANA. PLoS Pathog. 13, e1006555.

Hawley, R.J., Eitzen Jr., E.M., 2001. Biological weapons—a primer for microbiologists. Annu. Rev. Microbiol. 55, 235–253.

Henderson, D.A., Inglesby, T.V., Bartlett, J.G., Ascher, M.S., Eitzen, E., Jahrling, P.B., Hauer, J., Layton, M., Mcdade, J., Osterholm, M.T., 1999. Smallpox as a biological weapon: medical and public health management. JAMA 281, 2127–2137.

Henderson, D.A., Inglesby Jr., T.V., O'toole, T., Mortimer, P.P., 2003. Can postexposure vaccination against smallpox succeed? Clin. Infect. Dis. 36, 622–629.

Hindson, B.J., McBride, M.T., Makarewicz, A.J., Henderer, B.D., Setlur, U.S., Smith, S.M., Gutierrez, D.M., Metz, T.R., Nasarabadi, S.L., Venkateswaran, K.S., 2005. Autonomous detection of aerosolized biological agents by multiplexed immunoassay with polymerase chain reaction confirmation. Anal. Chem. 77, 284–289.

Hoffmann, E., Neumann, G., Hobom, G., Webster, R.G., Kawaoka, Y., 2000. "Ambisense" approach for the generation of influenza A virus: vRNA and mRNA synthesis from one template. Virology 267, 310–317.

Hoffmann, M., Hernandez, M.G., Berger, E., Marzi, A., Pöhlmann, S., 2016. The glycoproteins of all filovirus species use the same host factors for entry into bat and human cells but entry efficiency is species dependent. PLoS One 11, e0149651.

Horn, J.K., 2003. Bacterial agents used for bioterrorism. Surg. Infect. 4, 281–287.

Huang, C.Y.-H., Silengo, S.J., Whiteman, M.C., Kinney, R.M., 2005. Chimeric dengue 2 PDK-53/West Nile NY99 viruses retain the phenotypic attenuation markers of the candidate PDK-53 vaccine virus and protect mice against lethal challenge with West Nile virus. J. Virol. 79, 7300–7310.

Hugel, T., Michaelis, J., Hetherington, C.L., Jardine, P.J., Grimes, S., Walter, J.M., Falk, W., Anderson, D.L., Bustamante, C., 2007. Experimental test of connector rotation during DNA packaging into bacteriophage φ29 capsids. PLoS Biol. 5, e59.

Iversen, P., Warren, T., Wells, J., Garza, N., Mourich, D., Welch, L., Panchal, R., Bavari, S., 2012. Discovery and early development of AVI-7537 and AVI-7288 for the treatment of Ebola virus and Marburg virus infections. Viruses 4, 2806–2830.

Jacque, J.-M., Triques, K., Stevenson, M., 2002. Modulation of HIV-1 replication by RNA interference. Nature 418, 435.

Janse, I., Hamidjaja, R.A., Bok, J.M., Van Rotterdam, B.J., 2010. Reliable detection of *Bacillus anthracis*, *Francisella tularensis* and *Yersinia pestis* by using multiplex qPCR including internal controls for nucleic acid extraction and amplification. BMC Microbiol. 10, 314.

Jansen, H.-J., Breeveld, F.J., Stijnis, C., Grobusch, M.P., 2014. Biological warfare, bioterrorism, and biocrime. Clin. Microbiol. Infect. 20, 488–496.

Jhaveri, K.D., Rosner, M.H., 2018. Chimeric antigen receptor T cell therapy and the kidney: what the nephrologist needs to know. Clin. J. Am. Soc. Nephrol. 13, 796–798.

Jones, M., Schuh, A., Amman, B., Sealy, T., Zaki, S., Nichol, S., Towner, J., 2015. Experimental inoculation of Egyptian rousette bats (*Rousettus aegyptiacus*) with viruses of the Ebolavirus and Marburgvirus genera. Viruses 7, 3420–3442.

Kaneda, T., 1963. Biosynthesis of branched chain fatty acids I. Isolation and identification of fatty acids from *Bacillus subtilis* (ATCC 7059). J. Biol. Chem. 238, 1222–1228.

Kaptein, S.J., Neyts, J., 2016. Towards antiviral therapies for treating dengue virus infections. Curr. Opin. Pharmacol. 30, 1–7.

Karabatsos, N., 1985. International Catalogue of Arboviruses, Including Certain Other Viruses of Vertebrates, third ed. American Society of Tropical Medicine and Hygiene for the Subcommittee on Information Exchange of the American Committee on Arthropod-borne Viruses, p. 1147.

Keele, B.F., Van Heuverswyn, F., Li, Y., Bailes, E., Takehisa, J., Santiago, M.L., Bibollet-Ruche, F., Chen, Y., Wain, L.V., Liegeois, F., 2006. Chimpanzee reservoirs of pandemic and nonpandemic HIV-1. Science 313, 523–526.

Kemenesi, G., Kurucz, K., Dallos, B., Zana, B., Földes, F., Boldogh, S., Görföl, T., Carroll, M.W., Jakab, F., 2018. Re-emergence of Lloviu virus in Miniopterus schreibersii bats, Hungary, 2016. Emerg. Microbes Infect. 7, 66.

Komar, N., Langevin, S., Monath, T.P., 2009. Use of a surrogate chimeric virus to detect West Nile virus-neutralizing antibodies in avian and equine sera. Clin. Vaccine Immunol. 16, 134–135.

Krishan, K., Kaur, B., Sharma, A., 2017. India's preparedness against bioterrorism: biodefence strategies and policy measures. Curr. Sci. 113, 1675.

Kum, D.B., Mishra, N., Boudewijns, R., Gladwyn-NG, I., Alfano, C., Ma, J., Schmid, M.A., Marques, R.E., Schols, D., Kaptein, S., 2018. A yellow fever–Zika chimeric virus vaccine candidate protects against Zika infection and congenital malformations in mice. NPJ Vaccines 3, 56.

Kumar, V., Goel, R., Chawla, R., Silambarasan, M., Sharma, R.K., 2010. Chemical, biological, radiological, and nuclear decontamination: recent trends and future perspective. J. Pharm. Bioallied Sci. 2, 220.

Kuroda, M., Sekizuka, T., Shinya, F., Takeuchi, F., Kanno, T., Sata, T., Asano, S., 2012. Detection of a possible bioterrorism agent, Francisella sp., in a clinical specimen by use of next-generation direct DNA sequencing. J. Clin. Microbiol. 50, 1810–1812.

Lai, W., Lai, H., Kuo, S.P., Tarasenko, O., Levon, K., 2005. Decontamination of biological warfare agents by a microwave plasma torch. Phys. Plasmas 12, 023501.

Le Calvez, H., Yu, M., Fang, F., 2004. Biochemical prevention and treatment of viral infections—a new paradigm in medicine for infectious diseases. Virol. J. 1, 12.

Lee, Y.N., Bieniasz, P.D., 2007. Reconstitution of an infectious human endogenous retrovirus. PLoS Pathog. 3, e10.

Lefterova, M.I., Suarez, C.J., Banaei, N., Pinsky, B.A., 2015. Next-generation sequencing for infectious disease diagnosis and management: a report of the Association for Molecular Pathology. J. Mol. Diagn. 17, 623–634.

Lemoine, M., Nayagam, S., Thursz, M., 2013. Viral hepatitis in resource-limited countries and access to antiviral therapies: current and future challenges. Future Virol. 8, 371–380.

Lesser, I., Arquilla, J., Hoffman, B., Ronfeldt, D.F., Zanini, M., 1999. Countering the New Terrorism. RAND Corporation.

Levitt, N.H., Ramsburg, H.H., Hasty, S.E., Repik, P.M., Cole Jr., F.E., Lupton, H.W., 1986. Development of an attenuated strain of chikungunya virus for use in vaccine production. Vaccine 4, 157–162.

Li, W., Shi, Z., Yu, M., Ren, W., Smith, C., Epstein, J.H., Wang, H., Crameri, G., Hu, Z., Zhang, H., 2005. Bats are natural reservoirs of SARS-like coronaviruses. Science 310, 676–679.

Li, X.-F., Dong, H.-L., Wang, H.-J., Huang, X.-Y., Qiu, Y.-F., Ji, X., Ye, Q., Li, C., Liu, Y., Deng, Y.-Q., 2018. Development of a chimeric Zika vaccine using a licensed live-attenuated flavivirus vaccine as backbone. Nat. Commun. 9, 673.

Maartens, G., Celum, C., Lewin, S.R., 2014. HIV infection: epidemiology, pathogenesis, treatment, and prevention. Lancet 384, 258–271.

Madad, S.S., 2014. Bioterrorism: an emerging global health threat. J. Bioterr. Biodef. 5, 1–6.

Mangold, T., Goldberg, J., 1999. Plague Wars: A True Story of Biological Warfare New York. St. Martin's Press.

Manhart, W.A., Pacheco, J.R., Hume, A.J., Cressey, T.N., DeflubÉ, L.R., MÜhlberger, E., 2018. A chimeric Lloviu virus minigenome system reveals that the bat-derived filovirus replicates more similarly to Ebolaviruses than Marburgviruses. Cell Rep. 24, 2573–2580. e4.

Marlin, S.D., Staunton, D.E., Springer, T.A., Stratowa, C., Sommergruber, W., Merluzzi, V.J., 1990. A soluble form of intercellular adhesion molecule-1 inhibits rhinovirus infection. Nature 344, 70.

Maurin, M., Raoult, D.F., 1999. Q fever. Clin. Microbiol. Rev. 12, 518–553.

McBride, M.T., Gammon, S., Pitesky, M., O'Brien, T.W., Smith, T., Aldrich, J., Langlois, R.G., Colston, B., Venkateswaran, K.S., 2003. Multiplexed liquid arrays for simultaneous detection of simulants of biological warfare agents. Anal. Chem. 75, 1924–1930.

Metcalfe, N., 2002. A short history of biological warfare. Med. Confl. Surviv. 18, 271–282.

Monath, T.P., Vasconcelos, P.F.C., 2015. Yellow fever. J. Clin. Virol. 64, 160–173.

Murray, N.E.A., Quam, M.B., Wilder-Smith, A., 2013. Epidemiology of dengue: past, present and future prospects. Clin. Epidemiol. 5, 299.

Nan, Y., Zhang, Y., 2018. Antisense phosphorodiamidate morpholino oligomers as novel antiviral compounds. Front. Microbiol. 9, 750.

Narayanan, J., Sharma, M.K., Ponmariappan, S., Shaik, M., Upadhyay, S., 2015. Electrochemical immunosensor for botulinum neurotoxin type-E using covalently ordered graphene nanosheets modified electrodes and gold nanoparticles-enzyme conjugate. Biosens. Bioelectron. 69, 249–256.

Negredo, A., Palacios, G., Vázquez-Morón, S., González, F., Dopazo, H., Molero, F., Juste, J., Quetglas, J., Savji, N., De La Cruz Martínez, M., 2011. Discovery of an ebolavirus-like filovirus in europe. PLoS Pathog. 7, e1002304.

Neumann, G., Watanabe, T., Ito, H., Watanabe, S., Goto, H., Gao, P., Hughes, M., Perez, D.R., Donis, R., Hoffmann, E., 1999. Generation of influenza A viruses entirely from cloned cDNAs. Proc. Natl Acad. Sci. 96, 9345–9350.

Ni, T.-H., Zhou, X., McCarty, D.M., Zolotukhin, I., Muzyczka, N., 1994. In vitro replication of adeno-associated virus DNA. J. Virol. 68, 1128–1138.

Novina, C.D., Murray, M.F., Dykxhoorn, D.M., Beresford, P.J., Riess, J., Lee, S.K., Collman, R.G., Lieberman, J., Shankar, P., Sharp, P.A., 2003. Erratum: siRNA-directed inhibition of HIV-1 infection (Nature Medicine (2002) 8 (681–686)). Nat. Med. 9, 681–686.

Orr, R., 2001. Technology evaluation: fomivirsen, Isis Pharmaceuticals Inc/CIBA vision. Curr. Opin. Mol. Ther. 3, 288–294.

O'Toole, T., 1999. Richard Preston's The Cobra Event. Public Health Rep. 114, 186.

Pal, V., Sharma, M., Sharma, S., Goel, A., 2016. Biological warfare agents and their detection and monitoring techniques. Def. Sci. J. 66, 445–457.

Pardi, N., Hogan, M.J., Pelc, R.S., Muramatsu, H., Andersen, H., Demaso, C.R., Dowd, K.A., Sutherland, L.L., Scearce, R.M., Parks, R., 2017. Zika virus protection by a single low-dose nucleoside-modified mRNA vaccination. Nature 543, 248.

Paweska, J., Storm, N., Grobbelaar, A., Markotter, W., Kemp, A., Jansen Van Vuren, P., 2016. Experimental inoculation of Egyptian fruit bats (*Rousettus aegyptiacus*) with Ebola virus. Viruses 8, 29.

Quinn, S.C., Thomas, T., Kumar, S., 2008. The anthrax vaccine and research: reactions from postal workers and public health professionals. Biosecur. Bioterror. 6, 321–333.

Raabe, V., Koehler, J., 2017. Laboratory diagnosis of Lassa fever. J. Clin. Microbiol. 55, 1629–1637.

Raber, E., Jin, A., Noonan, K., McGuire, R., Kirvel, R.D., 2001. Decontamination issues for chemical and biological warfare agents: how clean is clean enough? Int. J. Environ. Health Res. 11, 128–148.

Rogers, P., Whitby, S., Dando, M., 1999. Biological warfare against crops. Sci. Am. 280, 70–75.

Rotz, L.D., Khan, A.S., Lillibridge, S.R., Ostroff, S.M., Hughes, J.M., 2002. Public health assessment of potential biological terrorism agents. Emerg. Infect. Dis. 8, 225.

Salmain, M., Ghasemi, M., Boujday, S., Pradier, C.-M., 2012. Elaboration of a reusable immunosensor for the detection of staphylococcal enterotoxin A (SEA) in milk with a quartz crystal microbalance. Sensors Actuators B Chem. 173, 148–156.

Salvo, M.A., Kingstad-Bakke, B., Salas-Quinchucua, C., Camacho, E., Osorio, J.E., 2018. Zika virus like particles elicit protective antibodies in mice. PLoS Negl. Trop. Dis. 12, e0006210.

Sapsford, K.E., Bradburne, C., Delehanty, J.B., Medintz, I.L., 2008. Sensors for detecting biological agents. Mater. Today 11, 38–49.

Sarwar, U.N., Sitar, S., Ledgerwood, J.E., 2011. Filovirus emergence and vaccine development: a perspective for health care practitioners in travel medicine. Travel Med. Infect. Dis. 9, 126–134.

Scott, L.J., 2016. Tetravalent dengue vaccine: a review in the prevention of dengue disease. Drugs 76, 1301–1312.

Seto, Y., 2009. Decontamination of chemical and biological warfare agents. Yakugaku Zasshi 129, 53–69.

Shan, C., Muruato, A.E., Nunes, B.T., Luo, H., Xie, X., Medeiros, D.B., Wakamiya, M., Tesh, R.B., Barrett, A.D., Wang, T., 2017. A live-attenuated Zika virus vaccine candidate induces sterilizing immunity in mouse models. Nat. Med. 23, 763.

Sharma, N., Hotta, A., Yamamoto, Y., Fujita, O., Uda, A., Morikawa, S., Yamada, A., Tanabayashi, K., 2013. Detection of Francisella tularensis-specific antibodies in patients with tularemia by a novel competitive enzyme-linked immunosorbent assay. Clin. Vaccine Immunol. 20, 9–16.

Sharma, M.K., Narayanan, J., Upadhyay, S., Goel, A.K., 2015. Electrochemical immunosensor based on bismuth nanocomposite film and cadmium ions functionalized titanium phosphates for the detection of anthrax protective antigen toxin. Biosens. Bioelectron. 74, 299–304.

Sharma, M.K., Narayanan, J., Pardasani, D., Srivastava, D.N., Upadhyay, S., Goel, A.K., 2016. Ultrasensitive electrochemical immunoassay for surface array protein, a *Bacillus anthracis* biomarker using Au–Pd nanocrystals loaded on boron-nitride nanosheets as catalytic labels. Biosens. Bioelectron. 80, 442–449.

Sleator, R. D. 2010. The story of Mycoplasma mycoides JCVI-syn1. 0: the forty million dollar microbe. Bioeng. Bugs 1 (4), 231–232.

Smith, H.O., Hutchison, C.A., Pfannkoch, C., Venter, J.C., 2003. Generating a synthetic genome by whole genome assembly: φX174 bacteriophage from synthetic oligonucleotides. Proc. Natl Acad. Sci. 100, 15440–15445.

Sobel, J., 2005. Botulism. Clin. Infect. Dis. 41, 1167–1173.

Spencer, J., Scardaville, M., 1999. Understanding the bioterrorist threat: facts & figures. US Army 163, 18.

Sumathy, K., Kulkarni, B., Gondu, R.K., Ponnuru, S.K., Bonguram, N., Eligeti, R., Gadiyaram, S., Praturi, U., Chougule, B., Karunakaran, L., 2017. Protective efficacy of Zika vaccine in AG129 mouse model. Sci. Rep. 7, 46375.

Sun, S., Rao, V.B., Rossmann, M.G., 2010. Genome packaging in viruses. Curr. Opin. Struct. Biol. 20, 114–120.

Suter, K., 2003. The troubled history of chemical and biological warfare. Contemp. Theatr. Rev. 283, 161.

Takehisa, J., Kraus, M.H., Decker, J.M., Li, Y., Keele, B.F., Bibollet-Ruche, F., Zammit, K.P., Weng, Z., Santiago, M.L., Kamenya, S., 2007. Generation of infectious molecular clones of simian immunodeficiency virus from fecal consensus sequences of wild chimpanzees. J. Virol. 81, 7463–7475.

Tang, X.B., Hobom, G., Luo, D., 1994. Ribozyme mediated destruction of influenza A virus in vitro and in vivo. J. Med. Virol. 42, 385–395.

Taubenberger, J.K., Morens, D.M., 2008. The pathology of influenza virus infections. Annu. Rev. Pathol. 3, 499–522.

Taubenberger, J.K., Reid, A.H., Krafft, A.E., Bijwaard, K.E., Fanning, T.G., 1997. Initial genetic characterization of the 1918 "Spanish" influenza virus. Science 275, 1793–1796.

Taubenberger, J.K., Reid, A.H., Lourens, R.M., Wang, R., Jin, G., Fanning, T.G., 2005. Characterization of the 1918 influenza virus polymerase genes. Nature 437, 889.

Taubenberger, J.K., Hultin, J.V., Morens, D.M., 2007. Discovery and characterization of the 1918 pandemic influenza virus in historical context. Antivir. Ther. 12, 581.

Thavaselvam, D., Vijayaraghavan, R., 2010. Biological warfare agents. J. Pharm. Bioallied Sci. 2, 179.

To, A., Medina, L.O., Mfuh, K.O., Lieberman, M.M., Wong, T.A.S., Namekar, M., Nakano, E., Lai, C.-Y., Kumar, M., Nerurkar, V.R., 2018. Recombinant Zika virus subunits are immunogenic and efficacious in mice. MSphere 3. e00576-17.

Touret, F., Gilles, M., Klitting, R., Aubry, F., Lamballerie, D., X. & NougairÈDE, A., 2018. Live Zika virus chimeric vaccine candidate based on a yellow fever 17-D attenuated backbone. Emerg. Microbes Infect. 7, 1–12.

Trombley, A.R., Wachter, L., Garrison, J., Buckley-Beason, V.A., Jahrling, J., Hensley, L.E., Schoepp, R.J., Norwood, D.A., Goba, A., Fair, J.N., 2010. Comprehensive panel of real-time taqman™ polymerase chain reaction assays for detection and absolute quantification of filoviruses, arenaviruses, and new world hantaviruses. Am. J. Trop. Med. Hyg. 82, 954–960.

Tucker, J.B., 1999. Historical trends related to bioterrorism: an empirical analysis. Emerg. Infect. Dis. 5, 498.

Ursic-Bedoya, R., Mire, C.E., Robbins, M., Geisbert, J.B., Judge, A., Maclachlan, I., Geisbert, T.W., 2013. Protection against lethal Marburg virus infection mediated by lipid encapsulated small interfering RNA. J. Infect. Dis. 209, 562–570.

Van Aken, J., Hammond, E., 2003. Genetic engineering and biological weapons: new technologies, desires and threats from biological research. EMBO Rep. 4, S57–S60.

Van Zandt, K.E., Greer, M.T., Gelhaus, H.C., 2013. Glanders: an overview of infection in humans. Orphanet J. Rare Dis. 8, 131.

Vijayanand, P., Wilkins, E., Woodhead, M., 2004. Severe acute respiratory syndrome (SARS): a review. Clin. Med. 4, 152–160.

Wang, E., Volkova, E., Adams, A.P., Forrester, N., Xiao, S.-Y., Frolov, I., Weaver, S.C., 2008. Chimeric alphavirus vaccine candidates for chikungunya. Vaccine 26, 5030–5039.

Wang, D.-B., Yang, R., Zhang, Z.-P., Bi, L.-J., You, X.-Y., Wei, H.-P., Zhou, Y.-F., Yu, Z., Zhang, X.-E., 2009. Detection of *B. anthracis* spores and vegetative cells with the same monoclonal antibodies. PloS One 4, e7810.

Wang, E., Weaver, S.C., Frolov, I., 2011. Chimeric Chikungunya viruses are nonpathogenic in highly sensitive mouse models but efficiently induce a protective immune response. J. Virol. 85, 9249–9252.

Weaver, S.C., Ferro, C., Barrera, R., Boshell, J., Navarro, J.-C., 2004. Venezuelan equine encephalitis. Annu. Rev. Entomol. 49, 141–174.

Weis, C.P., Intrepido, A.J., Miller, A.K., Cowin, P.G., Durno, M.A., Gebhardt, J.S., Bull, R., 2002. Secondary aerosolization of viable *Bacillus anthracis* spores in a contaminated US Senate Office. JAMA 288, 2853–2858.

Welch, P., Tritz, R., Yei, S., Leavitt, M., Yu, M., Barber, J., 1996. A potential therapeutic application of hairpin ribozymes: in vitro and in vivo studies of gene therapy for hepatitis C virus infection. Gene Ther. 3, 994–1001.

Welch, P., Tritz, R., Yei, S., Barber, J., Yu, M., 1997. Intracellular application of hairpin ribozyme genes against hepatitis B virus. Gene Ther. 4, 736.

Wimmer, E., Mueller, S., Tumpey, T.M., Taubenberger, J.K., 2009. Synthetic viruses: a new opportunity to understand and prevent viral disease. Nat. Biotechnol. 27, 1163.

Wright, J., Qu, G., Tang, C., Sommer, J., 2003. Recombinant adeno-associated virus: formulation challenges and strategies for a gene therapy vector. Curr. Opin. Drug Discov. Devel. 6, 174–178.

Wu, H., Zuo, Y., Cui, C., Yang, W., Ma, H., Wang, X., 2013. Rapid quantitative detection of brucella melitensis by a label-free impedance immunosensor based on a gold nanoparticle-modified screen-printed carbon electrode. Sensors 13, 8551–8563.

Xu, K., Song, Y., Dai, L., Zhang, Y., LU, X., Xie, Y., Zhang, H., Cheng, T., Wang, Q. & Huang, Q., 2018. Recombinant chimpanzee adenovirus vaccine AdC7-M/E protects against Zika virus infection and testis damage. J. Virol. 92. e01722-17.

Yang, X.-L., Tan, C.W., Anderson, D.E., Jiang, R.-D., Li, B., Zhang, W., Zhu, Y., Lim, X.F., Zhou, P., Liu, X.-L., 2019. Characterization of a filovirus (Měnglà virus) from Rousettus bats in China. Nat. Microbiol. 1, 390–395.

Ye, X., Ku, Z., Liu, Q., Wang, X., Shi, J., Zhang, Y., Kong, L., Cong, Y., Huang, Z., 2014. Chimeric virus-like particle vaccines displaying conserved enterovirus 71 epitopes elicit protective neutralizing antibodies in mice through divergent mechanisms. J. Virol. 88, 72–81.

Ye, Q., Liu, Z.-Y., Han, J.-F., Jiang, T., Li, X.-F., Qin, C.-F., 2016. Genomic characterization and phylogenetic analysis of Zika virus circulating in the Americas. Infect. Genet. Evol. 43, 43–49.

Yousaf, M.Z., Qasim, M., Zia, S., Ashfaq, U.A., Khan, S., 2012. Rabies molecular virology, diagnosis, prevention and treatment. Virol. J. 9, 50.

Yu, M., Ojwang, J., Yamada, O., Hampel, A., Rapapport, J., Looney, D., Wong-Staal, F., 1993. A hairpin ribozyme inhibits expression of diverse strains of human immunodeficiency virus type 1. Proc. Natl Acad. Sci. 90, 6340–6344.

Zhou, X., Muzyczka, N., 1998. In vitro packaging of adeno-associated virus DNA. J. Virol. 72, 3241–3247.

Zimmer, S.M., Burke, D.S., 2009. Historical perspective—emergence of influenza A (H1N1) viruses. N. Engl. J. Med. 361, 279–285.

Genome information of BW agents and their application in biodefence

13

Anoop Kumar, S.J.S. Flora

National Institute of Pharmaceutical Education and Research-Raebareli, Lucknow, India

Introduction

Biological warfare agents (BWAs) are high-risk weapons and considered the weapon of low-income countries due to low production cost as compared to nuclear weapons. These agents can be manufactured easily. The raw material such as bacterial and viral strains and plasmids can be easily obtained from microbiology scientists or repositories for the production of BWAs. Further, the availability of microbial genome databases will be helpful for the identification of potential genes which are involved in virulence and manufacturing the most lethal combinations. Biological warfare agents may be bacteria, viruses, and toxins. During the 20th century, various programs related to bio-weapons were carried out by most of the developed countries (Szinicz, 2005). In modern times, the actual use of biological warfare agents in military conflicts has never been confirmed (Martin, 2002). However, an attack by using these agents is much easier to plan and execute than was the recent attack on Pulwama, Jammu, and Kashmir, India.

Modern genome sequencing technology has started to disclose the subsets of genes in the genome of each pathogen that are essential for infection, virulence, and antibiotic resistance (Schuster, 2007). This information will be helpful to design new diagnostics, new chemical entities (NCEs), and vaccines. Unfortunately, this information can also be misused for the creation of new types of biological warfare agents (Ainscough, 2004), enhancement of antibiotic resistance (Lowy, 2003), and the creation of more virulent agents (Fraser and Dando, 2001). Such "tailoring" of classical biological warfare agents could make them harder to detect, diagnose, and treat, and make BWAs more serious. Thus, advances in genetic research may play a significant role in the advancement of biological warfare agents.

Fortunately, the same information could also be used for biodefense against them (Minogue et al., 2019). It can be achieved by development of vaccines, novel antimicrobial compounds, and detection methods for BWAs, etc. This chapter summarizes genomic information of potential BW agents including bacteria, viruses, and toxins, and their application in biodefense.

Genome information of BWAs

Genome information and subsets of genes in the genome of BWAs will be helpful for the detection of bio-threat or even unknown threats. The researchers who are working in the field of genome around the world are now revealing the complete sequences of the most common bacterial, fungal, and parasitic pathogens of humans, animals, and plants. This information is useful in the creation of bio-weapons as well as countermeasures against them. The potential BWAs are bacteria, fungus, and virus. The genomic information of individual BWAs are described below.

Bacteria

Bacillus anthracis

Bacillus anthracis is an endospore-forming gram-positive bacterium, which causes anthrax—one of the lethal diseases. It is considered as one of the potential biological warfare agents (Turnbull, 1999). Bacillus anthracis bacteria has a single circular chromosome of 5.2 kb size and circular double-stranded two virulent plasmids, named pXO1 and pXO2 (Chun et al., 2012). These two virulence plasmids are responsible for the virulence of the organism (Bourgogne et al., 2003). The size of pXO1 is 182 kb forms binary products either with *lef* and *pag* or with *cya* and *pag* that encodes a toxin named tripartite toxin (Little and Ivins, 1999; Goel, 2015). These binary toxins (with lef (lethal factor, LF) and pag (protective antigen, PA) or with cya (edema factor (EF) and pag (protective antigen, PA) together target the multiple functions of the host, which results in the immune-suppression. Finally, death can happened due to a variety of interrupted systems.

The plasmid pXO2 (94.8 kb) encodes for 5 gene operons that synthesize polyglutamate, a capsular polysaccharide (CPS), which protect it from phagocytosis (Goel, 2015). The G+C content of the genome is 35.4% in a chromosome and 32.5% and 33.0% in pXO1 and pXO2, respectively. The chromosome of *Bacillus anthracis* has 2762 genes with assigned function; whereas, pXO1 and pXO2 have 65 and 38 genes with assigned functions, respectively (Liang et al., 2017). The number of genes with unknown function is 657 in a chromosome, 8 in pXO1, and 5 in pXO2 plasmid.

Brucella melitensis

The genome size of *Brucella melitensis* strain 16M is 3.2 kb distributed over two circular chromosomes. No plasmid is present in these bacteria (Delvecchio et al., 2002; Rao et al., 2014; Azam et al., 2016). Chromosome I is composed of 2.1 kb whereas chromosome II is composed of 1.2 kb. A total of 3198 open reading frames (ORFs) are predicted, out of which, 2059 ORFs are present on chromosome 1 and remaining 1138 on chromosome number 2 (Wang et al., 2011). The chromosome origins of replication (ORI) are similar in almost every bacterium. Genes (housekeeping genes) that encode for DNA replication, protein synthesis, and cell-wall biosynthesis are found on both chromosomes. Adhesins, invasions, and hemolysinsencoding genes are also identified. The G+C content of *Brucella melitensis* is 57%. *Brucella melitensis* genome has 78% genome with assigned function, however, 22% genome is without assigned functions (Cao et al., 2017).

Yersinia pestis

Yersinia pestis is a nonmotile, gram-negative, facultative anaerobic, nonspore forming, rod-shaped coccobacillus bacteria. The genome size of *Yersinia pestis* strain CO92 is 4.65-megabase (Mb); it has a single circular chromosome and three virulent plasmids. The sizes of plasmids are 96.2, 70.3, and 9.6 kb (Parkhill, 2001). *Y. pestis* genome is strangely rich in insertion sequences and its G+C content in a chromosome is 47.64%, whereas, plasmids have G+C 45.27%, 44.84%, and 50.23% in pPst, pYV1, and pFra, respectively (McDonough and Hare, 1997). Many genes encoding for adhesins, secretion systems, and insecticidal toxins have been developed from other bacteria and viruses. The genome of *Yersinia pestis* comprises around 150 pseudo genes (Parkhill, 2001). Average gene length of the chromosome is 998 bp. The plasmids of *Yersinia pestis* pPst, pYV, and pFrais are 611, 1643, and 835 bp, respectively.

Brucella abortus

Brucella abortus is a gram-negative, spore-forming, heterotrophic, rod-shaped bacterium, which causes brucellosis (Malta fever) in humans. *Brucella. abortus* biovar 1 genome is 3.3 Mb in size and sequence determined by the shotgun method. The genome is composed of two circular chromosomes. The first chromosome is of 2.1 kb (Chr I) and second is 1.1 kb (Chr II) in size (Halling et al., 2005; Georgi et al., 2017). The genome possesses high G+C percentage; the G+C contents of Chromosome I and Chromosome II are similar, i.e., 57.2% and 57.3%, respectively. *B. abortus* genome contains 3296 ORFs annotated as genes, 2158 ORF on Chromosome I and 1138 ORF on Chromosome II (Michaux et al., 1993; Michaux-Charachon et al., 1997). Genomic islands (GIs) are present in the genome, which encode metabolic pathways and/or virulence factors, and also transform nonpathogenic into pathogenic (Hacker and Kaper, 2000). Nine islands are present in Brucella genome: out of which GI-1, GI-5, and GI-6 do not contribute to Brucella virulence, whereas GI-3 is pathogenic for humans, and contains 29 genes (Rajashekara et al., 2008).

Burkholderia mallei

Burkholderia mallei is a bipolar, gram-negative, aerobic bacterium, which causes glanders disease in animals and humans. The genome of *Burkholderia mallei* consist of a single circular chromosome of size 3.5 Mb and a plasmid of 2.3 Mb. The genome was sequenced in the United States by The Institute of Genomic Research. Genomes of *B. mallei* have insertion sequences and phase-variable genes (Losada et al., 2010). The genes present on the chromosome are responsible for metabolism, capsule formation, and lipopolysaccharide biosynthesis. The pathogenic potential of *B. mallei* is due to its polysaccharide capsule. The mega plasmid encodes for the secretion systems genes and virulence-associated genes (Table 1).

Table 1 Genomic information of common bacterial BW agents.

Bacteria	Genome size (kb)	Chromo-somes	Plasmid number	No. of genes	G+C content	Genes with known function	Unknown genes	Coding
Bacillus anthracis	5.2	1	2	5838	35.4	2762	657	84.3%
Brucella melitensis	3.2	2	–	3197	57	2487	710	87%
Y. pestis	4.6	1	3	4262	47.6	4012	149	83.8%
B. abortus	3.2	2	1	3227	57.2	3018	208	93.5%

Virus

Viruses like Ebola virus, Japanese encephalitis virus, Marburg virus, variola virus, etc. are highly infectious and lethal in nature. If these agents were used in attacks, a large number of causalities would potentially result. Thus, these agents are considered as potential BWAs.

Ebola virus

Ebola virus is one of the main viruses of genus *Ebola* and causes a severe hemorrhagic fever in humans and animals (Kumar, 2016). Ebola virus is a negative sense single-stranded RNA of ~18 kb in size. The 3′ end of the genome is not polyadenylated (-AAAAA) and similarly, the 5′ terminus is not capped. Ebola virus genome requires only 472 nucleotides from the 731 nucleotides from the 5′ terminus and 3′ terminus for replication, but these nucleotide sequences are not sufficient for infection. The gene order in the Ebola virus genome is 3-leader-NP-VP35-VP40-GP/sGP-VP30-VP24-L-trailer-5′ (Fig. 1). The leader and trailer sites are nontranscribed regions, which play a role in signals transfer for control replication, transcription, and packaging of the viral genomes into new variations during the infection process.

Ebola virus genome consists of the precursor, a glycoprotein precursor (GP0), which is cleaved to GP1 and GP2 (Volchkov et al., 1998). These two molecules assemble, first into heterodimers, and then into trimers to give the surface peplomers. Secreted glycoprotein (sGP) precursor is cleaved to sGP and delta peptide, both of which are released from the cell (Lee et al., 2008; Singh Jadav et al., 2015). When

FIG. 1

Ebola genome that encodes seven structural proteins with a gene order of: 5′, tailer polymerase L protein, VP24, VP30, glycoprotein (GP), virion protein (VP) VP40, VP35, nucleoprotein (NP), 3′ leader.

the viral protein levels rise, a switch occurs from translation to replication into the genome. As the replication starts, it synthesizes a complementary +ssRNA using the negative-sense genomic RNA as a template. The newly synthesized +ssRNA strand is then used as a template for the synthesis of new genomic (−) ssRNA, which is rapidly encapsulated. These newly encapsulated and envelope proteins are attached at the plasma membrane of the host cells and replicate. Finally, budding occurs, which results in the destruction of the whole cell.

Japanese encephalitis virus

Japanese encephalitis virus (JEV) is a flavivirus, which mainly affects the brain, and is spread by mosquitoes. The genomic RNA of Japanese encephalitis virus (JEV) is a positive-sense; single-stranded RNA (+ssRNA) virus diameter and genome size of the virus is of ~11 kb. The genome of Japanese encephalitis virus possesses a 5′ cap but on the other side, it lacks a 3′ poly tail (Vashist et al., 2011). The sequenced genome of JEV-Y contains 10,976 nucleotides. The genomic RNA contains a single open reading frame (ORF) and codes for a polyprotein of ~3400 amino acids. The full-length genome consists of 10,976 bp, which includes a 95 bp of 5′-Untranslated Regions (UTRs) and595 bp of 3′-untranslated regions. The polyprotein is cleaved by viral and host proteases into the 10 proteins. The genome is divided into two types of genes, i.e., structural and nonstructural genes. The only structural genes are involved in antigenicity. In the genome, three structural genes are present. These 3 structural genes are core (C), premembrane (prM), and envelope (E), which are involved in capsid formation of the virus. Gene E is the most significant and is the most studied one among all three genes. The genes that are involved in virus replication are 7 nonstructural genes (NS1, NS2a, NS2b, NS3, NS4a, NS4b, and NS5) as presented in Fig. 2 (Saxena et al., 2011; Yang et al., 2011).

The C gene code for C protein is ~12 kDa in size and, along with the RNA, it fuses to form the nucleocapsid of the virus. The heterodimer is formed by the PrM protein along with the E protein, which acts as a "chaperone," hindering its function. The protease cleaved the PrM protein into mature M protein form before the release of the virion. This modification in the protein passes signals to the E protein, which results in the formation and activation of E protein homodimers. The E protein, along with approximately 500 amino acids, forms the largest structural protein of the genome. This protein is supposed to play a dynamic role in the entry of viral proteins into the host cells and the main target for the humoral immune response (Solomon, 2003).

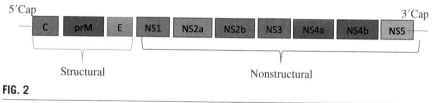

FIG. 2

Japanese encephalitis virus three structural (C, prM, E) and seven nonstructural proteins (NS1, NS2A, NS2B, NS3, NS4A, NS4B, NS5).

The nonstructural proteins are involved in the replication, transcription, and regulation of the innate immune response (Li et al., 2012; Zhang et al., 2012). The NS3 and NS4 proteins code for serine protease and RNA dependent RNA polymerase (RdRp) (Lu and Gong, 2013). All the nonstructural (NS) proteins are important for viral transcription, regulation, and replication, any one of them can be modified as a target for therapeutic intervention (Anantpadma and Vrati, 2011; Mastrangelo et al., 2012).

Marburg virus

Marburg virus (MARV) is considered an extremely dangerous virus and causes viral hemorrhagic fever in humans and animals. The genome of MARV is negative sense single- stranded (-ssRNA) noninfectious, linear nonsegmented, and possess inverse-complementary 3′ and 5′ termini. The genome is not polyadenylated at 3′ end, and does not possess a 5′ cap. The genome size of Marburgvirus is ~19 kb. The total of seven genes are present in the genome, which are 3′-UTR-NP-VP35-VP40-GP1-GP2-VP30-VP24-L-5′-UTR (Fig. 3) (Kiley et al., 1982).

Each gene of the Marburg virus is composed of highly conserved transcription start and stop codon sites. Each gene possesses 3′OH and 5′ untranslated regions, and the open reading frame (ORF) (Pringle, 2005). Short intergenic regions are present in between these genes, which range from 4 to 97 nts. The transcription stops the signal of the upstream gene and starts the signal of the downstream gene site overlap, sharing five highly conserved nts (Pringle, 2005).

Marburg virus genome contains seven structural proteins and each of them plays a different function in the genome. Nucleotide (NP) proteins play a role in the encapsulation of RNA genome, nucleocapsid formation, and budding, and are also essential for replication and transcription (DiCarlo et al., 2007). Viron protein 35 acts as polymerase cofactor, IFN antagonist, and helps in nucleocapsid formation, whereas viron protein 40 helps in budding and antagonist of IFN signaling. Viron protein 30 plays a role in nucleocapsid formation, whereas viron protein 24 plays a role in the maturation of nucleocapsids. The budding glycoprotein present on the surface helps in attachment, receptor binding, and fusion. The tail L gene plays a role in the RNA-dependent RNA polymerase activity (Mühlberger et al., 1999).

Variola

Variola virus (VARV) is a member of the genus Orthopox virus, which causes the acute human disease, smallpox, which is highly lethal and contagious. The genome of variola is linear the dsDNA is 130–375 kb in size. The linear genome is flanked by inverted terminal repeat (ITR) sequences, which are covalently closed at their extremities (Fig. 4) (Smithson et al., 2017). The virus is large, brick-shaped, and

FIG. 3

Marburg virus (−) stranded RNA genome organization of seven genes.

5′ITR   3′ITR

FIG. 4

Virola virus gene organization in the genome.

consists of single linear double-stranded DNA. The genome of variola consists of 186 kb and contains a hairpin loop at each end. The hairpin loop at each end consists of 530 bp fragments. Virola virus has 187 open reading frames (ORF) specifying putative major proteins containing more than or equal to 65 amino acids (Smithson et al., 2017).

Genome lacks C'site polyadenylation tail, which results in circularization of the genome by interaction with 50 region. The conserved elements in the 30 regions have various functions such as virus replication, host-cell tropism, vector specificity, pathogenicity, and virulence. The basic protein helps in the capside formation, which interacts with RNA. It induces its assembly into nucleocapsid particles. Membrane glycoprotein is part of the nucleocapsid and assists the Eprotein for the mature virion formation. Envelope protein (E) is exposed on the viral surface and is mainly involved in the attachment process of the virus to the cell through cell receptors. The E glycoprotein is the major antigenic determinant on the virus surface, involved in the binding of virus and fusion during virus entry in the cell. E protein has three domains. The domain I form the structural domain, domain II dimmers links structural and binding domains, and domain III is the major binding domain of the virus.

Dengue virus

Dengue virus causes a disease in humans called dengue fever, also known as breakbone fever. It is a mosquito-borne viral disease occurring in tropical and subtropical areas. The genome of dengue virus consists of a nonsegmented positive-strand (+) RNA of approx. 11 kb (Lindenbach and Rice, 2003). It encodes a single polypeptide protein, which is processed through co and posttranslationally by host and virus-encoded serine protease into the three structural and seven nonstructural proteins (NS): C-prM-M-E-NS1-NS2A-NS2B-NS3-NS4A-NS4B-NS5, as shown in Fig. 5.

The genome of dengue virus has structural and nonstructural genes. All structural genes (four) are involved in antigenicity, as they are coded by capsid protein and gene C is involved in capsid formation. Among them, gene PrM and M are responsible for membrane formation, whereas gene E helps in the envelope formation of the virus. Nonstructural genes are 7 in number: NS1, NS2a, NS2b, NS3, NS4a, NS4b, NS5 (Lindenbach et al., 2007) (Fig. 5). The gene NS1 plays a role in replication, and NS2a provides IFN resistance. NS2b, along with NS3, helps I serine protease secretion, whereas NS3 alone shows helicase activity (Lobigs and Lee, 2004). NS4a gives signaling to NS4b, and then together both block IFN type signaling. The NS5 gene shows RNA dependent RNA polymerase activity during the replication process (Lobigs and Lee, 2004; Roby et al., 2015).

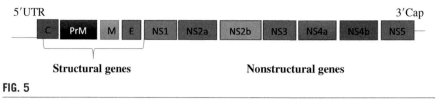

FIG. 5

Dengue virus gene organization in the genome.

Toxins

Trichothecene

Trichothecenes are a terpenoid toxin produced by different fungi, plants, and insects. The trichothecene gene cluster has been identified as a 26 kb length of DNA that consists of three loci, i.e., a single gene TRI101 locus and a two gene TRI1-TRI16 locus. The core trichothecene gene cluster comprises of 12 open reading frames (ORFs) that encode both structural and regulatory genes (Proctor et al., 2009; Brown et al., 2004). Gene cluster (TRI101 locus and the TRI1 locus) products from these three loci are required for the trichothecene biosynthesis (Villafana et al., 2019). These open reading frames consist of genes that play an important function in the biosynthesis, regulation, and transport of mycotoxins across the plasma membrane (McCormick et al., 2011, 2013). The genome has seven genes (TRI5, TRI4, TRI11, TRI3, TRI13, TRI7, TRI8), which are used in different biochemical steps in the biosynthetic pathways. TRI18 have deacetylase function, whereas TRI4, TRI11, and TRI13 are responsible for hydroxylation (P450 hydrolase activity). The function of TRI3 is in acetyl transferase and TRI7 gene acts as acetyl esterase. Gene TRI6 and TRIM help in the regulation, whereas gene TRI12 plays a self-protective role (Cardoza et al., 2011); however, the role of TRI9 and TRI14 is undefined (Proctor et al., 2009). Gene TRI6 and TRI10 encode for two Tri gene transcription factors sites (Rep and Kistler, 2010; Proctor et al., 2018). TRI101 gene is not involved in the second trichothecene biosynthetic gene cluster. The third trichothecenelocus genome includes a structural gene, TRI1, required for C-8 oxygenation during T-2 biosynthesis. The TRI1 is not located within 20–40 kb of the core cluster and any loss of these genes may affect the type of trichothecene mycotoxin, e.g., loss of TRI7 or TRI13 results in 3-ADON or 15-ADON mycotoxin production (Villafana et al., 2019).

Cholera toxin

The bacterium *Vibrio cholerae* secretes an AB5 multimeric protein complex, Cholera toxin, and causes watery diarrhea in humans. The *Vibrio cholerae* genome was sequenced by the clone or whole genome random sequencing method (Fraser et al., 1997; Heidelberg, 2000). Two circular chromosomes are present in the genome of *Vibrio cholerae*, which are of 2.9 (chromosome 1) and 1.0 kb (chromosome 2) in size. The G+C content of these chromosomes is 46.9% and 47.7%, respectively (Yamaichi et al., 1999; Trucksis et al., 1998). It has 3885 predicted open reading frames (ORFs) and 792 predicted Rho-independent terminators, with 2770 and 1115 ORFs, and 599 and 193

Rho-independent terminators on the individual chromosomes, respectively. The main genes are present in chromosome 1, which are required for growth and viability. The genes located on chromosome 2 are required for normal cell function (for example, dsdA, thrS, and the genes encoding ribosomal proteins L20 and L35). Additionally, many intermediaries of metabolic pathways are encoded only on chromosome 2 (Heidelberg, 2000). Protein with a known function is 58% in chromosome 1, which is required for proper functioning of the cell and 48% in chromosome 2. The proteins with unknown functions are 6 in both chromosome 1 and 2. The genomic sequence of *V. cholerae* is confirmed by the presence of a large integron island (a gene capture system) located on chromosome 2 (125.3 kbp) (Rowe-Magnus et al., 1999; Hall et al., 1991). The three genes encode products that may be involved in drug resistance (chloramphenicol acetyltransferase, fosfomycin resistance protein and glutathione transferase). The plasmid uses three virulent genes (hemagglutinin and lipoproteins) and DNA metabolism enzymes (MutT, transposase, and an integrase), which encode gene products similar to the host, which are used for their maintenance in host cells (Heidelberg, 2000).

Tetanus toxin

Clostridium tetani is a soil born bacteria that produces tetanus toxin. The vegetative cell of *C. tetani* under anaerobic conditions produces a tremendously potent neurotoxin that causes tetanus in humans. The *Clostridium tetani* genome is 2.7 kb in size and its chromosome encodes 2372 open reading frames (ORFs). The tetanus toxin is encoded by its plasmid of 74,082 bp, containing 61 open reading frames ORFs (Brüggemann et al., 2003). The chromosome of *C. tetani* contains a G+C content of 28.6%, whereas the pE88 plasmid exhibits G+C content of 24.5%. The replication origin of the chromosome started with the presence of characteristic replication proteins such as DnaA (Brüggemann et al., 2003). ORFs on the chromosome are transcribed in the $5' \rightarrow 3'$ direction, i.e., the same direction as DNA replication occurs. An array of surface and adhesion proteins (35 ORFs) virulence-related factors could be identified, some of them unique to *C. tetani* (Andkvist et al., 1997). The toxin causes disruption of the inhibitory mechanisms of the CNS, and the strongest toxins known to mankind.

Application of genomic information in biodefense

Genomic information of BWAs plays an important role in biodefense. This information will be helpful for the development of vaccines, understanding of resistance pattern or virulence, environmental detection of BWAs, development of new detection technology, etc. The various applications of genome information of BW agents are described below.

Design and development of vaccines against BW agents

The genome information of a particular pathogen will be helpful for the perfect identification of targets, as well as for the design and development of new vaccines. The genetic pattern related to the virulence can be identified with the help of genome sequencing. Pizza et al. (2000) identified new vaccine candidates against Neisseria meningitides (Men B) with the help of genome sequence information. Currently, researchers are using various immunoinformatic tools for the identification of good candidates for the development of vaccines. Immunoinformatic tools involve computational screening of the entire genome of the particular pathogen to find genes that encode proteins with the attributes of good vaccine targets (Rezaei et al., 2019; Unni et al., 2019). Thus, genomic information will play a major role in the timely development of vaccines against potential BW agents.

Understanding of virulence and resistance patterns of BW agents

The understanding of the virulence and resistance pattern of the individual pathogen is necessary for the development of new chemical entities (NCEs) against them. The genome information and subsets of genes in the genome of each BW agent could be obtained with the help of DNA arrays and proteomic techniques, and this information will be helpful to understand the mechanism of virulence and resistance pattern of BW agents. Baba et al. (2002) have reported the virulence potential of *S. aureus* strains with the help of cloning and sequencing techniques.

Environmental detection of BW agents

An environmental detector is required to sense potential BW agents. However, there is no real time environmental detector available. The major challenge in the development of such a detector is to distinguish the presence of BW agents from environmental contaminants. This can be done with the help of genomic information of individual BW agents. It is hoped that in the future, such a detector will be created.

Identification and characterization of potential BW agents

The verification of attack and identification of BW agent should be done quickly in a single assay to provide rapid and highly effective responses. The first step in the development of a single assay is the collection of all coding sequences from multiple isolates of important human, animal, and plant pathogens. Thus, genomic information of individual BW agents will play a major role in the identification and characterization of potential BW agents.

Design and development of new chemical entity (NCE) against BW agents

The new targets can be identified with the help of modern sequencing techniques. New chemical entities (NCE) can be identified and developed against the identified targets for the destruction of potential BW agents.

Understanding of infectious disease process

The cascade of signaling molecules after infection in human cells and the microbial cell could be analyzed with the help of DNA array and proteomics approaches. The identification of new proteins or pathways will be helpful for the detection of BW agents, and the designing of novel vaccines or NCE against BW agents. The development of a generalized diagnostic system based on searching for characteristic host responses among individuals potentially exposed to traditional or advanced biological warfare (ABW) agents would be a major step forward for biodefense.

Conclusion

Genomic information of BW agents will be helpful in detecting bio-threats, unexpected pathogens, or even the completely novel, previously unknown threat. This information can lead to better diagnostics, therapeutics, and prophylactics. Unfortunately, this information can also be used with malevolent intent to design and engineer combinatorial pathogens that do not exist in nature. We hope that misuse of science and technology should not be done as it is for peaceful purposes.

References

Ainscough, M., 2004. Next Generation Bio-weapons: Genetic Engineering and BW. The Gathering Biological Warfare Storm, 255.

Anantpadma, M., Vrati, S., 2011. siRNA-mediated suppression of *Japanese encephalitis* virus replication in cultured cells and mice. J. Antimicrob. Chemother. 67 (2), 444–451.

Andkvist, M., Michel, L.O., Hough, L.P., Morales, V.M., Bagdasarian, M., Koomey, M., DiRita, V.J., Bagdasarian, M., 1997. General secretion pathway (eps) genes required for toxin secretion and outer membrane biogenesis in *Vibrio cholerae*. J. Bacteriol. 179, 6994–7003.

Azam, S., et al., 2016. Genetic characterization and comparative genome analysis of *Brucella melitensis* isolates from India. Int. J. Genomics. 13 pages. 3034756.

Baba, T., Takeuchi, F., Kuroda, M., Yuzawa, H., Aoki, K.I., Oguchi, A., Nagai, Y., Iwama, N., Asano, K., Naimi, T., Kuroda, H., 2002. Genome and virulence determinants of high virulence community-acquired MRSA. Lancet 359 (9320), 1819–1827.

Bourgogne, A., Drysdale, M., Hilsenbeck, S.G., Peterson, S.N., Koehler, T.M., 2003. Global effects of virulence gene regulators in a *Bacillus anthracis* strain with both virulence plasmids. Infect. Immun. 71, 2736–2743.

Brown, D.W., Dyer, R.B., McCormick, S.P., Kendra, D.F., Plattner, R.D., 2004. Functional demarcation of the Fusarium core trichothecene gene cluster. Fungal Genet. Biol. 41, 454–462.

Brüggemann, H., et al., 2003. The genome sequence of *Clostridium tetani*, the causative agent of tetanus disease. Proc. Natl. Acad. Sci. U. S. A. 100 (3), 1316–1321.

Cao, X., et al., 2017. Whole-genome sequences of *Brucella melitensis* strain QY1, isolated from sheep in Gansu, China. Genome Announc. 5 (35), 17 pages. e00896.

Cardoza, R., Malmierca, M., Hermosa, M., Alexander, N., McCormick, S., Proctor, R., Tijerino, A., Rumbero, A., Monte, E., Gutiérrez, S., 2011. Identification of loci and functional characterization of trichothecene biosynthetic genes in the filamentous fungus Trichoderma. Appl. Environ. Microbiol. 2011, 4867–4877.

Chun, J.H., et al., 2012. Complete genome sequence of *Bacillus anthracis* H9401, an isolate from a Korean patient with anthrax. J. Bacteriol. 194 (15), 4116–4117.

Delvecchio, V.G., Kapatral, V., et al., 2002. The genome sequence of the facultative intracellular pathogen *Brucella melitensis*. Proc. Natl. Acad. Sci. U. S. A. 99 (1), 443–448.

DiCarlo, A., et al., 2007. Nucleocapsid formation and RNA synthesis of Marburg virus is dependent on two coiled coil motifs in the nucleoprotein. Virol. J. 20074, 105.

Fraser, C.M., Dando, M.R., 2001. Genomics and future biological weapons: the need for preventive action by the biomedical community. Nat. Genet. 29 (3), 253.

Fraser, C.M., et al., 1997. Genomic sequence of a Lyme disease spirochaete, *Borrelia burgdorferi*. Nature 390, 580–586.

Georgi, E., et al., 2017. Whole genome sequencing of *Brucella melitensis* isolated from 57 patients in Germany reveals high diversity in strains from Middle East. PLoS One. https://doi.org/10.1371/journal.pone.0175425.

Goel, A.K., 2015. Anthrax: a disease of biowarfare and public health importance. World J. Clin. Cases 3 (1), 20–33.

Hacker, J., Kaper, J.B., 2000. Pathogenicity islands and the evolution of microbes. Annu. Rev. Microbiol. 54, 641–679.

Hall, R.M., Brookes, D.E., Stokes, H.W., 1991. Site-specific insertion of genes into integrons: role of the 59-base element and determination of the recombination cross-over point. Mol. Microbiol. 5, 1941–1959.

Halling, S.M., Peterson-Burch, B.D., Bricker, B.J., Zuerner, R.L., Qing, Z., Li, L.L., Kapur, V., Alt, D.P., Olsen, S.C., 2005. Completion of the genome sequence of *Brucella abortus* and comparison to the highly similar genomes of *Brucella melitensis* and *Brucella suis*. J. Bacteriol. 187 (8), 2715–2726.

Heidelberg, J.F., 2000. DNA sequence of both chromosomes of the cholera pathogen *Vibrio cholera*. Nature 406, 477–483.

Kiley, M.P., Bowen, E.T., Eddy, G.A., Isaäcson, M., Johnson, K.M., McCormick, J.B., Murphy, F.A., Pattyn, S.R., Peters, D., Prozesky, O.W., et al., 1982. Filoviridae: a taxonomic home for Marburg and Ebola viruses? Intervirology 18 (1–2), 24–32.

Kumar, A., 2016. Ebola virus altered innate and adaptive immune response signalling pathways: implications for novel therapeutic approaches. Infect. Disord. Drug Targets 16 (2), 79–94.

Lee, J.E., et al., 2008. Structure of the Ebola virus glycoprotein bound to a human survivor antibody. Nature 454 (7201), 177–182.

Li, Y., Counor, D., Lu, P., Duong, V., Yu, Y., Deubel, V., 2012. Protective immunity to *Japanese encephalitis* virus associated with anti-xlink antibodies in a mouse model. Virol. J. 9, 135.

Liang, X., et al., 2017. The pag gene of pXO1 is involved in capsule biosynthesis of *Bacillus anthracis* Pasteur II strain. Front. Cell. Infect. Microbiol. https://doi.org/10.3389/fcimb.2017.00203.

Lindenbach, B.D., Rice, C.M., 2003. Molecular biology of flaviviruses. Adv. Virus Res. 59, 23–61.

Lindenbach, B.D., Thiel, H.J., Rice, C.M., 2007. Flaviviridae: the viruses and their replication. In: Knipe, D.M., Howley, P.M. (Eds.), Fields Virology. 5th ed. Lippincott-Raven Publishers, Philidelphia, pp. 1101–1152.

Little, S.F., Ivins, B.E., 1999. Molecular pathogenesis of *Bacillus anthracis* infection. Microbes Infect. 1, 131–139.

Lobigs, M., Lee, E., 2004. Inefficient signalase cleavage promotes efficient nucleocapsid incorporation into budding flavivirus membranes. J. Virol. 78, 178–186.

Losada, L., et al., 2010. Continuing evolution of *Burkholderia mallei* through genome reduction and large-scale rearrangements. Genome Biol. Evol. 2, 102–116.

Lowy, F.D., 2003. Antimicrobial resistance: the example of *Staphylococcus aureus*. J. Clin. Investig. 111 (9), 1265–1273.

Lu, G., Gong, P., 2013. Crystal structure of the full-length *Japanese encephalitis* virus NS5 reveals a conserved methyltransferase-polymerase interface. PLoS Pathog. 9 (8), e1003549.

Martin, S.B., 2002. The role of biological weapons in international politics: the real military revolution. J. Strateg. Stud. 25 (1), 63–98.

Mastrangelo, E., Pezzullo, M., De Burghgraeve, T., Kaptein, S., Pastorino, B., Dallmeier, K., de Lamballerie, X., Neyts, J., Hanson, A.M., Frick, D.N., Bolognesi, M., Milani, M., 2012. Ivermectin is a potent inhibitor of flavivirus replication specifically targeting NS3 helicase activity: new prospects for an old drug. J. Antimicrob. Chemother. 67 (8), 1884–1894.

McCormick, S.P., Stanley, A.M., Stover, N.A., Alexander, N.J., 2011. Trichothecenes: from simple to complex mycotoxins. Toxins 3, 802–814.

McCormick, S.P., Alexander, N.J., Proctor, R.H., 2013. Trichothecene triangle: toxins, genes, and plant disease. In: Phytochemicals, Plant Growth, and the Environment. Springer, Berlin, Germany, pp. 1–17.

McDonough, K.A., Hare, J.M., 1997. Homology with a repeated *Yersinia pestis* DNA sequence IS100 correlates with pesticin sensitivity in *Yersinia pseudotuberculosis*. J. Bacteriol. 179, 2081–2085.

Michaux, S., Paillisson, J., Carles-Nurit, M.J., Bourg, G., Allardet-Servent, A., Ramuz, M., 1993. Presence of two independent chromosomes in the *Brucella melitensis* 16M genome. J. Bacteriol. 175, 701–705.

Michaux-Charachon, S., Bourg, G., Jumas-Bilak, E., Guigue-Talet, P., Allardet-Servent, A., O'Callaghan, D., Ramuz, M., 1997. Genome structure and phylogeny in the genus *Brucella*. J. Bacteriol. 179, 3244–3249.

Minogue, T.D., Koehler, J.W., Stefan, C.P., Conrad, T.A., 2019. Next-generation sequencing for biodefense: biothreat detection, forensics, and the clinic. Clin. Chem. 65 (3), 383–392.

Mühlberger, E., et al., 1999. Comparison of the transcription and replication strategies of marburg virus and Ebola virus by using artificial replication systems. J. Virol. 73 (3), 2333–2342.

Parkhill, J., 2001. Genome sequence of *Yersinia pestis*, the causative agent of plague. Nature 413 (6855), 523–527.

Pizza, M., Scarlato, V., Masignani, V., Giuliani, M.M., Arico, B., Comanducci, M., Jennings, G.T., Baldi, L., Bartolini, E., Capecchi, B., Galeotti, C.L., 2000. Identification of vaccine candidates against serogroup B meningococcus by whole-genome sequencing. Science 287 (5459), 1816–1820.

Pringle, C.R., 2005. Order mononegavirales. In: Fauquet, C.M., Mayo, M.A., Maniloff, J., Desselberger, U., Ball, L.A. (Eds.), Virus Taxonomy—Eighth Report of the International Committee on Taxonomy of Viruses. Elsevier/Academic Press, San Diego, USA, pp. 609–614.

Proctor, R.H., McCormick, S.P., Alexander, N.J., Desjardins, A.E., 2009. Evidence that a secondary metabolic biosynthetic gene cluster has grown by gene relocation during evolution of the filamentous fungus *Fusarium*. Mol. Microbiol. 74, 1128–1142.

Proctor, R.H., McCormick, S.P., Kim, H.S., Cardoza, R.E., Stanley, A.M., Lindo, L., Kelly, A., Brown, D.W., Lee, T., Vaughan, M.M., 2018. Evolution of structural diversity of trichothecenes, a family of toxins produced by plant pathogenic and entomopathogenic fungi. PLoS Pathog. 14 (4), e1006946.

Rajashekara, G., Covert, J., Petersen, E., Eskra, L., Splitter, G., 2008. Genomic island 2 of *Brucella melitensis* is a major virulence determinant: functional analyses of genomic islands. J. Bacteriol. 190, 6243–6252.

Rao, S.B., Gupta, V.K., et al., 2014. Draft genome sequence of the field isolate *Brucella melitensis* strain Bm IND1 from India. Genome Announc. 2 (3), 14 pages. e00497.

Rep, M., Kistler, H.C., 2010. The genomic organization of plant pathogenicity in *Fusarium* species. Curr. Opin. Plant Biol. 13, 420–426.

Rezaei, M., Rabbani-khorasgani, M., Zarkesh-Esfahani, S.H., Emamzadeh, R., Abtahi, H., 2019. Prediction of the Omp16 epitopes for the development of an epitope-based vaccine against brucellosis. Infect. Disord. Drug Targets 19 (1), 36–45.

Roby, J.E., et al., 2015. Post-translational regulation and modifications of flavivirus structural proteins. J. Gen. Virol. 96, 1551–1569.

Rowe-Magnus, D.A., Guerout, A.M., Mazel, D., 1999. Super-integrons. Res. Microbiol. 150, 641–651.

Saxena, S.K., Tiwari, S., Saxena, R., Mathur, A., Nair, M.P.N., 2011. *Japanese Encephalitis*: an emerging and spreading arbovirosis. In: Ruzek, D. (Ed.), Flavivirus Encephalitis. InTech, Croatia (European Union), pp. 295–316. ISBN: 979-953-307-775-7.

Schuster, S.C., 2007. Next-generation sequencing transforms today's biology. Nat. Methods 5 (1), 16.

Singh Jadav, S., Kumar, A., Jawed Ahsan, M., Jayaprakash, V., 2015. Ebola virus: current and future perspectives. Infect. Disord. Drug Targets 15 (1), 20–31.

Smithson, C., et al., 2017. Re-assembly and analysis of an ancient variola virus genome. Viruses 9 (9), 253.

Solomon, T., 2003. Recent advances in *Japanese encephalitis*. J. Neurovirol. 9 (2), 274–283.

Szinicz, L., 2005. History of chemical and biological warfare agents. Toxicology 214 (3), 167–181.

Trucksis, M., Michalski, J., Deng, Y.K., Kaper, J.B., 1998. The *Vibrio cholerae* genome contains two unique circular chromosomes. Proc. Natl. Acad. Sci. U. S. A. 95, 14464–14469.

Turnbull, P.C.B., 1999. Definitive identification of *Bacillus anthracis*—a review. J. Appl. Microbiol. (2)237–240.

Unni, P.A., Ali, A.M.T., Rout, M., Thabitha, A., Vino, S., Lulu, S.S., 2019. Designing of an epitope-based peptide vaccine against walking pneumonia: an immunoinformatics approach. Mol. Biol. Rep. 46 (1), 511–527.

Vashist, S., Bhullar, D., Vrati, S., 2011. La protein can simultaneously bind to both 30- and 50-noncoding regions of *Japanese encephalitis* virus genome. DNA Cell Biol. 30 (6), 339–346.

Villafana, R.T., et al., 2019. Selection of *Fusarium Trichothecene* toxin genes for molecular detection depends on TRI gene cluster organization and gene function. Toxins (Basel) 11 (1), 36.

Volchkov, V.E., et al., 1998. Processing of the Ebola virus glycoprotein by the proprotein convertase furin. Proc. Natl. Acad. Sci. U. S. A. 95 (10), 5762–5767.

Wang, F., Hu, S., Gao, Y., Qiao, Z., Liu, W., Bu, Z., 2011. Complete genome sequences of *Brucella melitensis* strains M28 and M5-90, with different virulence backgrounds. J. Bacteriol. 193 (11), 2904–2905.

Yamaichi, Y., Iida, T., Park, K.S., Yamamoto, K., Honda, T., 1999. Physical and genetic map of the genome of *Vibrio* parahaemolyticus: presence of two chromosomes in Vibrio species. Mol. Microbiol. 31, 1513–1521.

Yang, Y., Ye, J., Yang, X., Jiang, R., Chen, H., Cao, S., 2011. *Japanese encephalitis* virus infection induces changes of mRNA profile of mouse spleen and brain. Virol. J. 8, 80.

Zhang, T., Wu, Z., Du, J., Hu, Y., Liu, L., Yang, F., Jin, Q., 2012. Anti *Japaneseenceph litis* viral effects of kaempferol and daidzin and their RNA-binding characteristics. PLoS One 7 (1), e30259.

Further reading

Biedenkopf, N., et al., 2016. RNA binding of Ebola virus VP30 is essential for activating viral transcription. J. Virol. 90 (16), 7481–7496.

Couesnon, A., et al., 2006. Expression of botulinum neurotoxins A and E, and associated non-toxin genes, during the transition phase and stability at high temperature: analysis by quantitative reverse transcription-PCR. Microbiology 152, 759–770.

Lobigs, M., 1993. Flavivirus premembrane protein cleavage and spike heterodimer secretion require the function of the viral proteinase NS3. Proc. Natl. Acad. Sci. U. S. A. 90, 6218–6222.

Planning for protection of civilians against bioterrorism

14

V. Nagaraajan

VN Neurocare Center, Madurai, India

Planning against bioterrorism is normally focused on soldiers fighting on the battlefield, but now the concept is changing to include protection of the civilians from bioterrorism, to prevent calamities caused by toxins and microbes. The change in the attitude of protecting common people may involve additional financial involvement, but it becomes mandatory. The approach toward bioterrorism is multifocal. When the enemy applies bioterrorism on the battlefield, simultaneously they have a lateral approach to attack the civilians also to divert the total attention of the country. This also would lead to a financial crisis in the government, which will in turn impact and crash the defense system. So it becomes mandatory that a good amount of attention is focused against bioterrorism toward civilians.

Since bio threat instruments are easily available to the terrorists, laboratories should work on the basis of constant vigil with an advanced detective system to provide earliest detection, which would include identification of the contaminated areas, where the population is at risk, and also identify whether it is natural or induced. It has to be kept in mind that not only the rural areas but also the major cities are easily penetrable to the bioterrorist. Adequate forensic technologists should work on constantly identifying the newly emerging biological agents and their origins. Decontamination technologies should be constantly monitored and enforced without causing any additional environmental changes. For this, the antibioterrorism forensic people should co-opt themselves with the public health authorities, law enforcing authorities, district administration authorities, and relevant stakeholders. Availability of a national pool of vaccines and stock piles should be allowed to penetrate to all the cities with the cooperation of state and district officials, which is much lacking now (JCVI, n.d.)

Development of vigilance

Even though several biovigilance systems are available, the following are the modern mandatory additions that need to be incorporated:

(a) Presence and detection of aerosolized biological agents before the onset of epidemic. This system is highly significant and mandatory because it comprises a highly proactive response to an epidemic outbreak rather than the passive management and struggle after the epidemic.

(b) Generation 3 automated detection system. This system is an automatic response system within a few hours (4–6 h), to detect the incidence of bio attack (NIH, n.d.). The control system can extinguish the infection at its earliest stage of virulence.

(c) Enhancing the scientific potential to identify the first responders. For this, numerous strategies are required, especially constant surveillance with the objective of probing for dangers in suspicious powders and even cosmetic powders, which may gain contamination by many dangerous biological agents such as anthrax that gel with easily with common powders. Constant sampling of such random public utility powders in the commercial system, consumed by the first responders, is mandatory. This surveillance will create fear and doubt in the minds of bioterrorists and their handlers of getting detected when they attempt to mix the biologically harmful powders and thus it can be rendered ineffective (Buller, 2003).

(d) Enhanced anti bioterrorism equipment toward the first responders. The U.S. government has commercialized a new form of Tyvex armor which provides high protection from the chemical and biological contaminants to the first responders and patients. The development of self-contained breathing apparatus (SCBA) offers a high range of detection of bioterrorism agents that are aerosolized.

The need for the "Bio-shield" system

The accumulation of vaccine and treatment potentials for biological threats or medical counter measures looks conventional and they are termed a Bio-shield system (Kelle, 2007). The system is very viable and adequate stock piling of vaccines and efforts to counteract bioterrorist attacks are to be provided in every nook and corner of the country. National systems of chain drug stores to identify the availability of vaccines should be enforced and participation of private sectors in distribution of many counter measures should be encouraged. Adequate financial allocations and budget should be legalized through the parliament in every country in sectors of both defense and health. Novel approaches in the bio and health technology like "synthetic biology" require placement to design and detect various new types of biological warfare agents (Kahn, 2015; Anon, n.d.-a; Basulto, 2015; Wagner et al., 2001).

Future dangers in bioterrorism

Future dangers arise out of activities such as rogue research toward experiments on how to render vaccine ineffective, namely:

(a) research toward development of therapeutic resistance to the use of antibiotics and antiviral agents;

(b) development of biothreat robotics;

(c) research toward the enhancement of virulence of a pathogen and making a nonpathogen more virulent;

(d) incremental in the transmissibility of a pathogen;

(e) altering the host range of a pathogen;

(f) development of ability of a biothreat agent to evade possible diagnostic and detection tools; and

(g) ability to develop the weaponization of a biological agent or biological toxin.

Synthetic biology concerns or affects biosecurity and it is focused on the role of DNA synthesis and production of genetic material of lethal viruses (e.g., Spanish flu, polio) by producing these risk materials in laboratories (Anon, n.d.-b; Pellerin, 2011; Chen et al., 2010).

Use of CRISPR technology/Cas system

This technology has now emerged recently as an advanced tool technique of gene editing and development of new chromosomes of unnatural occurrence. The CRISPR technology has brought down the time and speed in editing the gene, from years to weeks. This technology raised number of ethical concerns with special reference to not to use technology in the field of biohacking space (Heitz, 2013; Locker, 2013; Cohen, 2014). CRISPR technology has been used to the benefit of society in curing and identifying the chromosomal aberrations in rare and inherited diseases. Now this is being misdirected by rogue science research to develop new unidentified organisms with high lethal value, producing lethal toxins, which has potential to become an unmanageable biothreat. Adequate legalization of this technology is highly warranted for preventing such technology and research from being misused against mankind.

Existence of biosurveillance

Biosurveillance is a recently evolving disciplinary science of real time disease outbreak detection, which can detect both natural and manmade epidemics, especially biothreats. Apart from the existing first automated bioterrorism detection system—real-time outbreak disease surveillance (RODS)—which has been designed to draw and collect data from many data sources and use them to provide signal detection regarding the possible bioterrorism event at the earliest possible time, other attempts include the following:

(a) Tiny electronic chips containing live nerve cells have been designed to warn of the presence of a broad range of bacterial toxins.

(b) Fiber optic tubes coated with antibodies coupled with light emitting molecules to identify the specific pathogens especially to identify anthrax, botulinum, and ricin.

(c) Ultra violet avalanche photo diodes offer the great gain and absolute reliability that is needed to detect anthrax and other toxic pollutions in the air (Tavernise, 2014).

Creation of novel vaccines

Development of newer vaccines for any type of newer encoded biothreat is possible now from various research centers (UT, Dallas, USA) that are mutating the DNA encoding with active A chain of the toxin, taking out the site that inhibits protein synthesis as well as the site responsible for inducing vascular leak. Such centers have created three recombinant version ricin A chains, and a couple of them were found to be effective as vaccine in experimental mice. Human application is still a work in progress. It is pertinent to note that ricin is not only dangerous but also a cost effective bioterrorism weapon and stockpiles of ricin toxin are found to be stored in certain pockets of the world.

Potential threats in force

Threats include:

(a) *Bacteria*: newer versions of nuclear chain mutated anthrax, plague, brucellosis, Q fever, shigellosis, typhus, and of late, beta-lactamase resistant staphylococci.
(b) *Viruses*: smallpox, Ebola, Marburg, dengue, H1N1 virus.
(c) *Toxins*: botulinum A to G, staph enterotoxin, ricin toxin, marine neurotoxin, clostridium perfringens toxins, mycotoxins.

Challenges

The future challenges will be the identification of pathogenesis and mechanisms involved in generating various diseases, making corresponding, adequate, appropriate animal experimental models, surrogate markers, and assessment of effect of appropriate reagents, which would antagonize the disease spreading system.

Technology development

Technology development requires:

(a) development toward confirming the identity of newer pathogens;
(b) gene probe polymerized chain reaction;

(c) amplification of select genetic sequences; and

(d) development of reagent sets for agent identification.

Toward improvement in vaccine production

For this, the following are necessary fields of research:

(a) development and qualification of assays for vaccine candidates;

(b) identification of surrogate markers of efficacy;

(c) aerosol challenge studies in animal models;

(d) vaccines—cell culture derived vaccine FY94;

(e) Venezuelan equine encephalitis virus vaccine (V 3526) AND FY00, live attenuated VEE vaccines derived by site directed mutagenesis of a full length infectious cDNA clone; and

(f) recombinant botulinum toxin vaccine, especially Type A and B - FY 00.

Biodefense mechanism and doctrine

Biodefense mechanisms and doctrines include the following:

(a) Education and training of the public and defense personnel on the ongoing newer concepts of rogue research and biothreat mechanisms.

(b) Intelligence, which comprises organization, delivery system to prevent the attack of biowarfare system initially to the war personnel, and also applying urgent appropriate therapeutics, prevention of death due to biowarfare, urgent detection of counter mechanism against biowarfare is needed. The intelligence requires assessments of bio threats and also may define what should be the subsequent mission against the modified bio warfare.

(c) Intelligent medical and research countermeasures comprising the development of antidote vaccines, newer diagnostic strategies against modern biothreats, and finally, an efficient and good diagnostic system and therapeutic system will fulfill the future need of impending biothreats in modern warfare.

References

Anon, n.d.-a. CRISPR, the Disruptor. Nature News & Comment (Retrieved 24.01.2016).

Anon, n.d.-b. Avalanche Photodiodes Target Bioterrorism Agents Newswise (Retrieved 25.06.2008).

Basulto, D., 2015. Everything you need to know about why CRISPR is such a hot technology. The Washington Post. ISSN: 0190-8286 (Retrieved 24.01.2014).

Buller, M., 2003. The potential use of genetic engineering to enhance orthopox viruses as bioweapons. In: Presentation at the International Conference 'Smallpox Biosecurity. Preventing the Unthinkable' (21–22 October 2003), Geneva, Switzerland).

Chen, H., Zeng, D., Pan, Y., 2010. Infectious Disease Informatics: Syndromic Surveillance for Public Health and Bio-Defense. (XXII, 209p. 68 illus., Hardcover).

Cohen, B., 2014. Kadlec says biological attack is uncertain, imminent reality. Bio Prep. Watch. (Retrieved 17.02.2014).

Heitz, D., 2013. Deadly bioterror threats: 6 real risks. Fox News. (Retrieved 17.02.2014).

JCVI, n.d. Research/Projects/Synthetic Genomics | Options for Governance/Overview. www.jcvi.org. (Retrieved 24.01.2016).

Kahn, J., 2015. The Crispr Quandary. The New York Times. ISSN: 0362-4331 (Retrieved 24.01.2016).

Kelle, A., 2007. Synthetic Biology & Biosecurity Awareness in Europe. (Bradford Science and Technology Report No. 9).

Locker, R., 2013. Pentagoseeking vaccine for bioterror disease threat. USA Today. (Retrieved 17.02.2014).

NIH, n.d. Office of Biotechnology Activities | Office of Science Policy (PDF). osp.od.nih.gov. (Retrieved 24.01.2016).

Pellerin, C., 2011. Global Nature of Terrorism Drives Biosurveillance. American Forces Press Service.

Tavernise, S., 2014. U.S. backs new global initiative against infectious diseases. New York Times. (Retrieved 17.02.2014).

Wagner, M.M., Aryel, R., et al., 2001. Availability and Comparative Value of Data Elements Required for an Effective Bioterrorism Detection System (PDF). Real-time Outbreak and Disease Surveillance Laboratory. (Retrieved 22.05.2009).

Further reading

Bell, L., 2013. Bioterrorism: a dirty little threat with huge potential consequences. Forbes (Retrieved 17.02.2014).

Tumpej, T.M., et al., 2005. Characterization of the reconstructed 1918 Spanish Influenza Pandemic Virus. Science 310 (5745), 77–80.

Index

Note: Page numbers followed by *f* indicate figures and *t* indicate tables.